ADVANCED CONTROL STRATEGIES FOR SOCIAL AND ECONOMIC SYSTEMS
(ACS'04)

A Proceedings volume from the IFAC Multitrack Conference,
Vienna, Austria, 2 – 4 September 2004

Edited by

P. KOPACEK

Institute for Handling Devices and Robotics,
Vienna University of Technology,
Vienna, Austria

Published for the

INTERNATIONAL FEDERATION OF AUTOMATIC CONTROL

by

ELSEVIER LIMITED

ELSEVIER Ltd
The Boulevard, Langford Lane
Kidlington, Oxford OX5 1GB, UK

Elsevier Internet Homepage
http://www.elsevier.com

Consult the Elsevier Homepage for full catalogue information on all books, journals and electronic products and services.

IFAC Publications Internet Homepage
http://www.elsevier.com/locate/ifac

Consult the IFAC Publications Homepage for full details on the preparation of IFAC meeting papers, published/forthcoming IFAC books, and information about the IFAC Journals and affiliated journals.

First edition 2005

Library of Congress Cataloging in Publication Data

A catalogue record for this book is available from the Library of Congress

British Library Cataloguing in Publication Data

A catalogue record for this book is available from the British Library

ISBN 978-0-08-044242-6

ISSN 1474-6670

Printed and bound in the United Kingdom
Transferred to Digital Print 2010

To Contact the Publisher

Elsevier welcomes enquiries concerning publishing proposals: books, journal special issues, conference proceedings, etc. All formats and media can be considered. Should you have a publishing proposal you wish to discuss, please contact, without obligation, the publisher responsible for Elsevier's industrial and control engineering publishing programme:

Christopher Greenwell
Senior Publishing Editor
Elsevier Ltd
The Boulevard, Langford Lane   Phone: +44 1865 843230
Kidlington, Oxford   Fax: +44 1865 843920
OX5 1GB, UK   E.mail: c.greenwell@elsevier.com

General enquiries, including placing orders, should be directed to Elsevier's Regional Sales Offices – please access the Elsevier homepage for full contact details (homepage details at the top of this page).

IFAC MULTITRACK CONFERENCE ON CONTROL STRATEGIES FOR SOCIAL AND ECONOMIC SYSTEMS 2004

Sponsored by
International Federation of Automatic Control (IFAC)
IFAC Technical Committees on:
- Supplemental Ways of Improving International Stability
- Cost Oriented Automation
- Economic and Business Systems
- Social Impact of Automation
- Developing Countries

Co-sponsored by
International Federation for Information Processing (IFIP)
International Federation of Operational Research Societies (IFORS)
Vienna University of Technology
IFAC Technical Committees on:
- Computers for Control
- Computers and Telematics
- Advanced Manufacturing Technology
- Manufacturing Modelling for Management and Control
- Enterprise Integration
- Control Education

Organized by
Vienna University of Technology
Institute for Handling Devices and Robotics

International Programme Committee (IPC)
Kopacek, P. (AUT) – Chair
Ederer, B. (AUT) – Vice-Chair from industry
Kopacek, B. (AUT) – Vice-Chair from industry

Vice-Chairs:
Cernetic, J. (SLO)
Dimirovski, G. (MAC)
Dinibütün, A.T. (TR)
Erbe, H. (GER)
Kile, F. (USA)
Neck, R. (AUT)
Stahre, J. (SWE)
Craig, I. (ZA)

Members:
Dumitrache, I. (ROM)
Groumpos, P.P. (GRC)
Hersh, M. (GBR)
Makarenko, A. (UKR)
Mansour, M. (CHE)
Paiuk, J. (ARG)
Thoma, M. (GER)

Shubin, T. (RUS)
Stapleton, L. (IRL)
Vamos, T. (HUN)

National Organizing Committee (NOC)

Han, M.-W. (AUT) – Chair

Members:

Prochazkova, D. (AUT)
Putz, B. (AUT)
Shirmbrand, E. (AUT)
Würzl, M. (AUT)

PREFACE

Automation and closely related technological changes have an increasing impact on society. This multitrack conference is a first trial to bring together scientists and industrial people to discuss emerging problems related to control and automation in the next triennium as well their impact on technological and economical development of society.

This Proceedings volume of the IFAC Multitrack Conference on Advanced Control Strategies for Social and Economic Systems (ACS '04) contains 5 survey papers and 29 technical papers selected from 61 submitted papers.

Therefore the contributions cover the whole field of more than 5 IFAC TC's, like managing the introduction of technical changes, technology and environmental stability, socially appropriate guidelines for control design, socio-technical systems as complex composite systems, ethics, effects of automation in engineering, upgrading the industrial control systems, decision and control in economics of developing countries, network economics and networking the national economics, e – Business, artificial intelligence for social/economic systems, control of chaos in social/economic systems.

Furthermore the results of a panel discussion are also included in this volume.

Summarizing the scientific technical output of this conference: ethics becomes more and more interest in control engineering, the classical IFAC topic on social effects is moving more and more to human machine cooperation, new automation technologies requires more interdisciplinary educated people and developing countries need the newest technology for an efficient improving of the industry.

On behalf of the International Program Committee (IPC) and the National Organizing Committee (NOC), we would like to thank all for participating in and contributing to this event and hope all enjoyed the familiar atmosphere, which stimulated intensive and successful discussions.

P. Kopacek

Vienna, October 2004

CONTENTS

EDUCATION

ETHICS AND AMAT IN A WORLD OF ILLUSION
MULTIPLE STRESSORS ASSOCIATED WITH THE CURRENT ENGINEERING TECHNOLOGY PROGRAMME

Stapleton, L.[1] **and Kile, F.**[2]

[1] *ISOL Research Centre, Waterford Institute of Technology,*
Republic of Ireland
[2]*Microtrend, 420 E. Sheffield Lane, Appleton, WI 54913-7181, USA*

Abstract: A post-structural analysis of current technological trajectories is developed which addresses ethical issues for engineering and technology research. It particularly concerns itself with two areas: environmental impact and social impact in terms of labour. This paper produces an analysis of current trends which has not been well developed within the social impact literature, demonstrates serious instabilities within current socio-technological systems at a global level and how ethical questions must be framed in a very wide context. It shows that questions of ethics and social impact go to the very heart of the work of engineers, technologists and AMAT researchers. *Copyright © 2004 IFAC*

Keywords: Automation, Economic systems, Labor dislocation, information technology.

1. INTRODUCTION

Advances in technology continue apace. However, there is an increasing concern at various levels of society, up to United Nations, that the developments in AMAT (Automation and Machine-Assisted Thinking) and ICT (Information and Communications Technology are creating significant problems for our global society. This paper argues that some of these problems go to the very heart of any ethical analysis of AMAT and ICT, and raise deep questions for the programme of engineering research and development currently underway. Specifically, this paper addresses two major issues:

1. Downward pressure on labor forces, in both developed and developing areas, resulting from Automation and Machine-Assisted Thinking (AMAT). This pressure creates an increasing need for training in applications of technology, resulting in replacing skilled persons with less skilled persons who are dependent on AMAT to function in their jobs. Pressure on labor forces displaces people at skill level "n" with machine-assisted people at skill level "n-1, and so on through a spectrum of labor expertise." The final consequence of this downward cascading of

pressure on labor results in underemployment among many in both developed and developing economies and total loss of employment opportunities among the least qualified in developing areas.

2. Evolution of "markets" from exchanges of products and goods to markets based on the sale and procurement of "signs" ("simulacra" in the idiom suggested by J. Baudrillard). This evolution of "markets" does not exclude exchange of tangible products and goods; rather, this transformation of markets stems from how tangibles and intangibles are blended in the minds of both sellers and buyers. Increasingly, "market exchanges" are moving from products apart from attributed images in the direction of exchanging "signs" through which both tangibles and intangibles are identified. Thus, one "purchases" status through possession of an expensive, carefully marketed high-end automobile, though this automobile may in no way be superior to a less highly "imaged" machine.

Although, superficially, these two issues may seem unrelated, recent analyses show that this is not the case. However, to date, the literature in engineering ethics, systems engineering development and related work, have not juxtaposed these two issues in one paper. Neither has the post-structural analysis presented here received very much attention, with one or two notable exceptions[1].

2. SIMULACRA, PASSIVITY AND IDLENESS

One analysis sees simulacra and the consumption of signs as a means by which social action can be derailed structurally within the post-capitalist system (e.g. Chomsky (1994)). Thus, rather than becoming *agent provocateurs*, a healthy opposition to the programme of global expansion of western capitalism, the socially marginalized elements of society are effectively neutered at source. If people feel uncomfortable by the image of an African child with flies in her eyes, they can change channels to "Friends" or an Australian soap opera.

All this may in fact be exacerbated by technology and systems engineering research, which assumes progress through the continuous creation of ever more advanced technology products (Douthewaite (1992)). This can be presented as a positive feedback loop, which self-perpetuates and increases the problem significantly. A major implication of the positive feedback loop is that, according to current techno-science thinkers, the model of technological "progress" may also contain within it a critical point of over-development at which the system itself reaches a sort of end point[2] and emerges as an entirely new process in which the previous assumptions no longer hold. In the reading of the post-structural analysis presented in this paper, "escape velocity" (a reference to a relationship between societies of humans and machines) is interpreted as a form of runaway system behaviour: in control theory the result of an accelerating positive feedback loop. According to the post-structuralist analysis, this "escape velocity" may already have been reached.

2. ENVIRONMENT

AMAT is leading to high levels of environmental stress at a global level. Globalization of technology appears inevitable, even if protesters against globalization should persuade decision makers that regional and national markets are preferable to a single global market. Barriers to dissemination of new technology remain ineffective.
Much of the rise in employment during the 1990s was

due to increased consumption. It is clear that increased consumption is de-stabilizing the global environment. A continued increase in consumption will, at some point, trigger widespread environmental catastrophes, migrations, ensuing wars, and massive loss of life. These assumptions may seem alarmist. However, note that the entire snow cover in the Alps of Switzerland (excepting snow which had fallen on existing glaciers) melted in 2003 for the first time in recorded history. Moreover, Alpine glaciers are receding at rapid rates. Even conservative estimates see the glaciers fully melted by 2020. Some observers estimate that, in terms of providing summer waterflow from glacial melt, no meaningful Alpine glaciers will remain by 2010.

In this view, nature is becoming insignificant and a sort of encumbrance upon technological advance. We continue to create more and more condensed technical systems, functions and models and we transform all the rest into waste - residues. In a sense we have left the earth, we have reached an escape velocity in the west so that we have entered a new reality: hyperreality. This hyperreality transforms the planet itself into a "marginal territory," a form of waste product. For example, building a freeway (or a toll road) or a shopping center transforms the natural landscape around into a wilderness. Even human interaction is subject to this rule. By creating ever faster communications networks, we reduce human exchange into a residue of "txt msgs." This view of society and environment sees a holistic system of environmental and human waste, (boat people, refugees, human trafficking, nuclear waste and so on) all moving around within this meta-system. In this analysis, nature and human society (if that term still has meaning) are reduced to residues within a hypothetically "managed" system, which is, in fact, a new and even more unmanageable entity. In this post-structural analysis the recycling of waste, the control of emissions and so on are signs of the hyperreality of the Western system. As Baudrillard says
"what is worst is not that we are submerged by the waste products of industrial and urban concentration, but that *we ourselves are transformed into residues*" (p. 78 Baudrillard (1992) italics in the original).

Disconnectivity and Illusion: The above argument results in a fatal situation in which the Western world, the world which creates so many of the environmental problems, is deeply disconnected from the real, the material. We recycle our glass bottles but do little for our discarded humans (human waste).

3. LABOR

Due to AMAT we have increasing unemployment, underemployment and less full employment positions.

---

[1] Some authors have attempted to tackle these issues in the literature but deep debate remains elusive.

[2] In this analysis notions like *end* and *beginning* no longer really apply (Baudrillard (1994))

3.1 Unemployment and Underemployment

Firstly this is a global stressor. The unemployed are becoming unemployable due to technological advances: for example, ICT literacy is a major issue in Europe. Underemployment occurs when people work below their skill level or with low pay or less than a desirable number of hours per week. These factors vary by funds needed, job satisfaction cultural expectations, difficulty of travel to work, etc. Underemployment is a social burden often overlooked in studies of total unemployment. Moreover, underemployment may be part of a continuum leading to complete unemployment.

3.2 Post-Structuralism and Catastrophe Theory

It should be noted that this post-structuralist scenario, though perhaps societally accurate, and accurate in the description of markets, assumes that this behavior (Western Behavior, if the reader prefers), can continue unchallenged by environmental forces. At some point disease, famine, or mass migrations under the pressure of extreme environmental change, will stress both the "developed world" (or "over-developed" world in the sense of a runaway system which has reached escape velocity) and the "developing world" so powerfully that social upheaval is capable of creating social chaos. This chaos can be viewed in terms of catastrophes, which entirely realign, or eliminate, underlying assumptions. Clearly, the attacks on New York and Washington, DC in the USA on September 11, 2001 were a warning sign of more dramatic, analogous future events. Indeed ,the escalation of similar (if less extensive attacks) in Turkey, Saudi Arabia and so on emphasise this from the perspective of the "developed" world. It is important to note that the "developing world" exists in large pockets even in the most advanced nations of the West. One need merely refer to poverty and unemployability among people of South Central Los Angeles or the vast "favelas" surrounding the glittering towers of Sao Paulo in Brazil.

3.3 Displacement and Underemployment

People unemployed because the work for which they are qualified has been displaced by AMAT may elect to accept work below their optimal skill level (underemployment). After a time, leads to loss of hard-won skills and also to diminished self-esteem.

Data regarding underemployment are scanty. It is common knowledge that many people are employed below their skill levels, suggesting that little serious effort has been made to quantify data reflecting the seriousness of this issue. During our travel to the Former Republic of Yugoslavia in May, 2000, highly educated citizens of one of its successor states told the authors that 50% of workers were unemployed. Some of the 50% unemployed were professional people working in a shadow economy, often below a suitable skill-level.

When a person displaced by AMAT accepts work below his/her skill level, another, less skilled person is displaced. The displaced person may accept a position beneath his/her skill level. In this way, underemployment cascades through a chain of skill levels at large costs to many people, each working at a lower level than previously.

An economy, with high productivity through use of AMAT, can create underemployment across many sectors of the economy. Underutilized human capacity increases further as potential customers earn too little to purchase products of the seemingly improved economy. This is a positive feedback loop with negative consequences for the economy.

3.4 Employment

In this context, effects on the workplace are primarily a "developed" world stressor. The employed are under increasing stress due to technological ▯advances.▯ For example, figure 1 shows a real wages fall in recent years. Furthermore, evidence gathered in modern organisations indicates that people are seeking medical attention for technology-induced stress-related illnesses (Stapleton & Murphy (2002)). By current standards more people in developed societies are overworked than at any time since the post-World War 2 recovery period.

For many in the "developed" world, the nature of work has changed with the increasing pace of technology change. Consequences include invasion of personal space by information and communications technologies (Carew & Stapleton (2004)), increased stress in the workplace as people must adapt more and more to globally constituted integrated information and automation systems (Stapleton (2003)), sweatshop-call centres and so on (Knight (2001)).

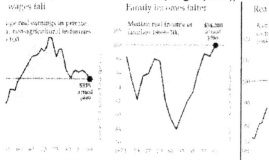

Fig. 1. Real Wages and Family Income.

4. DISENFRANCHISEMENT AND DESTABILISATION

The technology development programme, underpinned as it is by the profit motive and the dynamics of a post-structuralist consumer "society", may exacerbate the systemic problems which leave so many people disenfranchised from global wealth creation systems.

4.1 The Dislocation of the Disenfranchised

Enormous masses of rural poor in the developing world are never exposed to the products which comprise the globalization of technology. However, they are exposed through mass media to the supposed benefits of new technology. Most of these same areas are the source of rapid population growth, further increasing pressure for increased consumption and unrest following inability to attain the unrealistic consumption goals set forth in the mass media.

Furthermore, the disjunction between Western goals and methods on the one hand and the processes at work in developing societies are exacerbated by technological developments and systems engineering research based on the idea of progress as creation of ever more advanced technological products. This acts as a positive feedback loop which perpetuates and exacerbates the social and technical dissonance between the West and much of the developing world.

4.2 The Disappearance of the Disenfranchised

One analysis sees simulacra and the consumption of signs as a means by which social action can be derailed structurally within the post-capitalist system. Thus, rather than opposing expansion of Western industrial goals and methods into developing societies incapable of adapting to these goals and methods, the NGOs may actually facilitate change without analysis of its likely effects. Socially (as well as economically and educationally) marginalized sectors of society are thus excluded from changes occurring in their own nations or regions. As we noted earlier, if people feel uncomfortable by the image of an African child with flies in her eyes, they simply change channels a blandly entertaining program.

4.3 Revolution and Neutralisation: The absence of systemic corrective feedback

These stressors would normally lead to revolutionary activity which would destabilise elite power structures. The march of history shows us that these interventions ensure that societies remain stable. So, the peasants revolt (an event linked to environmental change as much as anything else) ensured that the feudal system incorporated social impact imperatives. The longer the stressors are allowed to accumulate pressure, the greater the ultimate devastation of the social earthquake that results (witness the collapse of the Roman Empire).

Modern media and consumption structures fill up peoples lives to the extent that many have neither the time nor the inclination to agitate. Indeed, where agitation arises, it can be subsumed into a swirl of media fragments, text messages, MMS and the internet, and thus be neutralised. Consequently, the voices of the disenfranchised are reduced to a tiny whisper, and rarely even merit a soundbyte on CNN. This is an extremely dangerous situation. To illustrate: in this analysis, the "surprise" of 9/11 (an event predicted by many analysts), and the consequential reverberations through social, financial and political systems, was due to a sudden shock of those outside our hyperreal world suddenly engaging with the West, reminding the West that the tectonics of globalisation are part of a positive feedback, rather than negative feedback, loop (the reader is reminded that a "positive" feedback loop is a technical definition of system self-amplification, which finally causes the system to self-destruct. In this sense, positive loop behavior has highly negative consequences) . The more subsumed we are into our hyperreality the more disengaged we become with the real, and so on unto infinity.

Globally, this is a highly unstable situation. A comparison with 1920s Germany (when inflation destroyed the economy, destroyed the middle-class, and paved the way for Nazi-ism) illustrates the implications of massive derailment of major discourse with disenfranchised groups within societies.

To a growing extent technology displaces both manual labor and skilled jobs, including some jobs formerly thought to be "intellectual work." Increased training is needed with each new generation of technology, dislocating many formerly employable people. Manual work continues to disappear, both in the developed world and in the developing world. Rural workers flock to mega-cities, which are unable to absorb new people at an appropriate rate. As this paper is being written, the political leadership structure of one of the poorest nations in Latin America is collapsing. Political chaos of this sort is so widespread that it can no longer be thought of as a series of collapses. This cascade of political collapse is a phenomenon in its own right. During the 1920s and 1930s similar collapses for largely economic reasons created the conditions for World War II. The chaos emerging in the early years of the new century are close to triggering global chaos. International events following the September, 2001 attacks on the United States suggest that developed nations lack the will and manpower to contain spreading chaos.

The lessons of Germany in the twenties and thirties lead us to the conclusion that our hyperrealities will become derailed, and come crashing back into the crushing materiality of the poor or the environmental chaos of global climatic change (which may operate according to a step function (Coxon (2000)). In this analysis, technology is a major driver behind 9/11, the terrorist acts of the disenfranchised, or the state behavior of Western powers (which some have interpreted as state terrorism). This raises pointed and disturbing ethical questions for current technology research trajectories.

5. RECOMMENDATIONS

The systems discussed in this paper are extremely complex in terms of several key dimensions including power relations, social impact, technological programmes of development and so on. This paper calls for corrective and/or adaptive responses to AMAT, ICT, and other technological forces for societal change. The paper is iontended more to stimulate debate for ethicists and social impact research, than provide conclusive solutions. Highly complex systems resist prescriptive solutions. Such systems are better dealt with through navigatory aids, and the identification of patterns and traces (Halpin & Stapleton (2003), Suchman (1987)).

5.1 AMAT Responses to Environmental and Social Impacts

With no caps on environmental loading, consequences cannot be predicted. This insight applies to all forms of loading, not merely greenhouse gases. But capping consumption is very politically charged: Many people in developed areas believe that they live in a global economy and are thus entitled to consume what they can afford. Additionally, some argue that increased consumption increases employment. These arguments overlook environmental loading. If loading must be capped, consumption must also be capped. Some "consumption" does not increase environmental loading; example: if a TV screen is viewed by five people instead of by four, environmental loading is not affected. "Consumption" of this type is environmentally negligible. It is essentially impossible to cap aggregate consumption unless both population and per capita consumption are capped. It is ironic that environmental limits call for capped consumption at the very time that people will have more free time.

It may be that through AMAT the gradual reduction of aggregate worker time needed for routine tasks would release a large pool of available labor to deal with emerging environmental issues. Two alternative scenarios are less promising:

1. Increasing machine-based entertainment, further distracting people from what have until now been consider core dimensions of community life, thus increasing forms of alienation from historically shaped social norms.

2. Increased control of "internet and related learning" by political and or religious movements, resulting in increasing polarization of society either through social splintering or mass movements resembling group hysteria. The recent phenomenon of "flash mobs" (instant, and very short-lived, i.e., as little as one minute, gatherings of people called together by anonymous email messages) suggests that mobs are easily created. Thus far these flash mobs have been created playfully by people with access to large networks of people in major urban centers. "Mobs" have gathered for minor events such as the opening of a new water fountain. These instant (and highly transitory) crowds (one "mob" dispersed after a planned 20-second duration) have thus far not exhibited classical mob behavior, but a move from playful hysteria to fear-based hysteria is not inconceivable.

Creation of nightmare societal phenomena through new technologies has a long historical pedigree. Clearly, ethicists and social theorists are called to develop new and attractive social theories moving ahead of technological developments. AMAT will undoubtedly migrate rapidly from its present form to new forms as technologies evolve.
This suggests that human behaviors and motivations must change. Motivations can be changed. Many of today's "wants" would not exist if advertising in mass media had not "created" these wants. But is a programme of "capping" going to deal with the disconnectivities described here? Clearly, we are not in a position to answer this question.

6. CONCLUSION

This paper explores the implications of recent post-structuralist theory for engineering and technology programmes of research and development. It merges certain aspects of post-structuralism with the current debate on ethics within automation and control engineering, indicating that this debate must (re)evaluate at a very deep level the core assumptions which underlie the current research trajectories, both in terms of social and environmental impact. It is self-evident that any discussion of ethics for engineers in the 21st century goes far beyond a "code" and goes to the heart of the global engineering research programme.

The analysis reveals that, as long been the case, development of ethics trails development of technology, often at staggering cost

REFERENCES

Baudrillard, J. (1994). *The Illusion of the End*, Stanford University Press.

Carew, P. & Stapleton, L. (2004). Privacy and Intrusiveness: The Legacy of the New Wave of Information Technologies, *Proceedings of the 2004 International Conference of Information Systems Development*, Kluwer Plenum: forthcoming.

Chomsky, N. (1994). *Keeping the Rabble in Line*, AK Press.

Coxon, P. (2000). Here Dr. Coxon uncovered evidence from Irish bog extracts that climatic change in the quaternary period is governed by step functions rather than slow, continuous changes.

Douthwaite, R. (1992). *The Growth Illusion*, Lilliput Press.

Stapleton, L. (2003). Information Systems and Automation Technology as Social Spaces , in Brandt. Et. Al., *Human Centred Issues in Advanced Engineering*, Elsevier

Halpin, L. & Stapleton, L. (2003). A Theoretical Framework Based on Complexity Theory For Evaluating Large-Scale Information Systems Development (ISD) projects, in *Proceedings of the European Conference on IT Evaluation (ECITE) 2003*, Madrid.

Knights, D., Noble,, Vurdubakis & Willmott (2001). Allegories of Creative Destruction: Virtual Progress in progressing the virtual?, *Proceedings on the 19th conference on Organisational Violence and Symbolism (SCOS XIX)*, Dublin.

Stapleton, L. & C. Murphy (2002). Revisiting the Nature of Information Systems: The Urgent Need for a Crisis in IS Theoretical Discourse, *Transactions of International Information Systems*, **vol. 1**, no. 4.

Suchman, L. (1987). *Plans and Situated Actions*, MIT Press.

ELSEVIER

IFAC
PUBLICATIONS
www.elsevier.com/locate/ifac

SCIENTIFIC AND TECHNOLOGICAL ETHICS, ITS EVOLUTION IN OUR TIME: A SURVEY

Jacques G. Richardson

Decision+Communication (Consultants)
Cidex 400
91410 Authon la Plaine, France
jaq.richard@noos.fr

Abstract. Ethics has come fairly late to scientific knowledge and its applications. It emerged concurrently with wartime excesses and the advent of "big science" six decades ago and subsequent fears stemming from modern industry, including that of nuclear energy. Ethics was further propelled by growing concerns with the state of the natural environment and further exacerbated by the emergence of the life-science technologies. States and worldwide organizations have reacted to all these developments, injecting increasingly ethical safeguards as to how we make use of innovative knowledge created by scientific research. *Copyright © 2004 IFAC*

Keywords: Science, technology, risk, ethics, moral judgements, social safeguards.

1. BACKGROUND

The application of science–based knowledge—technological advance—in countries experiencing the industrial revolution has been a fairly linear movement. In the industrializing countries (the poor nations of the world) the challenge has been to close the gap by what is, for practical purposes, nonlinear advance.

The inadequacy of education, specialized training, nutrition and health care, capital, accessible markets and economic opportunity is supplemened by other factors. These include social disparity, poor distribution of wealth, and lack of a sense of competition, entrepreneurial spirit and time.

The ethical applications of science-based knowledge and its resulting technologies began to be of social concern in the second half of the 20th century. Helping resolve such issues should contribute, in turn, to improved economic and political stability at the local, national, regional and global levels. The theoretical and practical flow of ethical impacts resulting from scientific and technological advance, beginning with concept through application and consequences, may be shown as in Figure 1.

2. ETHICS ENTERS THE SCENE

In 1942 two separate but related events occurred. U.S. President Franklin Roosevelt named for the first time a national coordinator of scientific research and development, Dr. Vannevar Bush, appointed head of the Office of Scientific Research and Development (OSRD). This marked the beginning, in effect, of what became known as "big science". In Europe, the Third Reich adopted the "final solution" to the presence of unwanted minority elements in Germany's national population. The latter inspired, in turn, the protective "Nuremberg Code" of 1948, an outcome of the war-crimes trials held in the city of the same name during the months following the Second World War. The Code set standards for, among other things, various types of medical research and obtaining the consent of the human subjects involved. A highly innovative trend was under way.

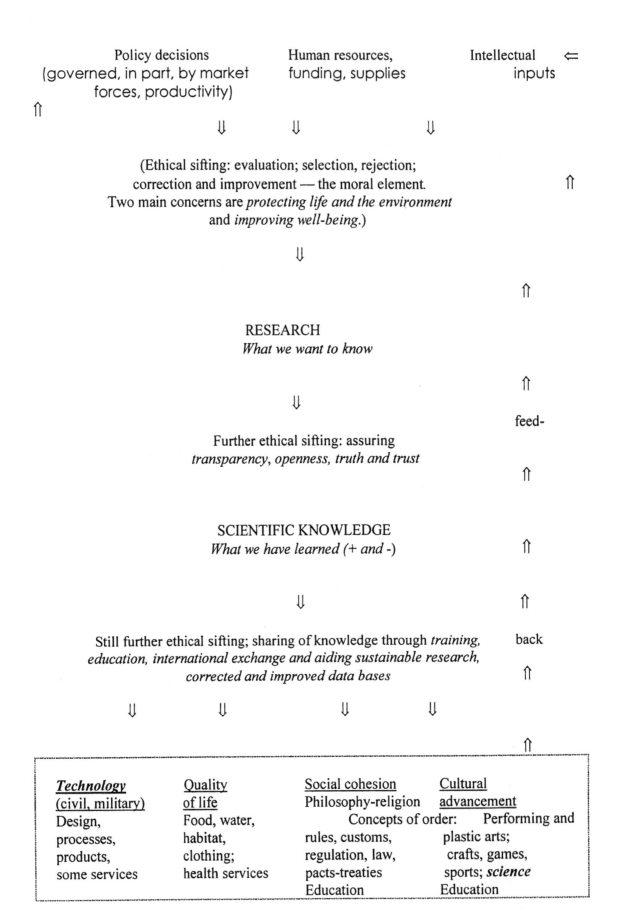

Policy decisions
(governed, in part, by market
forces, productivity)

Human resources,
funding, supplies

Intellectual
inputs ⇐

⇑

⇓   ⇓   ⇓

(Ethical sifting: evaluation; selection, rejection;
correction and improvement — the moral element.
Two main concerns are *protecting life and the environment*
and *improving well-being*.)

⇑

⇓

⇑

RESEARCH
What we want to know

⇑

⇓

feed-

Further ethical sifting: assuring
transparency, openness, truth and trust

⇑

SCIENTIFIC KNOWLEDGE
What we have learned (+ and -)

⇑

⇓

⇑

Still further ethical sifting; sharing of knowledge through *training,*
education, international exchange and aiding sustainable research,
corrected and improved data bases

back

⇑

⇓   ⇓   ⇓   ⇓

⇑

Technology (civil, military)	Quality of life	Social cohesion Philosophy-religion	Cultural advancement
Design, processes, products, some services	Food, water, habitat, clothing; health services	Concepts of order: rules, customs, regulation, law, pacts-treaties Education	Performing and plastic arts; crafts, games, sports; *science* Education

Fig. 1. Science, technology and ethics

Then an American biologist and a French engineer-physicist, Rachel Carson and Michel Batisse, first drew attention circa 1962 to the risks affecting the environment when their risk alerts concerning pollution and aridity were heard ten years before the UN's first conference on the environment (held in Stockholm, 1972). By 1980 the Congress of the United States adopted the Bays-Dole Act, setting strict humanitarian standards to be observed in life-science research funded (even in part) by Federal authorities.

During the 1980s several states (France and Denmark among them) created the first "bioethical" commissions intended to assure humane oversight of biomedical studies funded and undertaken within their national borders. The trend continued, with more specifics, and more numerics—mainly in terms of the numbers of humans involved and the levels of financing.

3. BIOENGINEERING

With the advent of biochemistry and molecular biology, the span of engineering has increasingly embraced the life sciences. Added to these, since the appearance of the transistor, are the technologies concerning the handling of information. Lest civil, mechanical, and electrical engineers think that the biological and informational technologies are "not really engineering", they need only recall that these technologies deal with structures and their interactions, how they form systems, and—as a dimension of complexity—affect other systems and subsystems as well as human involvement.

In 1990, for example, the United Kingdom decreed legislatively that embryonic research and technology were restricted to *in vitro* fertilization and to screening for genetic deyfects. The legal framework ramified further during the 1990s as Germany imposed severe limitations on some types of biotechnology, and more and more nations created national committees concerned with bioethics.

By 2004, the UK had organized its national Stem Cell Bank to provide free storage and distribution facilities of the (by now) controversial stem cells, but under specific conditions. Voters in California were to vote (November 2004), also, on the terms and conditions of publicly-funded biomedical research amounting to a total of $300 million.

4.. BIOETHICS GOES UNIVERSAL

Concomitantly with these developments, the International Bioethics Committee (IBC, 36 experts) was created in 1993 under the auspices of Unesco

and its then director-general, Spanish biochemist Federico Mayor Zaragoza. IBC's main concerns were to be those of scientific rigour, and its ethical and juridical interpretations throughout general biology, genetics and medicine.

This led, by 1998, to the founding of the Intergovernmental Bioethics Committee (also 36 experts) as a specialized ethical forum representing governmental positions. This official committee's work, which translates through diplomatic negotiations, into international agreements, was solidified in 1999 by a resolution adopted by the United Nations.

There came into existence, also in 1998, the International Commission on the Ethics of Scientific Knowledge and Technology (COMEST).
Headed since 2002 by the Norwegian mathematician Jens Fenstad, COMEST is responsible for
- facilitating technical dialogue between scientists and decision makers,
- providing special counsel to decision makers
- giving early notification to governments of risks in evolving scientific and technological endeavours.

The first areas of such undertakings include
- fresh-water supply and distribution
- energy
- new information and communication technologies, and the
- extra-atmospheric research and the manned/unmanned exploration of space. (Noteworthy in this respect is the work of Dr. Alain Pompidou, a physician, a study published in 2000 under the auspices of the European Space Agency.[1])

In 1997, largely at the behest of the aforementioned IBC, there was proclaimed the International Declaration on the Human Genome and Human Rights, adopted formally in 1998 by the UN General Assembly. This was followed in 1999 by the adoption of the International Declaration on Human Genetic Data. Both instruments were intended to protect, to the maximum possible, the rights of the individual as knowledge expanded on the human genome and its implications for both good health and disease.

5. CONCLUSIONS, CONSEQUENCES

The developments recounted above are, thus far, limited to the setting of norms and standards. They have few "teeth", indeed, beyond the laws and regulations that have been set at the level of the individual nation. The way to the future in the applications of ethics to our scientific knowledge and how we use it seems to be set, however and, with time, will become directly pertinent to how we conceive, plan, execute and evaluate new and and old technologies in daily life.

It is indicative, furthermore, that in 2002 the UN Educational, Scientific and Cultural Organization launched a novel programme called Ethics and Economics, one destined to bring, in time, a more humanistic order than was possible before to the worlds of applied economics: industry and commerce. Will applied science and the public-related technologies stay ahead?

> *... the fate of all complex adapting systems in the biosphere—from single cells to economies—*
> *is to evolve to a state between order and chaos, a grand compromise between structure and surprise.*
> — Stuart Kauffman, *At Home in the Universe, The Search*
> *for Laws of Self-Organization and Complexity,*
> New York, Oxford University Press, 1995

---

[1] Alain Pompidou, *The Ethics of Space Policy*, Paris, Unesco and the European Space Agency, 2000, An unusual perspective on the professional behaviour relating to astrophysical research and technology.

MANAGEMENT OF A MULTI-AGENT DEMAND & SUPPLY NETWORK

Agostino Villa and Irene Cassarino

*Dipartimento di Sistemi di Produzione ed Economia dell'Azienda
Politecnico di Torino, c.so Duca degli Abruzzi, 24, 10129 - Torino (Italy)
Phone: +39-011-564.7233; fax: -564.7299; e-mail: agostino.villa@polito.it*

Abstract: A Demand & Supply Network is a multi-agent "virtual enterprise" resulting from a temporary agreement among a number of firms, which decide to cooperate together into a given value chain for a limited time horizon. The network organization consist of a chain of production-and-service stages, each one being composed by either a firm or a set of firms, each one with proper individual management. The crucial point in the development of said network is "to assure collaboration among partners" and to organize the interactions among firms, accordingly. The organization of a *CO*llaborative *DE*mand & *S*upply *NET*work (*CO-DESNET[1]*) requires to apply robust criteria for designing collaborative management strategies and for analysing how collaboration among the component firms could be efficiently assured. The paper will present a new model of a multi-agent network of enterprises, specifically developed for an accurate description of the inter-firm interactions. Based on this model, conditions verifying how collaborative operations will occur, are derived. *Copyright © 2004 IFAC*

Keywords: Multi-agent networks; Supply Chains; Negotiation; Cooperation.

1. INTRODUCTION

A *Demand & Supply Network* is an "extended virtual enterprise", that means a temporary network of several firms which decide to cooperate together in a common given value chain for a limited time horizon. In practice, it consists of a set of different enterprises, able to producing different parts which could be utilized in a common family of final products, and to applying complementary production programs, planned together, for a common industrial goal. Several examples of these networks, structured mainly in the form of "supply chains" and operating in different industrial business lines, can be found ranging from automotive to furnitures as well as to paper production and other business lines (Tayur et al., 1999).

Recently, a new evolution of the supply chain organization has been characterized by "co-operation agreements" filled out by enterprises interested to cooperating together but for a finite time period, and in such a

way to involve only a part of their own core business. So, the resulting new supply chain can have a finite life and it does not completely reduce the autonomy of any component firm, because each one can still produce items for proper clients, and then operate in a proper market segment (Villa, 2001). More precisely, all the enterprises which agree to be included into a *Demand & Supply Network* and then be active inside the same supply chain, must sign an agreement to co-operating together in defining common production plans for specific products; obviously, they could also maintain their independence and autonomy for any other production.

In formal terms, a *Demand & Supply Network* can be viewed as a *virtually connected chain of service stages*, each one containing either a firm or a set of parallel firms, each firm with its proper autonomous decision-maker, denoted "*agent*" (Huang and Nof, 2000), who aims to cooperate but also wants to obtain the best profits for his own enterprise. Each service stage is connected to the upstream stages and to the down-

---
[1] *CO-DESNET* is the acronym of the Coordination Action (CA) project n° IST-2002-506673 / Joint Call IST-NMP-1, supported by the European Commission, Information Society Directorate-General, Communication Networks, Security and software, Applications, under the coordination of Politecnico di Torino, Prof. A. Villa, and with EC Official Dr. F. Frederix. This paper refers to research results developed for preparing the CA proposal at Politecnico di Torino.

stream ones through a virtual market place: it means that each firm can negotiate contracts for producing goods with downstream (buyer) firms as well as contracts for acquiring materials with upstream (supplier) firms. This *negotiation opportunity* is a qualifying character of a *Demand & Supply Network*. Since each firm aims to gain its own best income, it utilizes the *Demand & Supply Network* to which it belongs as a frame within which a "good negotiation" can be performed. Here "good negotiation" means that an agreement between each pair of "*consecutive agents*" (belonging to two consecutive stages of the chain) can be found such as to satisfy both agents, because both aim at being cooperative but, at the same time, want to make profits: then the desired agreement should assure a sufficient income to both of them.

This is the real problem now facing supply chain designers and organizers: under which conditions does an individual enterprise find convenient to sign a temporary agreement for becoming a partner within a multi-agent *Demand & Supply Network*? Industrial experience suggests that this is a multi-faced problem depending on economic, technological and managerial considerations. This paper will approach the problem according to the management point of view. And the goal is to develop a model of the multi-agent management problem in a *Demand & Supply Network* such as to analyse how cooperation of partners could be enforced.

The paper contents are as follows.
Section 2 introduces a model of the problem of managing interactions among the component firms of a *Demand & Supply Network*, by presenting a proper mathematical formulation in terms of a negotiation task, stated in therms of a multi-agent optimisation problem. Section 3 will discuss the above mentioned negotiation problem, in order to deriving conditions which can motivate cooperation of local agents together. Finally, the Section 4 will summarize conditions assuring a collaborative management of local autonomies into a format of practical impact.

2. MODELING THE DEMAND & SUPPLY NETWORK MANAGEMENT

The proposed model of a Demand & Supply Network is based on the idea that said virtual industrial system consist of an "open commercial system" within which two component firms interacts together by exchanging material and financial resources.
For each material resource (i.e., parts of a common final family of products), a proper market place exists on which this resource is negotiated through a proper monetary value, between the supplier (i.e., the resource producer) and a client (i.e. the purchaser who will utilize the resource in its own production sequence).
In the *Demand & Supply Network*, considering the complete production cycle of a final product, each buyer will purchase at least a resource to apply his own manu-

facturing operations in order to transform it into a new item with more added value, to be sold in a downstream market place. So, the principal functions of any component firm are: purchasing, transforming (through either manufacturing or servicing), and sale of items.

In a *Collaborative Demand & Supply Network*, this set of dynamic commercial interactions occur into a coordinated protected industrial system. Partners inteed are connected together through collaboration agreements, which reflect into effectiveness of the comercial negotiations. These greements should give each component firm a sufficient assurance of economic survival, by suitable management of usual "hard" business relations between large-scale and small-scale firms.

The following model has the main scope of allowing an easy but correct evaluation of costs and advantages of the above sketched regulated industrial system. The main goal of the model is to give a description of the costs of each partner, so that they could be measured, depending on some technical and financial parameters describing the partner itself and the connected market places.

To this aim, a *Collaborative Demand & Supply Network* composed by 4 enterprises, belonging to 2 different statges, is considered, together with a final client. Two market places, one internal and one between producers and the final client, define the interactions among the agents.

At the material input stage, two firm (denoted by index $n=1,2$) produce parts to be sold to the two firm beloning to the following stage (denoted by index $n=3,4$). These last ones will apply final operations to transform parts into products, to be sold to the final client, whose exogenous demand is denoted by A_t.

Each agent n (i.e. the management of a component enterprise) is modelled by a production optimisation problem which variables to be optimised are production volumes and price. This model describes the enterprise maximization of its own profit (i.e. the difference between sell return and production costs) under the constraints describing the inventory dynamics and the production capacity saturation.

The notations adopted for the first stage component firms $n=1,2$ are the following (all referred to time period t):

$P_{n,t}$ price to sell items from supplier n to the second-stage buyers at the internal market place;

$X_{n,t}$ volume sold from supplier n;

$Y_{n,t}$ volume produced by supplier n;

$I_{n,t,F}$ output storage of finished parts at supplier n;

c_n storage unitary cost;

r_n production unitary cost;

C_n production capacity of supplier n.

The resulting model for the first stage component firms $n=1,2$ is as follows:

$$\min_{X,P,Y} J_n = \sum_t (-P_{n,t} X_{n,t} + c_n I_{n,t,F} + r_n Y_{n,t}) \qquad (1)$$

$$I_{n,t,F} = I_{n,t-1,F} + Y_{n,t} - X_{n,t} \qquad (2)$$

$$Y_{n,t} \leq C_n \qquad (3)$$

$$X_{n,t}, I_{n,t,F}, Y_{n,t}, P_{n,t} \geq 0 \qquad (4)$$

Referring to the second stage component firms, $n=3,4$ (denoted as "buyers"), the adopted notations and related models are as follows:

$Q_{n,t}$ price to purchase items by buyer n from the first-stage suppliers at the internal market place;

$Z_{n,t}$ volume purchased by buyer n;

$D_{n,t}$ volume sold by buyer n to the downstream final client;

A_t external demand from final client;

$W_{n,t}$ volume produced by buyer n;

$I_{n,t,P}$ output storage of final products produced by buyer n;

$I_{n,t,M}$ input storage of parts to be used by buyer n;

S_t unitary price for selling final products during period t.;

$c_{n,P}$ unitary cost of final products storage;

$c_{n,M}$ unitary cost of input storage of parts.

$$\min_{Q,Z,W,D} J_n = \sum_t (Q_{n,t} Z_{n,t} + r_n W_{n,t} + c_{n,M} I_{n,t,M} + \\ + c_{n,P} I_{n,t,P} - S_t D_{n,t}) \qquad (5)$$

$$I_{n,t,M} = I_{n,t-1,M} + Z_{n,t} - W_{n,t} \qquad (6)$$

$$I_{n,t,P} = I_{n,t-1,P} + W_{n,t} - D_{n,t} \qquad (7)$$

$$W_{n,t} \leq C_n \qquad (8)$$

$$Z_{n,t}, I_{n,t,P}, I_{n,t,M}, W_{n,t}, S_t \geq 0 \qquad (9)$$

The four enterprises are assumed to belong to the same *Collaborative Demand & Supply Network*. That means that some conditions have to regulate interactions among partner firms, namely:

I. Conditions stating that production volumes must be balanced with internal and final demands:

    I-a. All parts produced by firms in the upstream stage have to be purchased by firms of the second stage:

$$\sum_{n=1,2} X_{n,t} = \sum_{n=3,4} Z_{n,t} \qquad (10)$$

    I-b. The final client demand has to be fully satisfied:

$$\sum_{n=3,4} D_{n,t} - A_t = 0 \qquad (11)$$

II. Conditions stating that prices of parts in the internal market place have to be balanced:

$$P_{1,t} - P_{2,t} = 0 \qquad (12)$$

$$P_{1,t} - Q_{4,t} = 0 \qquad (13)$$

$$Q_{3,t} - Q_{4,t} = 0 \qquad (14)$$

In addition, the interactions have to satisfy some co-operation conditions, which regulate the operations within the network, namely

III. Condition stating that the production volumes of the firms belonging to a same stage have to be balanced according to the respective efficiency:

$$\frac{1}{C_1} (\sum_t X_{1,t})^2 - \frac{1}{C_2} (\sum_t X_{2,t})^2 = 0 \qquad (15)$$

IV. Condition stating that the volumes of final products sold by the firms belonging to the final stage, to the client, have to be balanced according to the respective efficiency of the firms themselves:

$$\frac{1}{C_3} (\sum_t D_{3,t})^2 - \frac{1}{C_4} (\sum_t D_{4,t})^2 = 0 \qquad (16)$$

Note that these last two conditions state the effective collaborative rules for the enterprises belonging to a common *Demand & Supply Network*. In practice, the agreement to be a network partner states that each firm agrees in producing, for the network needs, by using its own production capacity at a rate which must be balanced with the capacities of the other network partners: no enterprise will receive a demand for products "unbalanced" with respect to the others.

In addition, conditions (12) to (14) will impose an equilibrium on the network internal prices.

The global management problem results from the whole set of conditions (1) to (16) above stated. It consists of a large-scale non-linear otimization problem: existence of an optimal solution can be proven according to standard optimisation theory results (Brandimarte & Villa, 1995).

By applying Lagrangian relaxation, the complete optimisation problem can be splitted into four optimization sub-problems, all interrelated together, each one linked to a component firm n, namely:

- for the firm $n=1$

$$\min_{X,P,Y} \Pi_1 = \sum_t (-P_{1,t} X_{n,t} + c_1 I_{1,t,F} + r_1 Y_{1,t} + \\ + \alpha_t X_{1,t} + \gamma_t \frac{1}{C_1} \left[\sum_t X_{1,t} \right]^2 + (\upsilon_t - \delta_t) P_{1,t}) \qquad (17)$$

- for the firm $n=2$

13

$$\min_{X,P,Y} \Pi_2 = \sum_t (-P_{2,t} X_{2,t} + c_2 I_{2,t,F} + r_2 Y_{2,t} +$$

$$+ \alpha_t X_{2,t} - \gamma_t \frac{1}{C_2}\left[\sum_t X_{2,t}\right]^2 - \delta_t P_{2,t}) \qquad (18)$$

- for the firm $n=3$

$$\min_{Q,Z,W,D} \Pi_3 = \sum_t (Q_{3,t} Z_{3,t} + r_3 W_{3,t} + c_{3,M} I_{3,t,M} +$$

$$+ c_{3,P} I_{3,t,P} - S_t D_{3,t} - \alpha_t Z_{3,t} \qquad (19)$$

$$- \eta_t \frac{1}{C_3}\left[\sum_t D_{3,t}\right]^2 + \beta_t Q_{3,t} + \omega_t D_{3,t})$$

- for the firm $n=4$

$$\min_{Q,Z,W,D} \Pi_4 = \sum_t (Q_{4,t} Z_{4,t} + r_4 W_{4,t} + c_{4,M} I_{4,t,M} +$$

$$+ c_{4,P} I_{4,t,P} - S_t D_{4,t} - \alpha_t Z_{4,t}$$

$$+ \eta_t \frac{1}{C_4}\left[\sum_t D_{4,t}\right]^2 + (\delta_t - \beta_t) Q_{4,t} + \qquad (20)$$

$$+ \omega_t D_{4,t})$$

In above stated formulations, the newly introduced variables (in Greek letters) denote the Lagrangian variables, by which the constraints have been included into the augmented global functional to be minimized.
All avove stated functional terms have to be maximized with respect to the Lagrangian variables: *this maximization will allow to obtain the co-ordination conditions.*

3. CONDITIONS TO ASSURING COOPERATION AMONG PARTNER FIRMS

Some conditions which can assure collaborative operations of all firms belonging to the *Demand & Supply Network*, can be obtained by analysing the Lagrangian variables, introduced in the relaxed formulation. The motivation of this analysis approach is that said Lagrangian variables denote co-ordination costs.

Let us first refer to variable α_t, namely to the co-ordination cost concerning the volumes exchanged between the two suppliers and the two buyers ("vertical co-ordination"), associated to constraint (10). In case the offered volume is greater than the requested demand, said cost is greater than zero. Then, from (17) and (18), both first-stage producers are pushed to reduce their throughput. For the second-stage buyers, from (19)-(20), the effet is reverse.

The Lagrangian variable γ_t denotes the co-ordination cost concerning volumes of production to be shared between the two first-stage suppliers ("horizontal co-ordination"), related to condition (15). The squared valued functions in (17) and (18) force a split of production volumes respectively depending on the rate between the production throughput $X_{n,t}$ and the firm capacity utilization, estimated by $\frac{1}{U_{n,t}} = \frac{X_{n,t}}{C_n}$, for $n=1,2$. The cost γ_t forces to minimize unbalances between concurrent producers.

The companion Lagragian variable η_t operates in a similar way as γ_t, on the two firms belonging to the second stage, as in constraint (16).

With reference to the Lagrangian variable υ_t, the "horizonal co-ordination" cost concerning a potential difference between prices offered by the concurrent first-stage suppliers is taken into account, as referred to constraint (12). In case said cost will increase, the first supplier ($n=1$) is compelled to pay an adjoint quantity of production for each unit sold by the second supplier ($n=2$), as shown in (17) and (18).

A similar effect is due to the second-stage "horizontal co-ordination" cost β_t, as in (19)-(20), now associated to constraint (14).

Note that this effect is a demonstration of how a *Collaborative Demand & Supply Network* should operate in order to regulate the financial transactions between concurrent producers: prices must be maintained as equal as possible otherwise a "dumping" attempt will be paid in terms of reduction of the planned throughput.

Again referring to price management within the network, the variable δ_t is referred to the "vertical co-ordination" between the price proposed by upstream producers ($n=1,2$) and the price which downstream buyers ($n=3,4$) could agree to pay, i.e. associated to constraint (13). As soon as δ_t reduces, buyer 3 should pay an additional cost per part to supplier 1.

Note that the price-coordinating costs are strictly interrelated together. Any price variation between two firm of the same stage should immediately reflect on price unbalances towards the other stage and between the firms beloning to the other stege, as well. This effect is evidenced by the Lagrangian variables associated to the "triangular conditions" (12-(14).

ω_t is the co-ordination cost associated to the equilibrium requirement between the volumes offered by downstream firms ($n=3,4$) and the demands of the final client in (11), the real exogenous input of the whole *Demand & Supply Network*. The sensitivity of the network management to ω_t can be interpreted as a cost paid by final products' producers to the final client if the volumes offered by themselves are greater than the demands (i.e., a discount to push sales).

4. SOME SUGGESTIONS FOR A STABLE NETWORK MANAGEMENT

The above sketched analysis of the effects of Lagrangian variables in solving the network management oprimization problem allows to make evidence of the most characterizing behaviour of the whole set of network-component agents.

First, the stability of the *Demand & Supply Newtork* operations can be assured, since:
a. each endogenous perturbation (i.e. due to a variation of some internal variables, as either produced volumes or prices) is counteracted by the Lagrangian variables (thus justifying their denomination as "co-ordination costs"), which force the set of interactions to come back to equilibrium;
b. each perturbation generates new additional costs to the whole set of component enterprises, thus forcing feedback effects.

As examples of these preliminary considerations, the reaction against perturbation of a production throughput and of an internal price are outlined.

1st case: Control of production volumes' perturbations.
Assume that the production volume $X_{1,t}$ increases at a certin time t. As a consequence, the "vertical co-ordination" cost variable α_t also increases, thus forcing an increase of the volumes to be purchased by second-stage buyers ($Z_{n,t}, t = 3,4$). But the other effect of the same co-ordination cost variable α_t is to force a reduction of the first-stage production rates ($X_{n,t}, t = 1,2$), as in (17) and (18). In addition, the Lagrangian variable γ_t denoting the co-ordination cost for the volumes of production to be shared between the two first-stage suppliers, also increases and the same should happen for the production rate $X_{2,t}$; this last evolution should counteract the dynamics driven by α_t.

In practice, any variation of the production throughput of a component firm should be paid by the same firm proportionally to its own production efficiency.

2nd case: Control of prices' perturbations.
On the other hand, assume that a variation of the price $P_{1,t}$ versus the other prices $P_{n,t}, n = 2,3,4$, can occur, in terms of an increase. In this case, the Lagrangian variable υ_t, named "horizonal co-ordination" cost concerning a potential difference between prices offered by the concurrent first-stage suppliers, should also increase, thus forcing the concurrent price $P_{2,t}$ to increase as well. But this effect will compel the original price $P_{1,t}$ to decrease again, towards an equilibrium condition. An additional effect is a reduction of δ_t, which has been referred to as the "vertical co-ordination" between the price proposed by upstream producers ($P_{n,t}, n = 1,2$) and the price which downstream buyers ($Q_{n,t}, n = 1,2$) could agree to pay. This forces $Q_{4,t}$ to increase, also driving $Q_{3,t}$ towards a

greater value, owing to the evolution of β_t. As a consequence, all the four prices should settle at a new equilibrium value, together.

Second, the effectiveness of a *Demand & Supply Network*, in terms of convenience of a firm to become a partner, can also be verified. To this aim, two complementary actions of industrial politics have to be taken into account:
a. the direct interest of each component firm to have, at its own output, a stable market place, i.e. a market which dynamics can be a priori forecast, at least for the major rate; this is what happens when the network operations can be maintained at equilibrium values, in terms of both volumes and prices;
b. the common interest of the whole network to have an increased strength for operating in the real market of the final products' families; this is what can be obtained by a large network of enterprises, even if each one is a SME.

Both above sketched conditions are clrealy verified by the network model above stated: the former, by noting that a *Demand & Supply Network* modeled by (1) to (16) should impose to each component firm agreement conditions which will assure co-operation; the latter, by noting that said model suggests to assure the exogenous demand satisfaction with the best possible internal equilibrium of price and volumes, the last ones assigned to partners according to their respective capacities. Then, a balanced income will be assured to all component firms.

5. FINAL CONSIDERATIONS

The performance evaluation of the *Demand & Supply Network* above presented have to be completed by better specifying profitability conditions. In practice, a potential manager of a network of this type should be able to prove that the required investment for the network organization should be so convenient as to promote a real collaboration among partners (Bullinger et al., 2002; and Supply Chain Council, 2001).
To this aim, some preliminary considerations will be outlined in the following, all motivated by the proposed *Demand & Supply Network* model (even if not completely proved, till now).
It should be approached, inteed, the problem of validating the economic convenience of an investment devoted to promoting a *Demand & Supply Network*, by investigating if there is evident interest to increase as much as possible its dimension, as it could be useful from the structural point of view of acquiring even larger production capacity and wider market coverage.
A practical answer is that complexity-induced costs cannot overcome advantages offered by the "protected market", advantages consisting of the reduced global transaction costs. Once the most relevant transactions costs are recognized as those due to marketing information procurement and to the partner selection, then enterprises can be convinced of the convenience to

stipulate multi-agent contracts, by specifying the contribution of each partner and the respective rate of the global income. This is the first suggestions that the presented network model gives in terms of "equilibium-assuring management".

A further problem concerns how to evaluate the investment for the development and innovation of a *Demand & Supply Network*. Several classic criteria have been introduced to perform an analysis of the return of investment, generally based on the balance between costs and return flows (Rossetto, 1999). Recently, a method based on the concept of "real options" (i.e. by noting that, when an investment is decided, the decision-maker still has the opportunity to choose at which time to invest as well as in which type of realization) became more and more interesting. Referring to a *Demand & Supply Network*, this method seems to be even interesting, because it allows a supply chain organizer (e.g., a coordinator of the whole set of individual managers) to analyse alternatives: if to enlarge the chain, or to improve the connections among agents, or to promote specific technologies at some partner. Obviously, a robust estimation of the return of investment, at any option time and for any potential option, is mandatory. The model proposed in the previous Sections can be used as a "planner" of the network operations over a future time horizon: thus, it can be applied as an "obserber" of the network evolution in front of modification hypotheses, and an "estimator" of the network innovations' convenience.

Since an investment in supply network innovation is usually very expensive, the "real options" approach calls for industrial policies which could give to managers sufficient confidence about the future of their investments (Williamson, 1975). In Italy, a recent government measure, oriented according to this idea, is the so-called "Dual Income Tax (DIT)", which has been introduced in order to promote re-capitalization of enterprises. In practice, DIT assumes that an investment will have a fixed return C_o, to which a reduced tax rate is attributed. Then, the effective tax rate for an investment of amount I will result from a fixed rate (due to the assumed return C_o), to which a reduced tax rate t_1 is applied, and a residual rate depending on the real value of the return R, accounting for the usual tax rate t_2 (Rossetto & Villa, 2003):

$$C_o \times I \times t_1 + (R - C_o \times I) \times t_2 \qquad (21)$$

Intuitively, this measure operates as an investment multiplier, and can greatly promote decisions dedicated to industrial development.

Reports of industrial applications of the ideas above reported will be presented during the development of the cited EC-funded Coordination Action project *CODESNET*.

Acknowledgements: This research has been partially supported by the Italian Ministry of Education, University and Research, within a national-interest research programme PRIN.

REFERENCES

Brandimarte, P., and Villa, A. (1995). *Advanced Models for Manufacturing Systems* Management, CRC Press, Boca Raton.

Bullinger, H-J., Kuhner, M., and Van Hoof, A. (2002). Analyzing supply chain performance using a balanced measurement method, *Int. Journal of Production Research*, **40**, 3533-3543.

Huang, C-Y., and Nof, S. Y. (2000). Autonomy and viability – measures for agent-based manufacturing systems, *Int. Journal of Production Research*, **33**, 4129-4148.

Rossetto, S. (1999), *Industrial Organization Manual*, UTET, Torino.

Rossetto, S., and Villa, A. (2003). Evaluating capital investment for worldwide supply chain organization, in Proc. IFAC Int. Workshop on Intelligent Assembly and Disassembly, T. Borangiu (ed), Bucharest, pp. 215-220.

Supply Chain Council (2001). *Supply Chain Operations Reference Model*, see the web site: http://www.supply-chain.org.

Tayur, S., Caneshan, R., Magazine, M. (eds) (1999). *Quantitative Models for Supply Chain Management*, Kluwer Academic, Boston.

Villa, A. (2001). "Introducing some supply chain management problems", *Int. J. Production Economics*, **73**, 1-4.

Villa, A. (2002). "Emerging trends in large-scale supply chain management", *Int. Journal of Production Research*, **40**, 3487-3498.

Williamson, O. (1975). Markets and Hiierarchies: Analysis and Anti-trust Implications, Free Press, New York.

ELSEVIER

IFAC
PUBLICATIONS
www.elsevier.com/locate/ifac

TOWARDS MUTUAL UNDERSTANDING IN GLOBAL MORAL CONFLICTS

Christina Rose

Postgraduate Research Programme on Global Challenges
University of Tuebingen
e-mail: christina.rose@uni-tuebingen.de

Abstract: Today, global challenges can be realized every day, and they appear to be relevant to almost all spheres of social and public affairs. One of them is social responsibility, e.g. in Control Engineering, as we are discussing it during this conference. Generally considered, in most cases difficulties root in interfering moral attitudes of two or more different groups, be it religions, cultures, or nations. Thus, the main issue of Ethics has become to establish means which improve mutual understanding through communication and which, ideally, help find moral consensus. Only then, these inter-cultural moral conflicts may be sorted out lastingly. Based upon the concept of Discourse Ethics, this presentation will introduce and discuss two examples of such moral discourse. Within the Community of Practice on 'Ethics in Economy' (CoP), members of the German-wide Service Network for Training and Continuous Education SENEKA come together to discuss ways of defining and implementing common moral standards within different kinds of organizations. Finding such solutions becomes increasingly difficult, the bigger the organization is and the more inter-culturally it works. Thus, it is essential to put emphasis on *inter-cultural understanding*, especially in global discourses. As an example, the *2nd Parliament of World's Religions* (1993) discussed and agreed upon the *Declaration toward a Global Ethic*, an inter-religious consensus on moral core commandments. These two examples may help to improve mutual understanding in discussing global responsibilities of entrepreneurship and engineering. *Copyright © 2004 IFAC*

Keywords: Ethics, responsibility, culture, religion, understanding, discourse, consensus, entrepreneurship, engineering.

1. INTRODUCTION: GLOBALISATION AND RESPONSIBILITY

As we are discussing emerging problems related to control and automation in the next triennium and their impact on technological and economical development of society, one of these challenges definitely is the need for taking over *social responsibility*. On the one side, the so-called 'globalisation' offers advantages for science and society, like joint research programmes, new markets, deregulation in law and politics etc.. On the other hand, disadvantages of global innovations even threaten lives, sometimes those of whole nations. Thus, the world is growing together not only because of new *possibilities*, be it in technology, communication, economy, or transport, but also because of newly realized *threats upon*

humankind. Looking at natural resources for example, it has become visible that not only partial, short-term win is essential, but also sustainability. Awareness of such global challenges has established itself during the last decades, and meanwhile, it is a widely *felt*, not only preached, responsibility to save life on earth in general. But still, questions remain *which* consequences *exactly* are to be taken. The need for solving problems such as hunger or poverty is obvious, but points of view differ on *how* to do it. Discussions on catchwords such as 'justice of distribution' (mainly concerning common goods such as water and oil) are highly controversial. Taking global social responsibility, thus, is not as simple as it may first seem. It requires the ability to decide *which* acting is to the benefit of humankind. The question behind this is the question of *ethical maxims or principles*: How, by which means, can such a process of moral decision-finding lead to results which can be universally accepted?

Now, *'ethics'* is a term which has become very popular and which is used quite frequently, but unfortunately it is very often confused with *'morals'* or with the Greek term *'ethos'*. Especially in the context of globalisation, a distinction between these three is important. It is therefore necessary to make clear the meanings of each of the terms.

2.   'MORALS', 'ETHOS', AND 'ETHICS'

'Morals' or the *'moral codex'* is a collection of all material norm and value ideas about a 'good and right behaviour' which are accepted within one community, as well as all those convictions which make these norms and ideas seem plausible, which are used to rectify them, or which lead to their modification. 'Morals' thus is the sum of customs, moral ideals or accepted moral rules of a social entity. They may have the character of moral norms, value judgements, institutions, as well as other kinds of traditional or conventional order and meaning which influence the behaviour within one group. Morals regulate the fulfilling of all needs within one community and describe the duties of its members. These value-judgements, however, may change when the group's image of itself changes. Contents of rules and regulations also differ from one community to the other. But it is constitutional for 'morals' that they claim *universal liability* within the community they belong to.

The Greek term *'Ethos'* (pl.: *'ethea'*) generally describes those aspects of a moral codex which an individual or a group of individuals have accepted as their *own, personal 'guidelines'* and see as their personal or *collective profile*. The ancient Greeks had two different spellings of 'ethos' in order to make a distinction between the two possible motives out of which it is accepted by a person or a group. Spelled with epsilon, 'ethos' indicates that an individual or a group of individuals have simply taken over those habits, customs and ways of life which were taught to them in their childhood. The term thus describes the junction between human and 'morals' rather as an incidentally acquired set of rules. Spelled with eta, though, 'éthos' refers to the individual person's conscience, virtues, and traits of character. 'Éthos' leads a person on not only to follow out of habit, the standards of a community but because of the personal *decision* to act in a way which he or she judges to be *'morally valuable behaviour'*.

The contents of an 'ethos' are clearly defined value orientations. They include a complete collection of attitudes, convictions and norms which are linked together as one coherent, within itself well-structured pattern. It is used by the individual or group as the instance for orientation concerning decision-finding for their own actions. The possible contents of an 'ethos' show a huge range of characteristics, as they may include attitudes (with special implicated value judgements and convictions), favoured ways of living, normative dispositions ('virtues'), role-giving with related sanctions, rules and norms with differing degrees of universality and liability, traditions, pure codices of behaviour, or simply modern trends. But every 'ethos' includes a minimum of inner structure concerning role-taking and -giving. This mainly is the starting point for the development of a 'corporate ethos', e.g. an ethos of all people from the same occupation (e.g. medics, engineers, etc.), social status (e.g. royality, workers, tramps, etc.), or within the same role in society (e.g. politicians, academics, members of an NGO, etc.). Such an 'ethos' is the overall pattern of moral attitudes and norms on which people of this occupation, status, or group orientate their actions at work and within society. In the same way as 'morals', 'ethos' is less often and less seriously modified, the more it belongs to a 'closed community', within which it can offer strong orientation and high moral safety standards. The more 'open' the community is, the less it becomes able to offer orientation, but the greater is the degree of individual or shared freedom within the 'ethos'.

'Ethics' now, as one special discipline within philosophy, is concerned with both, 'morals' and 'ethos', and is, thus, often also called 'Philosophy of Morals'. Its subject is the whole area of human acting, including all conditions concerning the members of a community. Its task is to theoretically and critically reflect the lived morals, the practically existing and valid moral convictions, and the lived ethea. 'Ethics' thus reflects and comments, by philosophical means, whether one specific moral statement (e.g. "reproductive cloning is immoral, because it neglects human dignity") can be generalized, is comprehensible and sound, and whether it is compatible with other existing morals or further convictions and reasons of judgement (e.g. religious or scientific ones). The judgements about such statements are normative, i.e., they claim to lead people's acting. The *application* of such statements which are judged positively is thus to be supported by ethics.

Today, however, it is no more sufficient to consider what is best for individuals or one particular community, but what is right and just for *humankind*. This need for *universality* of moral decisions has already been made out by Immanuel Kant very clearly. (see Rose, 2004:166-167) The pluralism of many different, partly contradicting concepts of 'good living', thus, has led to the distinction between *evaluative* questions of 'good life', and *normative* questions of 'morally right' actions in ethics. While the first one is mainly concerned with discussing *what we have got* within one community, the latter is concerned with the question of *what should be done*, i.e.: how to deal with conflicts of values and moral interests and how to sort them out peacefully and justly. The focus of these normative ethics of 'morally right behaviour' then is to generate, check and ground its normative statements. If such norms, maxims, or principles are grounded, they can claim universality and are guaranteed priority compared to particularistic concepts of a 'good life'.

The question from above was how to find moral decisions which can be accepted not only by one community, but by everybody to the benefit of humankind. It has led us to reconsider the term 'ethics'. On the background of these considerations, the meaning of 'ethics' in the context of this presentation can be described as the following two philosophical tasks: Firstly, ethics should lead to reconsidering carefully specific ethea, regarding questions such as: what its moral judgements are based upon, how they were made, and whether they might be universally accepted or not. This first step might be called 'irritation' or 'deconstruction' of a traditional, particularistic moral attitude. Some approaches within ethics regard this as the only task of the discipline. But in my eyes, a second step has to follow: Secondly, after establishing a kind of sensitivity for how relative and potentially conflict-bearing the own ethical point of view may be compared with others, ethics must provide, ground and make applicable new ways to sort out tradition-based moral conflicts. Very often, such suggestions have the character of an ethical 'maxim', or 'principle for acting'. This second step we may call 'dealing with the irritation', or 'construction'. 'Construction' here can of course *not* mean a *reconstruction* of the old, overcome traditions, but must aim at a new, global concept built up *together*.

It is obvious that these tasks cannot be solved by philosophers alone. Ethea are too complex and too much based upon human inter-relation that they could be evaluated, judged and improved by just looking at it from an outside point of view. On the contrary, the two steps of 'deconstruction' and 'construction' must be made by everybody him- or herself, as an individual and as a social group. Let us consider the example of ethics in medical engineering: Things are far too complex to allow a well-informed moral point of view, if experts from medicine, engineering, and people concerned (e.g. people who might be either cured or harmed by the new technologies) are not integrated into the discussion process. The role of a philosopher or of a philosophical ethical concept in this process is not to give advice in *what* to decide upon, but *how* to decide about it. This includes improving people's ability to ask themselves and others *questions* about

'why', 'what and whom for', etc. – and to answer them plausibly.

3. DISCOURSE ETHICS

Such discussions, of course, require the willingness of individuals or groups to reconsider their own ethea, to compare them to other concepts, and to discuss them with other individuals or groups. But as we have seen, the process of 'globally growing together' causes threats and challenges which cannot be ignored, if a peaceful life on earth should not be endangered gravely. Responsibility-taking by getting into dialogue with each other, thus, is a bare necessity of survival, and more and more people and groups are taking up this challenge and joining in. Besides politics, society has become an important actor in successful global interacting and building the 'global community'. Many examples (e.g. the multitude of NGOs, or the implementation of 'ethics committees') show this recent development as reaction to global ethical issues. With the need for communication and cooperation, the actual commitment is growing, too. This is reason enough to consider 'ethics of communication' a promising concept for sorting out tradition-based moral conflicts and to improve global co-operation. Such concept has, e.g., been introduced by Juergen Habermas and Karl-Otto Apel, both Frankfurt a. M., Germany, in the 70s of the last century and has been internationally discussed and developed further ever since.

Their approach has become widely known as 'discourse ethics'. It offers exactly one principle of universality for moral decision-finding, which is meant to help sort out any kind of moral disagreements. This principle contains the followings aspects: Every norm which claims liability must make sure that *all* consequences (direct or indirect) of its *universal* application must be of such a kind, that they can be *accepted willingly* by every single person *possibly concerned*. (see Habermas, 1983:103) Transformed into a 'maxim' or 'guideline' for moral decision-finding it reads: Our actions must only follow such maxims of which we have assured ourselves that all consequences (direct or indirect) of its *universal* application can be *accepted willingly* by every single person *possibly concerned*. This consensual acceptance can only be found out in *discourse* with those people themselves (or with their advocates), or, if a discourse is impossible, by trying to anticipate the others' opinions. (see Apel, 1988:123)

Even though Habermas and Apel tried to summarize their concept within that single maxim, it implies a complex set of fundamental rules for discourse which Habermas had already discussed previously in his Theory of Speech Act. (Habermas,

1981) These rules express such elements of communication which are necessarily to be accepted, if rational communication should work *at all*. Habermas subsequently analyses the speech act, and he discovers that any discourse mainly consists of situations in which the speaker uses language to convince others of what he thinks is *true*. Conversation thus includes the agreement on *logical rules of language* as well as the aim of inter-subjective *consensus*. But succeeding or failing within a discourse does not only depend upon whether the *present* partners in conversation agree or disagree: In addition to that, any claim of truth must consider the aspect of *universality*. Discourse Ethics thus applies Kant's claim of universality of moral argumentation to the act of discourse itself. The discourse should not only aim at convincing one special person, but it should as well be acceptable by the *ideal community in communication*, which includes all present and future members of humankind.

All these rules show that there are certain underlying values of discourse ethics. The three most prominent amongst them are *responsibility-taking* (e.g. by thinking through one's own and the others' opinions thoroughly etc), *equality* and *freedom* (both e.g. in entering and participating in the discourse). Regarding inter-cultural settings, these three aspects are essential for building the climate of mutual trust in moral discourse. Only if the participants gain the feeling of being respected as equals, they do not need to fear threats like cultural or religious 'imperialism'.

But unfortunately, such an ideal community of communication is hardly ever to be reached in everyday life. E.g. contradicting ethea, prejudices, or misunderstandings can lead to lack of trust. However, attempts are made to cope with these difficulties. Two such examples are presented in the following paragraphs.

4. 'ETHICS IN ECONOMY' WITHIN SENEKA

As during this IFAC conference here, within the Service Network for Training and Continuous Education SENEKA, the attempt has been successfully made to bring together scientists and practice-oriented people. The aim has been to establish a network for industry-focused applications of knowledge management across Germany. (see Henning and Schoeler, 2004). In this project, working groups have been established, amongst them some Communities of Practice (CoPs). These communities within SENEKA are open to all those who are interested in their topics. Beyond the borders of organizations and hierarchies, discussions are held on the level of equality. People can simply bring in and exchange their experiences and thoughts, but it is also possible to decide upon a certain *aim* of their co-operation, e.g. the development of a new joint product. One of these CoPs is the Community of Practice on 'Ethics in Economy'. Its focus is to discuss ethea of different organisations and compare them with each other. One of the aims behind this is to gain a better understanding of how to establish and support organisation values and ethea, which might be accepted within as well as beyond the borders of the own organisation. This attempt shows the commitment to take responsibility for the own organisation's actions and its impact on society. The participants regularly show great interest in how the others deal with strengthening and implementing ethea for the different kinds of organizations, but, it turned out to be extremely difficult to draw general conclusions from the specific cases which have been discussed.

As an example, during one meeting in Cologne in October, 2003, one experience was discussed which was from quite a small family enterprise. This publishing house focused on theological and societal topics and only had a few employees. It had very little need to doubt or modify its ethos, because the enterprise was strongly influenced by the personal value orientations of the family behind it. It regarded itself as a catholic organization with social commitment. The employees shared this tradition-based ethos without any problems. Family and enterprise value system were inseparable, and this ethos also dominated the content and commitment of the enterprise's work in publishing.

As a total contrast to this, at the same meeting the person in charge for 'ethics' from the big organization Siemens reported about the situation there. Since Siemens operates world-wide, it has many employees with most different cultural and moral backgrounds. Thus, the leaders felt the urgent need to deal with the topic of 'ethics' in order to define a clearly cut common set of values which should improve identification with the enterprise as well as solidarity among the people working there. As a second reason for working out their own ethos, the situation of Siemens as a global player was relevant. Since there is a great public interest in the organization to acting in a morally convincing manner, it had to be pointed out *what* exactly was meant by the term 'acting morally'. Siemens made great efforts to understand the fundamental value orientations of their employees world-wide. On this base, a new value profile was worked out and was condensed into five essential guidelines, the so-called 'Business Conduct Guidelines'. (see Franssen and Funk, 2004)

Which conclusions can be drawn from these experiences? Now, the SENEKA-CoP itself originally took into consideration to put together guidelines for 'best practice' concerning 'ethics in economy'. This turned out to be rather impossible. One essential outcome of the discussions is, thus, that every discourse about moral convictions and every effort to create a common ethos must take into consideration the *special situation* in which the group is, in this case the organization. The main emphasis here must be put on a) the question how large and heterogeneous the group of *people concerned* is, b) how to peacefully and lastingly sort out moral conflicts which might derive from those people's different traditional backgrounds, and c) which values and judgements all these people have in common, in order to strengthen corporate identity and responsibility-taking for each other. It is also necessary to include only such values which are really accepted and followed by the members of the group. Since for Siemens as a big organization it is of utmost importance that it sticks to their own promises and principles, it is meant to strictly control and guarantee its ethos to be *lived*. We may call this the needs for *authenticity* and *applicability*. If these needs are not fulfilled the ethos might well 'backfire': A group of people (such as an organization) who signed and announced values and norms which turn out to be empty phrases and which bear no consequences for everyday practice, loses trust of its partners (e.g. its customers etc.).

In this context, pluralism turns out to be the big challenge for authenticity and applicability. As we have seen when considering the term 'ethos', the more open a community is, the more difficult it is to find consensus and to offer safety and hold. The most widely cut ethea, thus, are always in danger to be too little concrete in what they say and thus, to lose power in offering applicable advice. This problem must be kept in mind also when discussing the following case.

5.   THE PARLIAMENT OF THE WORLD'S RELIGIONS: 'DECLARATION TOWARD A GLOBAL ETHIC'

Almost to the day 11 years ago, members of all religious communities of the world were invited to join in a meeting called the 'Parliament of the World's Religions'. Its aim was to find out whether there were some basic similarities within the ethea of all religions, which could help to establish mutual understanding, trust, and joint effort towards global peace. This 'overlapping consensus' was meant to name those values and norms which were elementary to *all* religions. The outcome of this discourse was the 'Declaration Toward a World Ethic', which was signed by 6500 partners in discourse. It includes the 'Golden Rule' as well as 'Four Irrevocable Directives'. (Parliament, 1993) The term 'ethic' was chosen in order to make clear that this collection of norms and values was not an

ethical concept, but that of common value judgements. Its German translation thus reads 'ethos', having the same meaning as we have already worked out above.

Finding this consensus was all but easy, though. The communication process was strongly influenced by the diversity of the members' moral points of view. Based upon religious traditional values, opinions were likely to differ very much. In addition, it was quite a challenge to overcome prejudices and age-old conflicting attitudes between the religions. Problems like these we may call challenges of *plurality*. They become more and more pressing, the bigger and the more heterogeneous a group is. Besides the fact that such an inter-particularistic communication requires a highly elaborated base of knowledge about the different underlying traditions, it can, for example, not be taken for granted that all partners in conversation using certain terms, are referring to exactly the same *meanings*.

Let me just mention one example which demonstrates these difficulties quite clearly. The 'Declaration' was meant to be a 'Declaration of the World's *Religions*'. It could, thus, easily be mistaken as a fact that the belief in 'God' or 'the Gods' as the ultimate grounding of all moral beliefs was something all these traditions had in common. But on the contrary, it turned out to be *impossible* to find inter-religious consensus about a 'Declaration' which includes the words, "In the name of God, the Almighty, the Creator of Heaven and Earth, ...", because in Buddhism, many Gods are known and praised, but 'God' or 'the Gods' are not believed to be the ultimate grounding, the creators, or the aim of life and reality, as is believed in other religions (e.g. in Christianity, Islam, or Judaism). In order not to provoke disagreements and not to stress differences between the traditions, the strong awareness of possible disagreements on such issues is, thus, as important as the ability of finding ways of dealing with them, wherever they appear despite all preliminary measures.

The concept of discourse ethics comes here to its limits. Too little attention is generally being paid to inter-cultural or inter-religious understanding and communication problems which ground in fundamentally differing traditions, and which result in lack of mutual understanding. Maybe partly because of this, Hans Kueng, who was an initial and driving force in getting the communication started, thus, did not call the process a 'discourse', but an 'inter-religious *dialogue*'. Besides implementing the fundamentals of a discourse – such as equality, freedom of participation, and responsibility-taking –, he put special emphasis on those aspects which should improve inter-religious *understanding*. This included, e.g., mutual studies by theologians from the world's religions in order to mutually gain better insights into the underlying beliefs, traditions and values, and in this way to find out about their *similarities*.

But the concept of a 'dialogue', as it is introduced e.g. by William Isaacs at MIT, in contrast to Habermas' and Apel's discourse concept mainly aims at implementing a common base of mutual *understanding*, rather than at finding consensus (Isaacs, 1999). Hans Kueng's idea of a 'Project of World Ethic' (Kueng, 1991) may, as a consequence, be seen as a mixture of both, intending as the *first step*, to implement mutual trust by mutual understanding through a *shared inter-religious knowledge base*, and as the *second step*, agree upon a *common moral core consensus*.

However, even after taking such special care on inter-particularistic challenges, problems of implementation remain for the 'Declaration' as an inter-religious consensus. Since the ethos (or, as Kueng calls it, 'Ethic') is meant to be a 'World Ethic', it must be acceptable to everyone, i.e.: to religious *as well as* to non-religious people. The great success of having overcome inter-religious struggles might, in this context, turn out to cause a new problem of particularism, since the consensus was found *exclusively* by *religious* people. The inter-religious consensus, thus, may, by *non-religious people*, just because of its inter-religiosity be seen as yet another (even bigger) particularistic threat to their identities, be it cultural or national ones. The problem here may be seen in the fact that originally non-religious people were not meant to take part in the decision-finding process of the World's Religions Parliament, and that their right of taking part in the discourse was thereby neglected. Kueng and the 'Foundation Global Ethic' (Tuebingen, Germany) are presently trying to prevent such discomfort and prejudices by implementing a moral discourse in all fields, mainly in education and science.

But still, I am slightly worried that this is a discourse about the *Declaration*, i.e. about the *manifested consensus* of those *religious* people. Implementing a really *universal* moral discourse which may help find *global* moral consensus, must, in my eyes, also allow a continuous discussion of the *contents* as well as the *phrasing*, integrating every one who wishes to take part in the conversation equally, disregarding his or her religious belief - or 'non-belief'. It would, thus, in my eyes be much more promising to institutionalise a 'Parliament of a global ethos (or World Ethic)', e.g. with annually meetings, in order to keep the discussion as open as it requires, and in order to abandon old and to prevent new prejudices and mistrust.

6. CONCLUSIONS

Why has nothing been written here about the *contents* of the 'business conduct guidelines' from Siemens, the 'Golden Rule', or the 'Four Irrevocable Directives' from the Declaration. Here the interactive part of this paper is meant to start: Those condensed directives and values mentioned were of course designed to be very far-reaching. Any reader would probably just nod and agree to them. This effect is not half as fulfilling as to take part in the *communication process itself*, be it a 'discourse' or a 'dialogue', about what *you personally* - and *those people you discuss with* – think might be (e.g.) the 'Four Irrevocable Directives' which could guarantee the peaceful survival of life on earth. It needs *experiencing* the development of such shared value judgements, because it requires the two steps of interactive 'irritation' and 'construction' mentioned before. Practice on 'ethics' means joining the 'Community of Communication' and entering the process "Towards Mutual Understanding in Global Moral Conflicts".

REFERENCES

Apel, K.-O. (1988): Kann der postkantische Standpunkt der Moralität ...? In: Apel, K.-O.: *Diskurs und Verantwortung*. 3rd ed. 1997, Frankfurt am Main: 103-178.

Franssen, M., Funk, G. (2004): Community of Practice 'Wirtschaftsethik'. *SENEKA Newsletter* 17, Feb., 2004.

Habermas, J. (1981): Theorie des kommunikativen Handelns. Vol I + II. Frankfurt am Main.

Habermas, J. (1983): Diskursethik – Notizen zu einem Begründungsprogramm. In: Habermas, J.: *Moralbewusstsein und kommunikatives Handeln*. Frankfurt am Main: 53-126.

Henning, K., Schoeler, G. (2004) (eds.): *Industry-Focused Applications of Knowledge Management across Germany*. Aachen.

Isaacs, W. (1999): *Dialogue and the Art of Thinking Together: A Pioneering Approach to Communication in Business and Life*. New York.

Kueng, H. (1991): *Global Responsibility. In Search of a New World Ethic*. London.

Parliament of the World's Religions (1993): Declaration. In: Kueng, H., Kuschel, K. (eds.): *A Global Ethic. The Declaration of the Parliament of the World's Religions*. London/ New York.

Rose, C. (2004): Considering Global Knowledge Networking from an Ethical Point of View. In: Henning, K., Schöler, G. (eds.): *Industry-Focused Applications of Knowledge Management across Germany*. Aachen: 164-171.

ELSEVIER

IFAC
PUBLICATIONS
www.elsevier.com/locate/ifac

ON HUMAN-HUMAN COLLABORATION IN NETWORKED AND EXTENDED ENTERPRISES
- AN OVERVIEW -

Heinz-H. Erbe

Center of Human-Machine Systems, Technische Universität Berlin

Abstract: Collaborative work over remote sites is a challenge to developers of information- and communication technology as well as to the involved workforce. New developments on cost-effective connections provid not only vision and auditory perception but also haptic perception are presented. Research fields for improving remote collaboration are discussed. Social aspects as new requirements on the employees of networked and extended enterprises are considered. *Copyright © 2004 IFAC*

Keywords: networked enterprises, e-/global-manufacturing, controlling remote collaboration, tele-presence, mixed reality, bond-graphs, hyper-bonds

1. INTRODUCTION

Collaboration is working together towards a common goal at different times, in different locations, at different companies in different functions. Principles are: support collaboration within the entire team including external suppliers and partners, support flexible team participation to minimize collocation as a team requirement, benefiting individuals by making their job easier and helping to achieve work-live

Applying LISI Levels to Collaboration

Level	Environment		Information Exchange for Decision Support
Enterprise	Universal	4	Advanced Collaboration Distributed Global Information Across Multiple Domains Simultaneous interactions with complex data Shared Data & Applications
Domain	Integrated	3	Sophisticated Collaboration (Common Op Picture) Shared Data for a Domain Separate Applications Business Rules & Processes Fused Information Products
Functional	Distributed	2	Basic Group Collaboration Across Programs Heterogeneous Product Exchange Separate Data & Applications Share Fused Information
Connected	Peer-to-Peer	1	Electronic Connection Homogeneous Product Exchange (voice, email, text) Separate Data & Applications Little Ability to Fuse Information
Isolated	Manual	0	Isolated / Standalone Manual (diskette, tape, hard copy)

Fig. 1. Levels of Information System Inter-operability (Mc Quay, B., 2003).

balances, providing a collaborative environment, allowing the team to tap into their inherent creativity and power of sharing ideas, focusing on people, process, communication and relationships in addition to technology (Fig. 1). This sounds like the intentions of sculpting a learning organization/enterprise as it was theoretically developed by Senge (1990) and Watkins & Marsick (1993). But now the challenge is to extend this theory to globally distributed companies. The benefits of collaboration are: reduced problems of resolution cycle time, increasing productivity and agility, reducing travel to remote sites, enabling more timely and effective interactions, faster design iterations, improving resource management and facilitate innovation.

At a recent workshop of the IFAC-Technical Board held in Rotterdam, The Netherlands, 2003, Human-Human Collaboration among others has been identified as an Emerging Area for Automatic Control. Do we need a collaborative control theory (Nof, 2004)?

Extended and networked enterprises distribute the design of products, planning of the production process, and manufacturing regionally if not globally. Employees are therefore confronted with collaborative work over remote distances. A cost effective collaboration depends highly on the organization maintaining a common understanding for this kind of work and a suitable support with information and communication technology. Developments providing not only vision and auditory perception but also haptic perception are very desirable.

Technological aspects: with the trend to extend the designing and processing of products over different and remotely located factories the problem arises how to secure an effective collaboration of the involved workforce. The usual face to face work is going to be replaced at least partly if not totally by computer mediated collaboration. Collaboration demands a deep involvement and commitment in a common design, production-process or service; i.e. to work jointly with others on a project, on parts or systems of parts. Information mediated only via vision and sound is insufficient for collaboration. In designing and manufacturing it is often desirable and in maintenance it is necessary to have the parts in your hands. To grasp a part at a remote site requires force (haptic)-feedback in addition to vision and sound.

Individual, social, and cultural aspects: Enterprises investing in the new information-technology and communication infrastructures have also to consider the important issues to develop a culture and shared values that can facilitate the adoption of such technologies. Investment in advanced technologies may not necessarily result in improved communication by and between the employees. Often managers and the developers of IC - technology assume too much about the anticipated use of the technology by the employees. For most employees, interacting in a virtual mode via mediating technologies may be totally new and may cause anxieties. The loss of human contact could be balanced by maintaining continuous communication as well as by holding occasional face-to-face meetings for information sharing and support.

The development and implementation of information- and communication-technology (ICT) suitably adapted to the needs of the workforce facilitating remotely distributed collaborative work is a challenge to engineers and the management of the involved enterprises. Most developments supporting remote service are sticking on transferring vision and sometimes sound as well. Demanded for are solutions to transfer also realistic haptic feedback.

After remarks on networked and extended enterprises where collaborative work is one of the backbones for a cost-effective design and production, the contribution deals with recent technological developments capable of supporting the work between remote sites and the individual and social aspects.

2. MODELS OF EXTENDED OR NETWORKED ENTERPRISES

2.1 Networks

Agile Manufacturing is build around the synthesis of a number of independent enterprises forming a network to join their core skills, competencies and capacities to be capable of operating profitably in a competitive environment of continually, and unpredictable, changing customer demands.

Networks of independent enterprises between suppliers and between suppliers and customers are designed as platforms, not as chains. A part of a chain can break and destroy or severely damage the whole network. The metaphor "platform" means that all enterprises in the network have access to all relevant information, regarding the order of a customer, the work-capacities of the partners in the network and their specific core competencies in order to discuss the distribution of tasks accordingly.

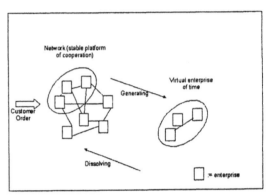

Fig. 2. Stable network and a virtual enterprise

At first the management of the individual enterprises have to understand or to analyze the advantages and/or disadvantages of networks. Secondly they have to understand that all personnel of the enterprise have to be involved. Then a lot of mental barriers must be overcome in order to make networks work effectively. Learning processes have to be developed wherein all employees are involved. It needs to be found how this will effectively work in view of the particular conditions confronting small enterprises.

Respecting customer orders not all partners are necessarily involved in processing them. That depends on the particular order regarding capacities and special equipment available. Therefore sometimes only few enterprises of the network are generating a "virtual" enterprise for processing the order of a customer. Figure 2 illustrates the generation and dissolution of a virtual enterprise based on a stable platform of networked enterprises (Schuh et al, 1998).

2.2 Extended enterprises

These enterprises have organized cross border activities, i.e. design and production sites remotely distributed, partly around the globe (Figure 3). The key objective of these enterprises is to cater to maximal agility: provide anything, anytime, anywhere, anyhow. Such organizations are using the maximal capabilities of both technology and humans to meet this objective. While technology is being used for accumulation, sharing and communication of information, the activities of judgment and decision-making are being entrusted more to the employees who are most familiar with the work situation.

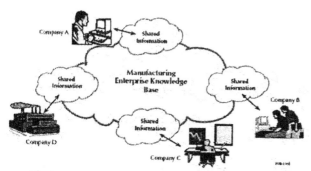

Fig. 3. Remote distributed factory sites (Center for e-design, Univ. of Pittsburgh).

bond graphs, first introduced by Paynter (1961) and later developed further by Karnop et al (1990).

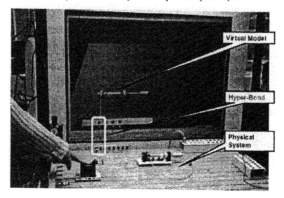

Fig. 4. Mixed reality demonstration (Bruns, 2001).

Hyper-bonds have been first used for discrete events creating learning environments through connecting

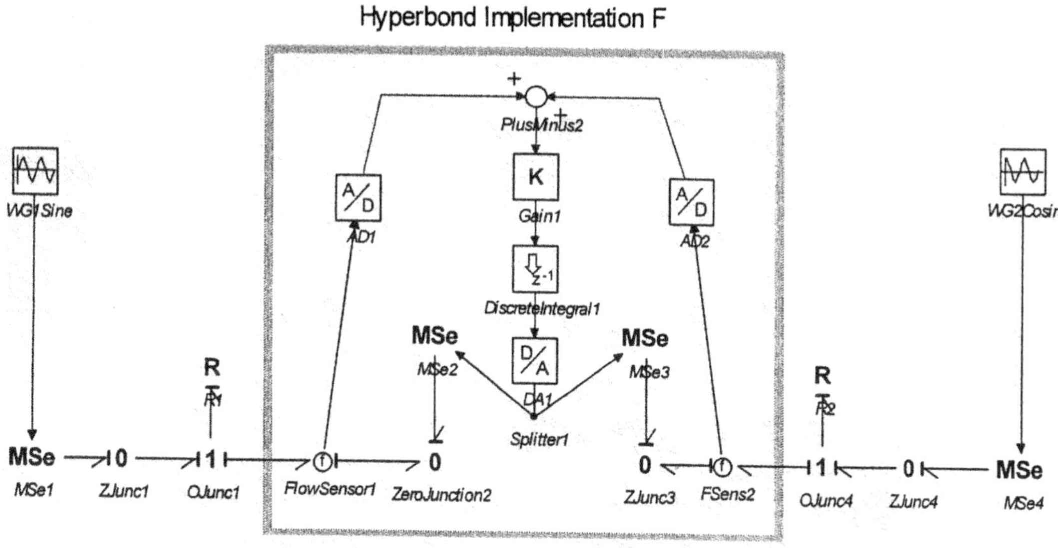

Hyperbond Implementation F

Fig. 5. Hyper-bond sensing flow, generating effort

3. TECHNOLOGY FACILITATING COLLABORATIVE HUMAN-HUMAN WORK

3.1 Hyperbonds and mixed reality

Bruns (2001) developed Hyper-bonds as universal interface type connecting reality with virtuality. It is a mechanism based on the translation between physical effort/flow phenomena and digital information like any other analog/digital and digital/analog conversion. But it aims at a unified application oriented solution connecting the physical with its virtual representation and continuation. The name Hyper-bond has been chosen because of its relation to the description of dynamic systems with

remote labs (Bruns, 2001). The figure 4 shows a simple example of pneumatics. A real pneumatic valve is connected to a pressure source (on the left side) and to a hyper-bond below the screen. The valve is under real pressure. If pushed, the tube connecting the valve and the hyper-bond is under pressure and triggers an input signal to generate virtual pressure in the virtual tube-continuation. The simulated cylinder drives out. If the pushbutton is released, the valve opens a pressure release, the real pressure drops and the virtual pressure driven by the back-spring of the cylinder causes an outflow of air through the hyper-bond. This demonstrates the bi-directional character of the hyper-bond. Figure 5 shows a more sophisticated hyper-bond. The (modulated) sources of efforts are wave generators on both sides. The difference of the efforts on both sides of the hyper-bond generates efforts in a control loop to preserve the overall behavior of the wave generators with two resistors R (Bruns, 2004).

Hyper-bond technology can be used for tele-collaboration in manufacturing and maintenance via the internet. Employees working at remote but real workplaces are connected via the internet. In a work application real objects at remote distributed sites can be handled with force feedback collaboratively through workers on both sites (Bruns & Yoo, 2004). While experiences with remote training of workers in control technique of electro-pneumatics are available, applications of this promising technology for tele-work are still under development (see 3.3).

3.2 Video conferences, tele-design and caves

Distributed design of products over remote sites is a strong demand of the automobile industry for saving time and therefore costs. Centers for e-design got established at different universities, just to name Pittsburgh University together with the University of Massachusetts, Amherst (Nnaji, 2004), among others. The communication is mostly done with data-transfer using distributed CAE-systems like CATIA, i.e. a visual access to the products under the design process. The designers have to negotiate a shared understanding. This is done partly through video conferences. But in the camera-monitor mediated world of videoconferencing, the limitations of communications bandwidth and equipment capability tend to place a severe handicap on the senses of sight and sound and eliminate the sense of touch (Xu & Cooperstock, 1999). As a result, even in state of the

Fig. 6. Model of remotely connected caves

Fig. 7. The "flying carpet" in a cave

26

art videoconference rooms using the highest quality equipment, the sense of co-presence enjoyed by individuals in the same room is never fully achieved. "Gaze awareness, recognition of facial gestures, social cues through peripheral or background awareness, and sound spatialization through binaural audio, all important characteristics of multi-party interaction, are often lost in a videoconference", as Xu & Cooperstock put it. Cooperstock and his team developed a "shared reality environment" to improve videoconference environments with television monitors and stereo speakers to immersive spaces in which video fills the participant's visual field and is reinforced by spatialized audio cues. Moreover computers play a more active role.

While this research and development is not focused to distributed design and manufacturing but to tele-medicine and training, a haptic feedback would help bridging the physical separation of remote persons.

This feedback should range from reproducing the floor vibrations in response to a user walking about to the tactile response of a surgeon's instrument as it moves through different tissue.

Cooperstone (2003) and his team use a certain kind of self-made cave-technology. Available caves at the market are very expensive. But cheaper versions with nearly the same performance are possible. Integrating the hyper-bond technology (mentioned in section 3.1) into this cave-technology (Fig. 6) to support tele-design and tele-maintenance could solve the problem of the presently missing feeling when touching a part, as believed to be a necessity for collaborative work over remote distances. The work-action of persons control via the hyper-bond signals the cameras and video-beamers of the cave. This synchronizes the force-feedback mediated via the hyper-bond. Some research and development have still to be done to make this approach applicable for tele-cooperation. Fig. 7 shows a development of Bruns and his group at the University of Bremen. A "carpet" controlled by a person soars over a simulated landscape in a cave.

3.3 Tele-operation, Tele-presence

A multi-modal tele-presence technology, i.e. the extension of human sensory and manipulative capabilities to remote environments, attracted research and development efforts in recent years (Buss et al, 2001, 2003). The application is focused to tele-operation in hazardous environments (space/nuclear), long-distance tele-maintenance and tele-service. Realistic tele-presence requires feedback of information in multiple modalities of human perception: visual, auditory and haptic. The intentions of tele-presence are different from remote collaboration. No person should be necessary at the remote site. But this could turn out as unrealistic, at least with respect to tele-maintenance. Then a bi-directional tele-presence is desirable. The application of Hyper-bond technology could transfer back the concrete force generated at the remote site.

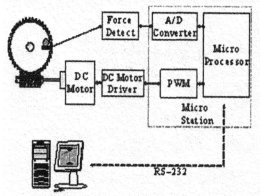

Fig. 8. Low cost momentum handle

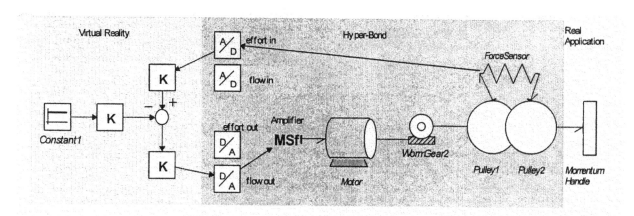

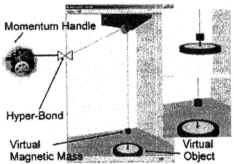

Fig. 9. Force feedback demonstration and hyper-bond

Yoo, and Bruns (2004) propose a low-cost momentum handle for force feedback (Fig. 8). The handle, actively driven by a motor, is always in a momentum-equilibrium through the wheel with a

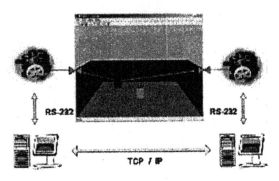

Fig. 10. and 11. Tele cooperation demonstration and model

virtual force/momentum. A pressure-sensor attached to the handle senses a force if the user applies a momentum. This analog signal is fed via an A/D-converter to a microprocessor. The microprocessor controls via a D/A-converter the motor driving the wheel. This micro controller is connected to a serial port of a PC, where it gets the virtual force/momentum from. As an example Yoo & Bruns (2004) demonstrated the lift of a virtual mass represented by a programmed algorithm running at a PC (Fig. 9). The respecting signals are transferred to

the micro controller generating a force at the handle via the motor moment.

Another interesting experiment regarding force feed back between remote sites has been shown by Yoo & Bruns (2004). A virtual mass at two computer screens of remote sites, connected through the internet, is moved through two momentum handles (Fig. 10 and 11). It is a consequent application of the concept of mixed reality. Research results until now could be enlarged for a remote human-human collaboration.

4. SOCIAL PROBLEMS OF REMOTE COLLABORATION

"...you cannot build network organizations on electronic networks alone...If so,... we will probably need an entirely new sociology of organizations."

(Nohria & Eccles, 1992)

One has to consider People, Process and Technology (Fig. 12). To extend their product development and the manufacturing globally, enterprises face increasing time compression, when using foreign based subcontracting labor. Global teams promise the flexibility, responsiveness, lower costs, and improved resource utilization necessary to meet ever-changing task requirements in highly turbulent and dynamic global business environments.

While the promises are laudable, a dark side to the new form also exists: such dysfunctions as low individual commitment, role overload, role ambiguity, absenteeism, and social loafing may be exaggerated in a virtual context. Moreover,

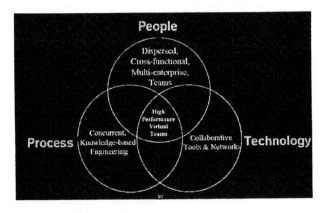

Fig. 12. The triangle of People, Process and Technology (Mc Quay, B., 2003)

customers might perceive a lack of permanency, reliability, and consistency in virtual forms. The question is whether global teams can function effectively in the absence of frequent face-to-face interaction.

4.1 Trust building and maintaining

Trust between the actors of virtual teams has been identified as a prerequisite for an effective collaboration cross-border of companies, countries and cultures. "trust needs touch" (Handy, 1995). However, paradox enough, only trust can prevent the geographical and organizational distances of global team members from becoming psychological distances. Trust allows people to take part in risky activities that they cannot control or monitor and yet where they may be disappointed by the actions of others (cited from Jarvenpaa & Leidner, 2000). In their case studies the authors questioned: Can trust exist in global virtual teams where the team members do not share any past, nor have any expectation of future, interaction? How might trust be developed in such teams? What communication behaviors may facilitate the development of trust? The global virtual teams had members who (1) were physically located

the global environment. Finally, it is a heavy reliance on computer-mediated communication technology that allows members separated by time and space to engage in collaborative work.

Findings: trust might be initially created, rather than imported, via communication behaviors in global virtual teams. The case studies portray marked variations in the levels of communication richness across teams, suggesting that the information richness is an interaction between the people, tasks, the 'organizational context, and perhaps familiarity with the technology in use. The studies also raised questions about how technology might obliterate, reduce, or delay the effects of culture and cultural diversity on communication behaviors when the setting is totally virtual. Some practical implications can be drawn from the studies. For the manager of a virtual team, one of the factors that might contribute to smooth coordination early in the existence of the team is a clear definition of responsibilities, as a lack

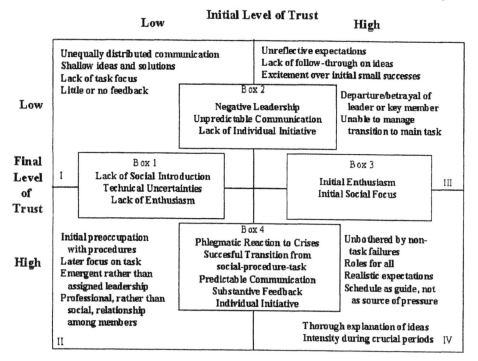

Fig. 13. Matrix of Low and High Level of trust in collaborative work (Jarvenpaa & Leidner, 2000)

in different countries, (2) interacted through the use of computer-mediated communication technologies (electronic mail, chat rooms, etc.), and (3) had no prior history of working together. A global virtual team was defined to be a temporary, culturally diverse, geographically dispersed, electronically communicating work group. The notion of temporary in the definition describes teams where members may have never worked together before and who may not expect to work together again as a group. The characterization of virtual teams as global implies culturally diverse and globally spanning members that can think and act in concert with the diversity of

of clarity may lead to confusion, frustration, and disincentive.

For the participants on virtual teams, there are some observations derived from the study which may be relevant to practice. Although it is not necessarily critical to meet in person, it is critical to engage in an open and thoughtful exchange of messages at the beginning of the team's existence. Cavalier attitudes that the virtual environment is no more challenging than a face-to-face environment prove to have ephemeral effects on participant enthusiasm and, once difficulties arise, the team lacks a substantive foundation upon which to overcome the real challenges imposed by the virtual context. Participants should also have an awareness of the importance of their providing to the others timely and detailed accounts of the work they are doing. Likewise, participants must be aware of the need to provide thorough feedback on the contributions of the other members. Finally, participants should be aware that it is not the quantity, but the quality and predictability, of their communication that is most critical to the effective functioning of the team. The research of Jarvenpaa & Leidner (2000) suggests that trust can exist in teams built purely on electronic networks. The study describes a number of communication behaviors and member actions that distinguished global virtual teams with high trust from global virtual teams with low trust (Fig. 13). Encouraging such behaviors and actions on the part of members of global virtual teams might help to foster a climate conducive to the existence of trust.

4.2 Companies responsibilities

Companies are investing in the new technology and communication infrastructures. However, the more important issue is the development of a culture and

- Socio-technical issue -- not just technology
  - People
  - Culture
  - Processes
  - Infrastructure -- framework & tools
- 19th-20th Century Technologies
  - Telephone
  - Automobile
  - Television
  - Computer/Internet
- 21st Century
  - Knowledge Based Society
  - Faster change

Fig. 14. Socio-technical issues (Mc Quay, B., 2003)

assume too much about the anticipated use of the technology by the employees. The crucial point is to realize that employees may need hand-holding in the beginning for realizing the anticipated benefits of technology. For most employees, interacting in a virtual mode via mediating technologies may be totally new and may cause anxieties in terms of issues such as performance and evaluation. Automobile companies are using virtual global teams

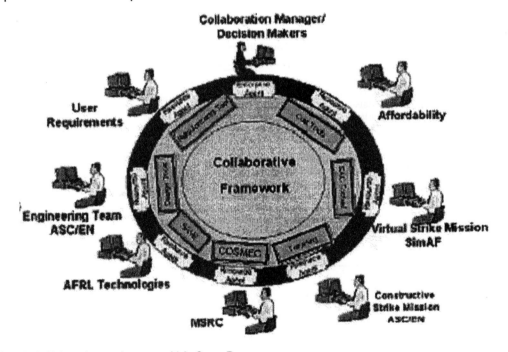

Fig. 15. Collaborative environment (Mc Quay, B., 2003)

shared values that can facilitate the adoption of such technologies and communication. Investment in advanced technologies may not necessarily result in improved communication by, and between, the employees (Fig. 14).
It is also crucial to understand how to combine the technology mediated collaboration with occasional face-to-face interactions with the employees. Often managers and the developers of IC- technology

for designing the latest car models. Their globally distributed employees are operating on a 24-hour day in which the design is communicated at the end of the work day to another part of the world where the work continues uninterrupted. The loss of human contact is being balanced by maintaining continuous communication as well as by holding occasional face-to-face meetings for information sharing and support. In several cases the human contact at the workplace is being replaced with increased contact with clients and customers in the field.

To provide their globally distributed workforce with suitable education and training for distributed virtual work companies are well advised to use action learning strategies as these are now well developed and tested (Yorks, L. et al, 1999). As innovation is not only the ability to combine existing experiences, knowledge and technologies effectively in order to develop products and services to put them on the market, but also the ability to change the organization appropriate to the demands of the market. Formal as well as informal learning within task-solving is not only to be understood solely as training for each individual staff member but also as a mean to develop the capacity for innovations within the organization. Continuous innovation requires a continuing education, continuing learning - this is where the leaning organization comes into play (Watkins & Marsick, 1993). Not only the management has to learn continuously, also at the shop floor level this continuous learning is particularly important, particularly as the knowledge about the manufacturing technology is present there.

5. CONTROLLING DISTRIBUTED COLLABORATIVE WORK

To make collaborative work between humans, humans and software agents, humans and automation systems (machines, robots) locally and globally effective it has to be controlled.

Protocols at the work application level, such as Task Administration Protocols, are defined as the logical rules for workflow control. Protocols enable effective collaboration by communication and resource allocation among production tasks.

Nof (2004) recommends active protocols, "active" implies that the protocol can, under coordination logic, trigger and initiate necessary, timely interaction tasks and decision-support functions to further improve performance. In smart distributed systems, e.g., multi-agent systems, coordination protocols are used to guide and support the automated interaction of each agent or an autonomous entity in order to achieve the common goal of a system. The efficient coordination process will lead to effective collaboration.

Auction-based protocols (Nof, 2004) have been developed and analyzed by parallel simulators for distributed task and resource allocation in multi-agent systems.

In the future, multi-participant negotiation processes, group decision mediation, and conflict management processes will be developed. In addition, the dynamic characteristics of the protocol which acts or reacts to the stochastic nature of distributed environments will be of interest.

To save manufacturing cost the integration of tasks are challenges for design and planning of new structures of decision making, conflict resolution and

task controlling. From islands of automation to global systems, from local optimization to global optimization are the demands fostering research and development. Integrated product and process development supported by concepts like the digital factory and through e- manufacturing and e-design is one of the outputs of this research (Fig. 15).

S. Nof (2004) states: "despite the advantages entailed with e-Work and e-Mfg., there are common challenges that emerge with collaboration: (1) Greater work complexity; (2) More limitations caused by increasing inter-dependence; (3) Issues of integrity and trust; (4) Greater need for coordination, cooperation, and synchronization; (5) Communication challenges and failures, complicating systems' requirements; (6) Problems of mismatch, e.g., inconsistent versions, cultural differences, etc.; (7) New users training requirements and associated costs. Failing to overcome these common challenges may explain why often, e-activities do not fulfill the expectation promised by technology".

"These concerns reflect the challenges of new work abilities to interface more complex systems, involving more information transfers, many more transactions, and the dawn of computer and robot servitude: beginning the delegation of control and cognitive tasks from humans to computers, machines, and robots. One of the main new concerns has been information overload. It means that humans, each

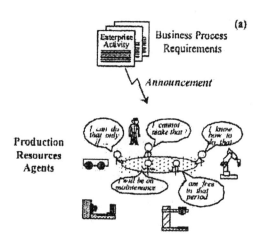

Fig.16. Software agents negotiating available resources (Rabelo et al, 1998)

with a single brain, have to interact and contend with computer-based systems and machines that may have certain superior skills not only in physical tasks, but also in information handling and computing. The objective is to reduce the information overload (only for the humans) by better design of interfaces and of computer systems, and by more training for the human users/participants" (Nof, 2004). Nof asks: will workers and managers be willing to trust the results obtained and delivered by autonomous systems like (multi) agent systems?

The application of software agents not only for information retrieving, but also for assisting interaction processes, such as business workflow, automated negotiation and error recovery have been developed (Fig.16).

The ability of a software agent to operate the same interface operated by the human user, and the ability of a software agent to act independently of, and concurrently with, the human user will become increasingly important characteristics of human-computer interaction. Agents will observe what human users do when they interact with interfaces, and provide assistance by manipulating the interface themselves, while the user is thinking or performing other operations. Increasingly, applications will be designed to be operated both by users and their agents, simultaneously. Figure 17 explains an e-collaboration continuum.

Shop floor control in an individual company is well developed (Scherer, 1998). The support system can be fully automated or can be a shop floor oriented planning and control software semi-automated based on electronic planning boards operated by the workforce. The goal is to achieve an optimum with respect to material and time resources regarding the tasks to be carried out. This leads to the normal work practice. Additional express orders or just not available resources causes a fluctuation around this normal work practice or performance of the manufacturing system. Inappropriate planning and control caused by time and costs pressure, tight resources on personal, or deficiencies by illness, all

support has to intercept such movements of the normal work performance against the limit of acceptable performance.

What is more or less feasible in an individual enterprise without information overload for the workforce does not work in a network. To achieve an optimum of the manufacturing process of networked enterprises regarding material and time resources to fulfill customer orders seems to be only possible with software support controlling the whole process. That means: cooperation requirement planning e-work parallelism, error and conflict handling and e-work effectiveness measures (Nof, 2004).

6. CONCLUSIONS

The contribution discussed requirements on supporting human-human collaboration in manufacturing. Until now, available Information- and Communication Technology do not properly support human-human collaboration. Tele-service developments restrict the information transfer to vision and sound. An immersion into the remote site to collaborate on solving maintenance or manufacturing problems is not supported. Recent results of research projects on hyper bonds and tele-cooperation promise a suitable support for the involved workforce if further developed. This would help to make tele-cooperation cost-effective by avoiding misinterpretation and therefore reducing the time-to-market, one of the drives for a global manufacturing.

Concluding with a statement of J. Cernetic (2003): "The perspective of using such relatively complex technological systems in all these applications is

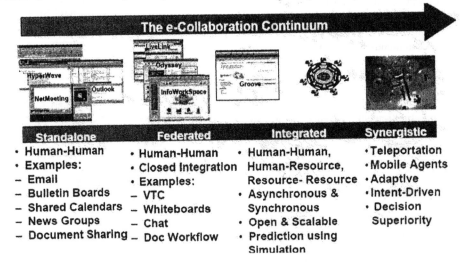

Fig. 17. © Center for e-design, U. Pittsburgh (2003)

unexpected issues are moving the system to the limit of the space of acceptable performance. This can occur unnoticed and sometimes are perceived only when the quality decreases, times of delivery are exceeded or through put time increases. The decision

leading to a much wider (and, at the same time, closer) exposure of advanced technology to a multitude of people, of which most of them will not be specially prepared for using it properly. If such products should be commercially successful, they will have to be well adapted to their intended users. In addition, these products and systems should not raise any doubts regarding their possibly negative social implications. This, in turn, means that these products or systems will have to be both, human-centered and socially appropriate. In this respect, the following statement from about 10 years ago is still fully true and deserves to be kept in mind whenever automation systems are being developed and implemented. *"Modern technology involves the application of science in social contexts. Automation technology, including digital computers and communication techniques which have become such vital elements and tools of automation, is an ever-growing technology with an enormous range of subjects and applications... However, as technological implementation becomes more application-oriented, consideration of the effects caused by automation technology becomes more important..." (Martin et al, 1991).*

However, under the rising pressure of global competitiveness, it is generally held that to make advanced-technology products and systems more appropriate to their users and to society, it requires additional knowledge, more time and more resources. This opens the question whether human-centered and socially appropriate automation can be considered cost-effective, or in other words, affordable."

REFERENCES

Bruns, F.W. (2001). Hyperbonds - Enabling Mixed Reality. artec paper 82, Univ. of Bremen ftp://artec-nt.artec.uni-bremen.de /pub/Field1/Publications/ artec-01-Bruns-hyperbonds.pdf

Bruns, F.W., Erbe, H.-H. (2003). Didactical Aspects of Mechatronics Education. In: *Proc. SICICA 2003 - Intelligent Components and Instruments for Control Applications.* L. Almeida, S. Boverie (eds.), Elsevier Ltd., Oxford

Bruns, F. W., Yoo, Y. (2004). Realtime collaborative mixed reality environment with force feedback. In: *Preprints of 7th IFAC Symposium on Cost Oriented Automation*, Ottawa, Canada.

Bruns, F. W. (2004). Hyper-Bonds-Application and challenges. Artec-paper 115, University of Bremen, Germany

Buss, M. et al. (2003). SFB 453, TU Muenchen

Cernetic, J. (2003). Affordable and socially appropriate automation. In: *Robotica*, vol 21, pp. 223-232, Cambridge University Press

Cooperstock, J. (2003). *CIM-McGill University*, Montreal

Handy, C. (1995). Trust and the virtual organization. *Harvard Business Review*, 73 (3), 40-50.

Jarvenpaa & Leidner, 2000. Communication and Trust in Global Virtual Teams. *Report*,

Graduate School of Business, The University of Texas at Austin.

Karnopp, D. C., Margolis, D. L., Rosenberg, R. C.(1990). *System Dynamics – A unified Approach.* John Wiley, New York.

Mc Quay, B. (2003). Distributed Environments Technology for e-design. *Report*, Air Force Research Laboratory, Wright-Patterson AFB OH, AFRL/IFSD

Martin, T., J. Kivinen, J.E. Rijnsdorp, M.G. Rodd and W.B. Rouse "Appropriate automation – integrating technical, human, organizational, economic, and cultural factors", *Automatica*, **27**, 901-917 (1991).

Nnaji, B. et al (2004). Cost effective Product Realization. In: *Preprints of 7th IFAC Symposium on Cost Oriented Automation*, Ottawa, Canada.

Nof, S. (2004). Collaborative e-Work and e-Mfg.: The state of the art and challenges for production and logistics managers. In: *Preprints of the 11th IFAC Symposium on Information Control Problems in Manufacturing – INCOM 2004* Salvador, Brazil; April 5 – 7.

Nohria, N., & Eccles, R. G. (1992). Face-to-face: Making network organizations work. In: N. Nohria and R.G. Eccles (Eds.), *Networks and organizations* (pp. 288-308). Boston, MA: Harvard Business School Press.

Paynter, H. M. (1961). *Analysis and Design of Engineering Systems.* MIT Press, Cambridge, MA.

Rabelo, R., Camarinha-Matos, L., Afsarmanesh, H. (1998). Multiagent Perspective to Agile Scheduling. In: *Intelligent Systems for Manufacturing* (Camarinha-Matos, L., Afsarmanesh, H., Marik, V. (Ed)) pp. 51-66. Kluwer Academic Publishers, London.

Scherer E. (Ed.) (1998). *Shop Floor Control - A Systems Perspective.* Springer-Verlag, Berlin.

Senge, P. (1990). *The Fifth Discipline.* Doubleday, New York.

Schuh, G., Millarg, K. and Göransson, A. (1998). *Virtuelle Fabrik – Neue Marktchancen durch dynamische Netzwerke.* München, C. Hanser Verlag.

Watkins, Karen E., Marsick, Victoria J.(1993). *Sculpting the learning organization.* San Francisco: Jossey-Bass Publishers, 1993.

Yorks, L., O'Neil, J., Marsick, V. J. (1999). *Action Learning. Advances in Developing Human Resources*, Swanson, R. (Ed.), Baton Rouge: Academy of Human Resource Development.

Xu A., J. Cooperstock (1999). CIM-McGill University, Montreal

BLOWING THE WHISTLE ON SOCIAL ORDER IN ENGINEERING

Dr. Larry Stapleton[1] and Dr. Marion Hersh[2]

*1.ISOL Research Group, Waterford Institute of Technology
Waterford, Republic of Ireland
Tel: +353 51 302059. Email: larrys@eircom.net
2. Department of Electronics and Electrical Engineering,
University of Glasgow, Glasgow G12 8LT, Scotland.
Tel: +44 141 330 4906. Email: m.hersh@elec.gla.ac.uk*

Abstract: Research ethics in science and engineering emphasises the importance of power structures and their relationship to an ethical program of research and practice. To date most analyses have focused primarily on social constructivitist approaches, which have the disadvantage of de-emphasising societal factors. This paper introduces two empirical studies of whistleblowing. It discusses agent network theory and the traditional nature of current engineering education programmes, which often act to discourage creativity and innovation. The results of these studies highlight the difficulties associated with whistleblowing and the importance of cultural factors in analysing socio-technical systems. *Copyright © 2004 IFAC*

Keywords: Ethics, social impact, ICT, engineering and automation systems, education

1. INTRODUCTION

Technology development is one of the most important factors in shaping modern society in all parts of the world. However many of the scientists and engineers who are involved in the research, development and implementation of these new technologies still consider themselves to be purely problem solvers and pay less attention to the nature of the problems they are solving, who has set them and whose interests the results will serve. Indeed, until quite recently, coherent, instrumentalist perspectives of interest networks in engineering and science have remained elusive. These interests are of importance from an ethical perspective, but need to be unpacked from the engineering and technology development discourse in order that the engineer can identify:

1. What those interests are
2. Where they (the engineer) stand

This unpacking is not a simple task, and recent analyses suggest that engineering education does little to prepare the engineer for such activities. One reason could be that, here, the instrumental world of the engineer collides with the soft, fuzzy world of social science. In particular, there is no preparation for working with system end-users who are not engineers, though much of engineering involves designing systems for people who are not engineers.

However there is growing awareness of the importance of ethical decision making in science and engineering and interest in the development of tools and codes to support it (Hersh, 2000). The focus of such codes is generally on the individual responsibilities of the scientist or engineer rather than the collective ethical responsibilities of the science and engineering communities. A mixture of this individualistic approach and an identification with management means that some codes even cut off engineers from sources of solidarity and support such as trade unions (Martin et al, 1996). These power dynamics generally acts to maintain the status quo and need to be taken into account for effective ethical decision making and action at all levels.

2. ENGINEERING EDUCATION, CONSENT, INNOVATION, HERESY & SOCIAL STABILITY

Many commentators have pointed to the ways in which mass consent is manufactured in apparently democratic societies in order to ensure political stability (Chomsky, 1994). This raises important questions for those involved in the education of technologists and engineers. Unfortunately the evidence suggests that modern western education and, in particular, engineering encourage students (Stapleton & O'Dowd Smyth, 2003) to be passive, non-questioning and acquiescent. However there is a often a close relationship between creativity, dissidence and innovation. For example, Brandt & Ihsen (1998) illustrate how highly creative thinking applied to robotics creates entirely new research trajectories within engineering and Court (1998) has shown how teaching creativity in engineering design can lead to many new product innovations. Holmes (1998) shows that creativity in engineering thought is 'not an optional extra', but goes to the very heart of engineering as a discipline. Platts (1998) argues that 'the inter-twining of technical and moral creative skills' is central to engineering advancement, and the creation of civilised society.

Lenschow (1998) argues for a paradigm shift in engineering education from 'teaching to learning', and he illustrates several ways in which this can be achieved. Research in engineering education recognises the need for more creative and innovative approaches which include the social responsibilities of engineers. This has been emphasised as a major issue for furthering research and practice in engineering ethics (van der Vorst, 1998). Brandt (1996) and Acar (1998) highlight the international nature of this issue citing similar experiences and imperatives in Germany, Slovenia and the USA. A number of evaluation models and frameworks for engineering education have been developed. (e.g. Atieh et al. (1991)). However such approaches rarely address deep structural issues in engineering. Consequently, there is no real debate or examination of the underlying values and structure of engineering education. The assumption seems to be that the overall educational process is appropriate in terms of creative thinking, reflection and ethos. The engineering education literature does not reflect the developments within education research more generally. This paper seeks to address these issues, with the overall objective of contributing to the creation of an engineering community which values divergent thinking, innovation and dissidence.

There are some innovative approaches, but engineering education in general is largely traditional. The focus is technology and questions about the role of engineers within society: who gains and who loses in the race for technological progress, are rarely discussed. Fundamental advances in educational philosophy and practice and the broader definitions of competence in the education research literature and practice-oriented disciplines, such as nursing and management, are generally ignored. The way in which engineers from traditional engineering programmes define the quality standards used to measure advances in engineering education typifies manufactured consent and how Foucauldian power structures are reinforced in society (Kuhn (1996); Foucault, 1965).

Thus engineering education is an example of the maintenance of a structural hegemony which is anti-innovation. Healthy discourse in any discipline needs to include the margins of that discipline. However people at the margins are often treated as 'heretics', a term more familiar from religion (Hersh and Moss, 2003). Scientific gate-keeping is used to ensure that only certain types of science and technology are given official sanction and that, if possible, proponents of heretical ideas are excluded from access to resources, including research grants, publication in respected journals and employment.

The prospect of paradigm shift can provoke very strong emotional reactions and a series of outraged objections, both relevant and irrelevant (Pugh, 1993). One of the factors underlying these emotional reactions is the perceived threat of a reduction in power (Johnson, 1988), with the most powerful people in an organisation deriving their influence at least in part from association with the 'constructs of power', making it very difficult for (other) members of the organisation to change or challenge paradigms accepted by the organisation. Such challenges are often seen as 'political' or 'cultural' rather than the subject of intellectual debate. Consequently responses may be 'political' in terms of action to preserve the status quo and existing power structures, rather than an analysis of the proposed new paradigm. In addition isolation and ostracism are very powerful tools for ensuring conformity and suppressing challenges to existing power structures.

A related issue is the status given to knowledge and technology on the basis of the identity, power and status of their proponents rather than a scientific analysis of their content and the process by which they have been obtained or developed. This is evidenced in Latour's (1999) analysis of Diesel's which led to the development of the diesel engine, discussed in the next section. In the modern era, we can see how small scale irrigation techniques used by local people in Africa are ignored and devalued, although they perform better than irrigation schemes constructed to fit a 'scientific' model (Ikkaracan and Appleton, 1995) and monocropping techniques suggested by 'expert' agronomists are still being promoted though traditional techniques of intercropping have been found to give much better yields throughout Africa (McCorkle, 1989). There is no explicit recognition of the networks of interests which lead to these scenarios. This understanding of the relationship between the technologies and tools and the power structures at work in society needs to be made explicit in engineering educational programmes, engineering research trajectories and set out as a core issue for engineering projects.

3. AGENT NETWORK THEORY

Thus there is often a relationship between power and identity factors and the ethical and other values used in decision-making. However, current analyses rarely pay very close attention to the power relations between materials, instruments and people involved in the scientific or engineering research community. What is needed in a theoretical trajectory to help make sense of these relations and explain how they work to undermine or support technological progress. One promising theory has emerged in the debates surrounding instrumental realism i.e. Latour's Agent Network Theory (ANT). Instrumental realism argues that scientific instrumentation interfaces between the worlds of science and technology, and that reality is approached through peoples relations with instruments (c.f. Ihde, 1991). Latour deals with the relationship between humans and non-humans and therefore provides a useful basis for considering some of the problems raised here. The focus is a new combination of the social and technological, the network of agents, rather than society or technology on its own. Latour develops a theory of agent networks in order to explain the relations between scientists (he uses Pasteur) and their instruments, as well as with other scientists and stakeholders. In this analysis the microbe or robot is not just an object of analysis, but an *interest* to be *managed and utilised* by scientists or engineers. Thus the viability of the socio-technical network is a primary success factor.

Adaptations of ANT can be used to theorise about engineering research as a set of socio-technical systems without losing sight of the non-human elements. ANT informs the debate on engineering ethics by identifying elements of these socio-technical systems, This is because ANT demonstrates the complex web of relations in engineering and how they work. ANT helps bring to light the often hidden (and ignored) social structures in engineering research. In this paper we make explicit the link between ANT and socio-technical systems research, a link heretofore generally ignored.

In order to understand ANT let us examine briefly the example of the great engineer, Diesel. According to Latour's theory of heterogeneous agent networks, to be successful as an inventor Diesel had to secure and manage the 'interests' of a wide variety of actors

including kerosene, air pumps, other engineers, financiers, entrepreneurs and consumer markets. The thermodynamic theory behind Diesel's work was that he could combust any fuel in his engine at low temperatures and high pressures. In Latour's view Diesel was 'let down' by the alliance he had formed with the fuels – they would not ignite and he therefore had to shift his alliances. He first had to identify the 'interests' of the fuels and from this he discovered that only kerosene ignited at low temperatures and high pressures. Now Diesel had to shift his set of alliances to take account of kerosene's interests without losing the support of the rest of his agent network. In the end Diesel lost the support of his financiers and was unable to build his engine, even though his basic theory was sound and his ideas solid. The above analysis, adapted from Latour's own work, shows the deep-seated connection between science and engineering on the one hand and social order on the other. Diesel failed because he was unable to manage the socio-technical structures which would help him construct his engine (his network of agent interests in ANT).

3.1 Problems with ANT

Like all theories, agent network theory also has its drawbacks. For instance, it does not adequately address social impact and ethical issues. This problem is a result of Latour's emphasis on instruments, materials and complex socio-technical spaces, rather than social systems and ethical considerations. Furthermore, although Latour's work can be linked to other work on knowledge and social order (Shapin & Schaffer, 1989; Foucault 1980) it does not always provide a very rich or multi-layered analysis of complex structures. For example, Latour's concern with instruments belies the fact that these are, themselves, socio-technical artefacts or, as Ihde has put it, techno-culture (Ihde, 1999). The technology remains a discrete component of the system, but essential connections between humans and machine are ignored (c.f. Stapleton, 2003).

This illustrates the need for a mature, far-reaching discourse on ethics at both at the individual and collective levels and including actors, such as financiers, stakeholders, and research laboratories, traditionally considered external to engineering and systems research. Ultimately this discourse must feed into research on tools which enable us to unpack engineering ethics at a practical level and small groups of engineers and their user communities to interact so that they can, together, expose hidden structures within both their specific projects, and background perspectives (Hersh & Moss, 2003). This can be achieved through techniques such as the Johari window (Stapleton & Hersh, 2003).

Agent networks are bound together by information and communications links which must be managed, maintained and, sometimes, re-configured. This gives a socio-technical network in which social order must be maintained in order for the network to survive. Faulty communications links can generate a positive feedback loop which can totally destabilise the system. Another possible source of instability is the voices of minority groups, which are so critical for innovation and creativity, within socially ordered structures like engineering and lead to the unveiling of agent networks. Unless appropriate mechanisms are available, engineering can become a very unstable or unethical system, as engineers attempt to forge new alliances to suppress the destabilising influences of the minority groups. Indeed, recently pharmaceutical research companies were linked to death threats and murder attempts against a world health organisation scientist whose work showed how the drug industry was unhelpful to medicine in poorer regions. Aspects of these relations can be explained in terms of ANT. It is critical that such systems have mechanisms for maintaining stability between actors. An actor is needed who can identify and communicate where faults in the network are appearing. The 'heresy' of whistleblowing is actually a negative feedback mechanism which can contribute to systems stability. Indeed, the greater the momentum and dynamism within the network, the greater the need for a negative feedback mechanism to halt any potential positive feedback processes.

4. WHISTLEBLOWING' AND 'THE HERETIC'

Whistleblowers (Hersh, 2002) can be vital to maintaining stability, particularly in large-scale technology projects such as ERP systems. However many organisational systems have a culture of blame and are intolerant of even the best intentioned criticism. Despite the importance of whistleblowing, the whistleblower is rarely considered a critical component of a successful agent network. Instead the whistleblower is depicted as a lone hero (and martyr) standing up against unacceptable practices in the organisation and paying the inevitable price in terms of isolation, ostracism, loss of promotion and possibly also loss of job, family, friends and health. One interesting feature of the treatment of whistleblowers is the shifting of responsibility from the organisation to the whistleblower and their 'false' views: the idea of the whistleblower as a heretic. For instance the 2nd and 3rd stages of a proposed 4-stage model of retaliation (O'Day, 1972) are isolation and defamation of character.

Over the years, a body of literature on whistleblowing has developed (Hersh, 2002). Whistleblowing can be seen as an important success factor in engineering projects by breaking groupthink (Pfeffer, 1982). However there are still problems with regards to both perceptions of whistleblowers and the very real negative consequences of whistleblowing. Unless these perceptions and consequences are addressed, many

engineering agent networks will fail. This paper will consider perceptions amongst engineering undergraduates and on a manufacturing engineering project, but not the consequences of whistleblowing.

4.1 A Short Empirical Study of Group Perceptions of Whistleblowing in Higher Education

The first author conducted a short study of perceptions of 'whistleblowing' amongst a group of final year undergraduates at an Institute of Technology in Ireland. The aim was to determine the group's perceptions of the term 'whistleblowing'. The study utilised a highly unstructured qualitative approach in which 47 students were organised into a single focus group, and (in this group context) asked to free-associate terms with the term 'whistleblowing'. Methodologically, this draws upon techniques used, for example, in clinical psychology research to gather unstructured, qualitative data from focus groups to assess perceptions of particular phenomena (Barker et. al., 2002). To try to dampen cultural effects, the group consisted of 20% non-Irish students (primarily Austrians and Asians). The researcher wrote the terms on a whiteboard in the front of the room and asked the group to discuss each term. This helped the researcher to interpret what the group meant by the term as they discussed it amongst themselves, and provided a way of openly recording discussions as they proceeded.

All the words that were written up were derogatory. The key terms that were widely agreed as representative were: *snitch, rat, grass, snake, dobber* (derogatory Irish colloquialism). These terms are interesting in that they are highly emotive (even inflammatory), especially in the Irish context. For example, for Irish people the term 'grass' has references to Northern Ireland where 'grasses' and 'supergrasses' betrayed people to the authorities. For this student group at least, the whistleblower is intensely disliked and not seen as a role to which one might aspire. Thus the results are influenced by the post-colonial psyche (75% of the group were Irish born and bred in the Irish Republic).

4.2 Whistleblowing in an Manufacturing Engineering Facility

Another study was conducted in a large scale engineering facility in Ireland, where a large scale enterprise resource planning system was being implemented across a large multi-national firm. This longitudinal study aimed to analyse the experiences of the organisation from the perspectives of the project team and key users. One aspect of the study assessed power relations and technological determinism. The study revealed how technical imperatives created a pressure upon the team not to blow the whistle when it became apparent that the system was deeply flawed. This made whistleblowing both very difficult and ineffective. Project momentum was extremely great and it had only become apparent quite late in the project that key functional areas would collapse once the system was in. It became evident to a senior finance manager that the problem was extremely serious but that it was being ignored. The following quote from the Finance Executive shows how a focus on system functionality as opposed to organisational readiness lead to serious conflicts...

'When I said I wasn't ready the comment was 'can you survive if the system goes live?' and I said 'I can survive but it will be bad' so they went live... if we got it wrong it's our responsibility, you're looking after a business function. How do you say 'it's not working well?' This is a reflection on you – we helped put the system in'

The above quote shows the difficulties of whistle-blowing even if the whistleblower is in a strong position within the firm. The interviewee above was a senior manager who had a deep knowledge of the system and was widely respected. However, even from this position it proved impossible for her to hold back the project. Subsequently, customer service collapsed, major production processes became unworkable and links to suppliers proved very hard to manage. In short, it was chaos, a chaos which could have been avoided if there had been a positive response to her whistleblowing

4.3 Implications of the findings

On the surface the student study indicates a worrying situation: i.e. that whistleblowers will be considered outsiders and that there will be considerable resistance to whistleblowing. This is possibly a consequence of the lack of course units on ethics in engineering and ICT degree programmes. Where ethics is taught it is generally relegated to sub-categories on syllabi, suggesting a significant problem in engineering education.

However a more in-depth analysis is required to understand whether negative views of whistleblowers really do indicate a lack of understanding of ethics and unwillingness to behave ethically. There is considerable discussion in the literature of the conflict of interests resulting from loyalty to the employer and/or organisation and the need to expose problematical or unethical behaviour (Hersh, 2002). However there is considerably less discussion of the conflicts of interests resulting from loyalty to colleagues. It could be postulated that such conflicts of loyalty are likely to be particularly strong amongst a group of largely Irish students. The descriptions of whistleblowers reported indicate a strong emotional reaction against betraying the group and collaborating with management, probably

symbolically identified with the landlord, resulting from the colonialist past. Thus raising the question of whistleblowing probably strengthens inter-group loyalty and the barriers against 'betraying' the group.

This also relates to the ethics of care, which is a context based approach to preserving relationships. It originates with Gilligan (1982), who found when applying Kohlberg's (1981) theory of human reasoning and moral development to women, that an ethics of care was more appropriate for women. The main aims include moral attention to the situation in all its complexities, sympathetic understanding and sensitivity to the wishes and interests of others, relationship awareness that the other person is in a relationship to you, accommodation to the needs of everyone, including yourself, and responses to need and showing caring. One of the problems of whistleblowing in this situation is that it is generally presented as an individual activity. Thus whistleblowing requires individuals to go outside the group and symbolically betray it to management (the landlord). This would result in deterioration and possibly a serious breakdown of relations between at least some members of the group and could lead to the destruction of the group as a meaningful entity.

Another important relationship is between individuals, the group and the technology. Technology is not an entity on its own and does not work in total isolation. It is sited in a social, cultural and political context and is used by people, sometimes as individuals, but frequently as part of teams or other groups. Thus the successful functioning of a particular technology is dependent on the group process. If this breaks down, then the technology is unlikely to work well or possibly at all. Thus if whistleblowing results in a dysfunctional group, then even a well designed technology may not function effectively. Conversely it may be possible to at least partially rescue (or subvert) a bad system through good group dynamics. In addition technology is produced by people and people work with it. This can lead to a sense of ownership and/or identification with the technology. Thus cricitism of the system through whistleblowing could be interpreted as criticism of individuals and the group.

The aim of this discussion is to highlight some of the possible reasons for the very negative reactions to whistleblowing. We are not suggesting that all the student respondents thought of all the arguments presented here. It may also be that the arguments were at the sub-conscious rather than conscious level. The discussion also indicates that appropriate solutions should be based on the group process and should include negotiation and consensus, as far as possible. Thus solutions could include a group redesign of the system to be more effective, co-operative group work to bypass or otherwise mitigate system inadequacies or representations to management agreed by and on behalf of the whole group about system problems. Another important issue is the importance of understanding the cultural context and drawing on the resulting strengths rather than seeing them as weaknesses or disadvantages. Thus the Irish context seems to have resulted in strong inter-group loyalties, as well as some suspicion of management and concern about group betrayal. However there may be suspicion of rather than identification with minorities and 'outsiders'.

5. CONCLUSIONS

This paper discusses two studies of whistleblowing. The first survey of Irish students found they had very negative perceptions of whistleblowing. The importance of understanding these results in the historical and political context has been shown, as well as the insights that have been gained from the applications of the ethics of care. The analysis has indicated that solutions based on group processes and negotiation may have the greatest chance of success in the Irish context. It also indicates the value of extending whistleblowing, which is often seen as the concern of a lone individual, to the group. The second study of a large scale manufacturing system bears out results in the literature both on the value of whistleblowing as essential in avoiding dysfunctional systems and the difficulties in getting it to be taken seriously. The authors hope to carry out more in-depth research at a number of locations to further investigate attitudes to whistleblowing and other ethical issues, as well as any differences based on gender, nationality or ethnic minority status.

REFERENCES

Acar, B. S. (1998). "Releasing Creativity in an Interdisciplinary Systems Engineering Course." *European Journal of Engineering Education* 23(2): 133-140.

Atieh, S. H., T. N. Al-Faraj, et al. (1991). "A Methodology for Evaluating College Teaching Effectiveness." *European Journal of Engineering Education* 16(4): 379-386.

Barker, C., Pistrang, N. & Elliott, R. (2002). *Research Methods in Clinical Psychology*, 2nd Ed., Wiley & Sons: NY.

Brandt, D. (1996). "Patterns and Challenges of Undergraduate Project Work in Germany." *European Journal of Engineering Education* 21(2): 197-204.

Brandt, D. and S. Ihsen (1998). "Creativity: How to Education and Train Innovative Engineers, or Robots Riding Bicycles." *European Journal of Engineering Education* 23(2): 131-132.

Chomsky, J. (1994). *Keeping the Rabble in Line*, Verso.

Court, A. W. (1998). "Improving Creativity in Engineering Design Education." *European Journal of Engineering Education* 23(2): 141-154.

Gilligan, C. (1982). *In a Different Voice*, Cambridge, Mass. Harvard University Press.Herman, E. &

Hersh, M.A. (2000). Environmental ethics for engineers, *Eng. Sci. & Education J.*, 9(1), 13-19.

Hersh, M.A. (2002). Whistleblowers – heroes or traitors?, *Ann Reviews in Control*, 26.

Hersh, M.A. and G. Moss (2004). Heresy and orthodoxy: *J. Int. Women's Studies*, 5(3).

Holmes, S. (1998). "There Must Be More To Life Than This." *European Journal of Engineering Education* 23(2): 191-198.

Ihde, D. (1999). *Expanding Hermeneutics: Visualism in Science,* Northwestern Univ. Press: Ill.

Ikkaracan, I. and H. Appleton (1995). *Women's Roles in Technical Innovation*, Intermediate Tech. Publ.

Johnson, G. (1988), *Processing of Managing Strategic Change*, Management Research News 11, 4/5, 43-6.

Kuhn, T. (1996), *The Structure of Scientific Revolutions*, 3rd ed., Univ. of Chicago Press.

Kohlberg, L. (1981). *The Philosophy of Moral Development*, New York, Harper and Row.

Latour, B. (1999). *Pandora's Hope: Essays on the Reality of Science Studies*, Harvard: MA.

Lenschow, R. J. (1996). "Industrial Development and Education." *European Journal of Engineering Education* 21(2): 149-160.

McCorkle, C. (1989). Price, preference and practice: In: *The Dynamics of Grain Marketing in Burinka Faso*, vol. II, University of Michigan Centre of Research on Economic Development.

Martin, M.W. and R. Schinzinger (1996). *Ethics in Engineering*, edition, McGraw Hill.

O'Day, R. (1972). Intimidation rituals, *J. of Applied Behavioral Science*, 10, 373-386.

Pfeffer, J. (1982). *Organisations and Organisation Theory*, Pitman: Boston.

Platts, J. (1998). "Participating in the Work of Creation." *European Journal of Engineering Education* 23(2): 163-170.

Pugh, D. (1993) Understanding and managing organisational change, In: *Managing Change*, Mabey, C. et al. (eds,), London, Paul Chapman

Stapleton, L. (2003). Information systems development as folding together humans & IT, In: Grundspenkis, J, et. al.(eds.), *Information Systems Development*, Kluwer/Plenum, pp.13-24.

Stapleton, L. and M.A. Hersh (2003). Exploring the deep structure of ethics in engineering technology in *SWIIS '03*, Waterford, Ireland, Elsevier.

Shapin, S. and Schaffer , S. (1989). *Leviathan and the Air-Pump*, Princeton Univ. Press: NY.

Stapleton, L. & C. O'Dowd-Smyth (2003). Anaesthetising ourselves, *SWIIS '03*, Waterford, Ireland, Elsevier.

Van-Der-Vorst, R. (1998). "Engineering, Ethics and Professionalism." *European Journal of Engineering Education* 23(2): 171-180.

ELSEVIER

IFAC

PUBLICATIONS

www.elsevier.com/locate/ifac

A METHODOLOGICAL APPROACH TO DIVERSITY ANALYSIS

P. Albertos*, I. J. Benítez*, J. A. Lacort*, J. L. Díez* and A. Ramos[#]

* Department of Systems Engineering and Control. (DISA)
Universidad Politécnica de Valencia , PO Box. 22012, E-46071 Valencia, Spain
E-mail: pedro@aii.upv.es

[#]Institut Universitari d'Estudis de la Dona, Universitat de Valencia
Avenida Blasco Ibáñez 32, 46010, Valencia, Spain
E-mail: amparo.ramos@uv.es

Abstract: Provided the changing framework for companies, a growing interest in diversity management styles has been noticed in organizations. In this paper, a methodology for diversity analysis at workplace is presented. When information about position characteristics and labour market for the studied job is available, any work team in an organisation can then be analysed. The diagnosis of the team from a diversity point of view is done and (should it be the case) suggestions of EO actions are given to improve the situation. *Copyright © 2004 IFAC*

Keywords: Simulation, Social and behavioural sciences, Work organization, Computer applications, Interactive programs, Knowledge-based systems, Recruitment. .

1. INTRODUCTION

Organizations are immerse, at the present moment, in a complicated environment of social transformation mainly due to the globalization process, the internationalization of markets and the irruption of IST (Information Society Technologies), among others. Novel values are appearing in society, leading to new client demands. Organizations have to change and adapt in order to keep their positions in the market. Companies need new management styles to cope and benefit from this situation (Alvesson and Billing, 1997; Asplund, 1988).

The diversity strategy (Barberá, *et al.*, 2000; Jacobson, 1999; Sarrió, *et al.*, 2003) takes advantage of heterogeneous groups, diverse in gender, age, race, nationality, etc. The key issue of diversity is to value people and their inter-individual uniqueness, since the team will be enriched from different experiences, point of views, abilities, etc. This strategy is based on equality (equalitarian access to positions) and difference (different ways of working) and it seems to be the solution for most companies, not creating drawbacks for any of them.

Although a plural labour force is available and advantages for companies of diversity strategies are clear, discrimination due to different reasons, such as gender, race or religion, can still be found at all levels of the labour environment, despite the norms and laws concerning equal opportunities (EO).

In this paper, a methodology is proposed for diversity analysis in organizations. First of all, a detailed explanation of a model for static analysis of diversity in organizations is done in section 2. A software tool called Divers@T that implements this methodology for analysis of work teams is presented in section 3, where an example of analysis of gender diversity in a real company is also included. Next, ongoing work related to dynamic analysis is presented in section 4. A final conclusions' section is also included.

2. DIVERSITY STATIC ANALYSIS

The proposed diversity analysis at a work team is based on data gathered by experts about the situation of the organization and the offer at a given time, from the point of view of the diversity aspect being analysed. First, from the requirements of the job, a diversity profile is generated, which suggests adequate percentages of diversity. Next, a study of the offer yields the available diversity for the analysed job. The resulting scenario is shown, suggesting optimal values for diversity according to the data. The user can then make a selection of applicants or enter current diversity percentages at the work team, to obtain an evaluation of the diversity, a diagnosis of the selection and the scenario, as well as, if needed, a collection of EO actions, generated from the analysis of the diversity constraints.

Two facts must be considered when analysing diversity at a work team. First, the number of possible diversity groups, according to the aspect of diversity under study. For instance, when coping with gender diversity there will be two groups, women and men. When age diversity is the issue being analysed, there can be as many groups as the user might want to include, such as young, middle-age and senior (three), or any other partition. The diversity analysis is, therefore a *multivariable* analysis, where every variable or diversity group must be user-defined prior to any analysis. It can also happen, and this is the case with age, that sometimes boundaries among diversity groups are not clearly established, acting as fuzzy membership functions (Fig. 1), while in other cases, such as gender, this will not be the case. The explanation of the whole analysis process in this paragraph will be made labelling the different diversity groups as A, B, C, etc.

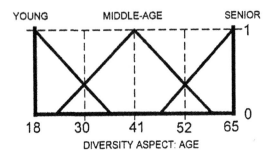

Fig. 1. Example of diversity groups for age analysis.

Once the different diversity groups have been defined, they must be somehow assigned a *discrimination degree*, since the analysis must take into account real disadvantages certain groups encounter on their way to be hired or promoted, called *diversity constraints*. This degree varies on user's decision for each group, from 0 (no discrimination) to 1 (total discrimination), admitting intermediate values.

When diversity groups have been established and discrimination degrees have been added, a study of the job is needed in order to set the diversity profile, which will yield the functional diversity.

2.1 Functional Diversity.

The functional diversity stands for the *requested* diversity, i.e., the optimal percentage of diversity groups for a specific work team, based on a previous study. This study is done by experts, based on a list of abilities or skills assumed to be needed for the job, and a classification of those abilities according to membership to a specific diversity group (group A, B, etc.), or to them all (being a *neutral* skill). This classification proves that most of the skills are neutral for most of the diversity aspects. However, there can be some skills classified as not neutral, belonging to a specific diversity group, allowing a slight deviation of functional diversity from the ideally mathematical and social standards. The overall distribution of skills into the different diversity groups or into the neutral category will yield the functional or demanded diversity for that job, that is, the expected percentages of workers from each diversity group to form the work team. Since the neutral factor is taken into account, the functional diversity is not a fixed number, but a *distribution* function, displaying a *range* of percentages where the diversity would be adequate, meeting the requirements of the job's diversity profile.

Another option would be to attempt to "fuzzify" each skill, stating a percentage of membership to specific sectors and to all of them ("neutral"), being 1 the sum of all percentages. The result would be again a distribution function, showing the range of percentages for an optimal adequacy to the studied job's diversity profile.

Moreover, the functional diversity takes into account percentages of people from the different diversity groups, either working at the organization and among the customers. These percentages receive the names of *internal* and *external* customer.

The functional diversity is used as a referent for the analysis. Along with the study of the offer, they build the scenario of the diversity prior to a selection or promotion.

2.2 Offer Diversity.

The offer diversity relates to the diversity *available* at the labour market, taking into account the requisites for the analysed job, and the limitation on number of applicants due to the diversity constraints.

Offer availability. Although the job's profile will be almost neutral in most of the cases, the situation at the labour market can be very different, having a specific diversity group as predominant in number and career, due to some reasons, being probably discrimination one of them.

Statistical data are needed in order to include the percentages of diversity groups available at the offer meeting the *minimum* requirements for the job. A further analysis can then be done, if a set of *desirable* extra requirements for the job is defined, and this set is clustered in different accumulative levels of knowledge or expertise, from the lowest level (the minimum) to reach, up to the level with the maximum specifications to accomplish. The percentages of offer composition per level must be entered. This process brings the possibility to choose different levels for the selection and observe the availability and the results.

The Analysis of Constraints. The diversity constraints gather all the known obstacles that certain diversity groups may find on their way to be hired or promoted, due to discrimination or prejudices. These constraints have been classified in three areas, which are the organization's *policies* and organization's *culture* and *values* considered both of them as external-organizational conditioners, and *personal* constraints as internal conditioners.

A constraints questionnaire, included as part of the analysis, captures the situation of the diversity constraints and the EO conditions at the organization. For the sake of veracity and the results' usefulness, this information, either introduced by the responsible or obtained from the workers, should be real information. From the constraints questionnaire, indices are obtained, which are a measure of how much these constraints affect the different diversity groups. The indices are used for two purposes: first, to reduce the availability of diversity groups at all knowledge levels, depending of the discrimination degree assigned to each group. Second, the indices will be used as part of the diagnosis to select specific EO actions to be suggested to eradicate the constraints.

Result. From the analysis of the available offer and the constraints, plus the required knowledge level, a new distribution function is obtained, called offer diversity. The offer diversity shows ranges of diversity groups available at the offer for a specific job with a specific level.

2.3 Resulting Scenario and Selection.

When putting together functional and offer diversity, an ideal situation would depict both functions very close to each other, but this situation will not be the usual. At this point, the user can observe which are the adequate percentages for both functional and offer diversity, and how far or close these values are from each other. Once the *scenario* has been displayed, the next step requires the user to enter the size of the work team being analysed and to make a selection of workers, or to enter current percentages of diversity groups at the team.

2.4 Results, Diagnosis and EO Actions.

The functional and offer distribution functions for each diversity group allow to visual and mathematically observe the adequacy of a selection to the ranges of diversity groups both requested for the job and available at the offer. These adequacy values are the *indices of diversity* and, along with the global index of constraints and other measures, will be used to generate a *diagnosis* of the situation and the selection made, by means of a *fuzzy inference system*, which will yield a global grade of diversity and EO. Furthermore, the analysis of the different constraint indices will be processed through an *expert system* to select and display specific actions, aimed to solve the problems of EO detected.

At this point, the analysis is concluded. Subsequent simulations can be done to observe and compare the different results and evaluations when changing the scenario or the study conditions, or attempting different selections.

3. GENDER DIVERSITY: THE DIVERS@ TOOL

The European project "Divers@: Gender and Diversity"[1] the authors are involved in, strives for the application of diversity and gender perspective to promote the women access to high positions of responsibility, contributing to the elimination of job segregation at the management levels of the labour hierarchy, in the University and in other working organizations. The basic idea is the need of changing of attitudes, values and general policies, through the application of the gender perspective and the diversity approach, specifically gender diversity.

A software simulation tool, called Divers@T (Fig. 2), was designed, with the aim to analyse gender diversity and EO conditions at work teams within an organization, especially at executive positions. It was developed blending the advantages of Visual Basic and Matlab^{TM}. In order to communicate both development platforms, a COM object has been built, allowing algorithms and functions designed in M files to be reused in standard applications, with minimum effort.

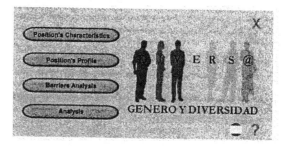

Fig. 2. Front-end of the Divers@T software.

The Divers@T software follows the methodology previously described in a step by step process, making a study of the job's profile, an analysis of the offer and constraints, an evaluation of the situation and a selection, and the obtention of EO indices and selection of EO actions.

When analysing gender diversity, the diversity groups object of study are two, namely, *women* group and *men* group. The discrimination degree in this case is maximum (1) for women and non-existent (0) for men.

[1]The project "Divers@: Gender and Diversity" is part of the EQUAL initiative, supported by the European Social Fund (Project ES296).

3.1 Job's Profile.

The list of abilities included are mainly related to managerial positions. A list of pre-defined job profiles, with the correspondent abilities, is included (Kite, 2001), although new profiles can be designed by the user.

For instance, an analysis of the Rectoral team of the University of Valencia, Spain, is carried out. The "Rectoral Team Member" profile is chosen, which comes with a set of pre-defined abilities or skills, classified into the different diversity groups (Feminine, Masculine, Neutral). For example, from the list there is an "Empathy" skill, classified as "Feminine", a "Capacity for individual leadership", classified as "Masculine", and a "Supervision Capacity", classified as a "Neutral" skill, among others (See Fig. 3).

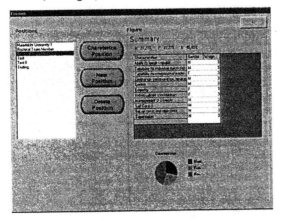

Fig. 3. Rectoral team member profile, with gender-classified abilities.

3.2 Barrier Analysis.

The diversity constraints concerning gender have been studied for a long time. They receive the name of *gender barriers*, and are present in the three areas previously referred (Section 2.2). The user of the Divers@T software enters the information about gender barriers through a barriers *questionnaire*. A set of affirmations are presented (Fig. 4), that the user has to value from 0 to 10, meaning 0 extremely negative or very poor, and 10 quite positive or excellent. From the assessments, EO indices are produced, that will be used to model the limitation that women suffer when accessing to the work team, as well as to select the specific EO actions.

3.3 Functional Diversity.

The functional diversity is the diversity being requested. It is based on the diversity job's profile, and also the percentages of women being *internal* and *external* customers (Section 2.1).

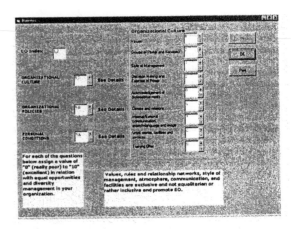

Fig. 4. Gender Barriers questionnaire.

The result is a distribution function, which is a modified generalized bell function, depicting the adequacy of percentages of women to the diversity job's profile (See Fig. 5).

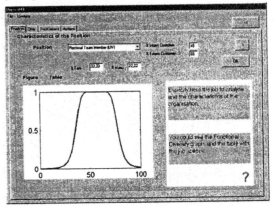

Fig. 5. Functional diversity.

In this example, the optimal adequacy (equivalent to 1 at the vertical axis) according to the functional diversity would be between 46.17% and 67.17% of women at the work team, being the centre of this range at a 56.67% of women.

3.4 Offer Diversity.

The user must enter data concerning the availability of women at the offer. After entering the total number of possible candidates and the percentage of women, the number of knowledge levels with percentages of women and men available at each one, and the desired level, along with the limitation on access due to the gender barrier index previously obtained, a figure is generated, depicting the evolution of the availability of women and men at the different levels (Fig. 6, a). From the required level, the offer diversity is obtained (Fig. 6, b). In this example, the resulting offer diversity shows the maximum adequacy at a 17.10% of women, which is highly bounded due to the gender barriers and the very low women availability in this case.

3.5 Scenario and Selection.

Next, the user enters the size of the team, 13 people in this case. The scenario is shown, suggesting the adequate composition of the team, in percentage and number, according to functional and offer diversities. These numbers are, from 6 to 9 women according to functional diversity, and only 2 according to the offer, due to the conditions previously mentioned. At this point, the user will make a selection of women for the team, either entering a new selection or the current number of women at the existing team.

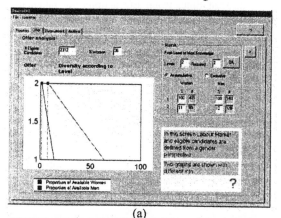

(a)

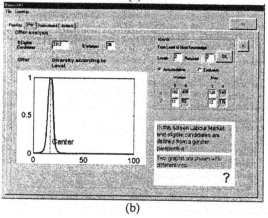

(b)

Fig. 6. Evolution of availability of women and men according to knowledge level (a) and the resulting Offer diversity (b).

3.6 Results, Diagnosis and Actions.

When a selection is made, being 2 women and 11 men in this case, the situation is shown (Fig. 7) and the diversity indices, for both functional and offer, are obtained (Fig. 8). These indices are the adequacy values of the selection to both functions, and are 9.08 for the offer diversity and 0.09 for the functional diversity. The global index of gender barriers, which is 6.93, ends up the collection of diversity indices.

A fuzzy inference system analyses the values obtained and yields three different EO degrees, varying from 0 (very bad) to 10 (very good), being

one for the scenario or situation of the diversity, obtaining a 4.02, another one for the selection made, obtaining a 2.31, and a third global degree, evaluating the situation and the selection at a whole, with the value of 3.09.

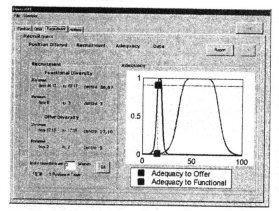

Fig. 7. The adequacies of the selection to both diversities are displayed.

The numeric information is completed with sentences describing and evaluating the results.

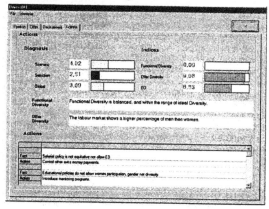

Fig. 8. A diagnosis of the situation and EO actions are displayed.

Finally, the analysis of all the indices obtained through the gender barriers questionnaire, are used to make a selection of negative facts detected and a set of actions suggested to solve them (Fig. 8).

All the information being generated can be arranged in a text format, and a report can be printed.

4. THE DYNAMIC ANALYSIS

A next step in the analysis of diversity at workplace takes into account not only the situation of the diversity at a given time, but its evolution through time, measuring and evaluating the trend of specific indices, such as general satisfaction, diversity indices, productivity, constraints indices, and global EO grade (Benítez, et al., 2003).

The objective is to be able to forecast the impact that the current environment and organization policies will have in both its productivity and diversity. It is important to understand that diversity is highly desirable from a fairness perspective as much as to achieve the maximum levels of productivity and quality of services or products from the organization.

In order to predict the situation in a period of time, we consider some variables that will indicate the changes in the personnel and therefore the composition of the staff in the future regarding the diversity aspect under study. Some of these variables are satisfaction, wages and policies, among others.

Given the satisfaction variable mentioned before, its variation along the time may be represented like in Fig. 9. This variation may occur as a result of different circumstances or events that may alter in one direction or in the other, as well as take a period of time to show the effect.

It is easy to foresee that a satisfaction of a member of a group dropping below an established threshold will lead to the person leaving the organization, and therefore the proportion of the diversity groups will be modified. The latter study should be repeated for as many variable as the user may understand are involved in the process.

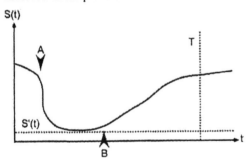

Fig. 9. Evolution of variable Satisfaction through time.

Now a simulation can be attempted, considering two scenarios, the ideal one in which the organisation has implemented the actions suggested, and the opposite, or the result of the evolution without any corrective action. In Fig. 10 the dotted line represents the evolution of the diversity along time when actions are not taken to improve EO, while the solid line depicts how diversity is improved by the actions. The graph can give us an idea of how conditions and quality will vary along time, since we understand both diversity and quality are related and the latter highly depend on the former.

The simulation may be repeated as many times as needed, and the static analysis will be based on the scenario obtained from the previous simulation.

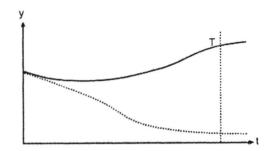

Fig. 10. Two cases of evolution of the diversity index through time, when EO actions are taken into account (solid line), and not (dotted line).

5. CONCLUSION

A methodology for diversity analysis at workplace has been presented. The method of analysis of diversity in a work team requires a detailed previous description of positions' characteristics (functional diversity) and labour market (diversity of the offer) for the studied job. Results include a comprehensive diagnosis and suggestions of EO actions if necessary.

REFERENCES

Alvesson, M. and Y.D. Billing (1997). *Understanding gender and organisations*. Sage, London.

Asplund, G. (1988). *Women managers: changing organisational cultures*. John Wiley & Sons, New York.

Barberá, E., M. Sarrió and A. Ramos (2000). Mujeres y estilos de dirección: el valor de la diversidad. *Revista de Intervención Psicosocial*, **vol. 9 (1)**. (In Spanish)

Benítez, I. J., P. Albertos, E. Barberá, J.L. Díez and M. Sarrió (2003). Equal Opportunities Analysis in the University: The Gender Perspective. In: *Computer Aided System Theory - Eurocast 2003. Revised Selected Papers* (Springer-Verlag Berlin Heidelberg New York), 139-150. Germany, ISBN 3-540-20221-8.

Jacobson, B. (1999). Diversity management process of transformational change. In: *Total E-Quality Management Conference*. Nuremberg, April 29th., 1999.

Kite, M. (2001). *Gender stereotypes. An Encyclopedy of women and gender. Sex similarities and differences, and the impact of society on gender*. Academic Press, **vol. 1 (A-K)**, 561-570.

Sarrió, M., A. Ramos and C. Candela (2003). Género, trabajo y poder. In: *Psicología y género* (E. Barberá, I. Martínez), Pearson Education (In press). (In Spanish)

ELSEVIER

IFAC

PUBLICATIONS
www.elsevier.com/locate/ifac

REVISING THE THEORY OF SOCIALLY INCLUSIVE SYSTEMS ENGINEERING

SOCIAL IMPACT CONSIDERATIONS IN DISTRIBUTED ASSISTIVE SYSTEMS FOR THE LEARNING DISABLED

Stapleton, L.[1], Duffy, D.[1], Lakov, D.[2], Jordanova, M.[2], Lyng, M.[1]

[1]*ISOL Research Centre, Waterford Institute of Technology, Republic of Ireland*

[2]*Bulgarian Academy of Sciences, Sofia, Bulgaria*

Abstract: the relationship between humans and advanced technology can be viewed as a network of interests of technical and non-technical agents. Drawing upon instrumental realist approaches as set out in agent network theory the paper describes a project currently underway in Ireland and Bulgaria which delivers comprehensive, assistive *systems* for people with learning disabilities. These systems address many of the difficulties associated with current assistive technology (AT) programmes, problems typically associated with the narrow focus of AT upon technology solutions. Whilst limited, it delivers a sound ethical basis for technology-centred programmes, and new trajectories for engineering research. *Copyright © 2004 IFAC*

Keywords: social impact of automation, assistive technology, ICT, artificial intelligence, ethics.

1. INTRODUCTION

Current theories of systems engineering often adopt a 'one solution fits all' approach (Siddiqi (1994), Stapleton & Murphy (2002)). Consequently, many engineering methodologies do not take sufficient account of local context issues, and especially ignore the difficulties that socially marginalized people face in working in contemporary organisations. The functional rationalism that underpins the one-solution-fits-all paradigm has recently come under significant pressure from systems and engineering theorists who argue that it is far too unsophisticated for the kinds of complex organisational and social information spaces that are now so common in both business and education (Stapleton (2001), Clarke & Lehaney (2000)). This functionally rationalistic approach reflects a scientific rationalism which has often excluded the marginalized from the centre of scientific discourse. Consequently, we have mobile phones that the elderly or visually impaired find difficult to use and large scale information systems which traumatise their user community (Stapleton (2003), (2002)).

2. HUMAN-TECHNOLOGY HYBRID SYSTEMS

In this paper the systems design process is seen as the folding together of humans and technology into a single, coherent system. Adapting Latour's instrumental realist approach yields a model which illustrates how such a folding process might work.

Latour argues that the twin mistake of functionally rationalism, on the one hand, and sociological approaches on the other, is that they both try to understand the relationship between humans and non-humans is their focus upon essences (artefact or human). In Latour, both are transformed into something new, as illustrated in figure 1. This illustrates to the software engineer and the information technologist one way in which social impact is created. The technology is no longer an essential thing, nor is the human. It is *both together* i.e. Human and artefact are folded into each other. They are transformed into something new, a composite of social and artefact (e.g. Ihde (1998)).

This shifts attention away from 'technology' or 'society' or 'social context' to a new combination of social and technological: the 'hybrid system'. Once we do this, we can see that goals (or functions) change from those of the individual components (human and non-human) to the goals/functions of the hybrid actor. This is a very important philosophical step in our base assumptions. In systems engineering we must now focus upon a whole new array of actors and actions – the hybrid systems and their functions. This opens a new research trajectory for the social impact of technological artefacts. We notice that we are now dealing with, not the goals of humans or technologies, but the new, distributed, mediated and nested set of practices whose sum may be possible 'to add up' but only if we respect the importance of mediation (*interference*) in the relationship.

As this process of *interference* and *folding* develops we note how the original (perhaps explicit) goals can be lost in a maze of new goals as the entire system becomes more and more complex. For example, an early human discovers the stick, and we have a stick-human hybrid. Perhaps the human initially uses this stick to plough the ground. However, the human becomes frustrated with the stick and sharpens it thus creating a whole new set of goals and functions, such as the stick as a defensive or offensive weapon. This whole new set of goals or functions could not have been foreseen at the outset when the stick was originally discovered and deployed. It illustrates how technology deployment must recognise that, as humans enter into and develop new relationships with the technology, goals and functions shift. This rationale directly implies that researchers of social impact must now introduce learning and adaptation theory into their armoury. Simultaneously, they emphasise design and re-design principles for the technical component. We have not been 'made by our tools' as indicated by Marx and Hegel (homo faber fabricatus). Rather the 'association of actants' is the important thing for the researcher of social impact associated with IT deployment.

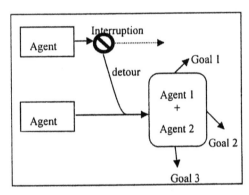

Figure 1. Interference & Goal/Function Transition (adapted from Latour (1999))

By implication systems engineering researchers must understand how new goals and functions appear, new goals and functions can be understood and (re)directed appropriately

It is apparent that this requires the application of a social theory that includes organisational learning and decision making. Any revised theory of technology deployment must emphasise the human element of the new human-machine system and cater for humans as they attempt to make sense of the new world into which they are thrust: an inter-subjective, shifting space in which they are intricately bound with a new information technology artefact, and which often makes little sense to them (Stapleton & Murphy (2002)). Systems (re-)design and deployment principles must be enhanced, or augmented, so that they can be folded into the overall management of the hybrid system. Furthermore, these approaches must be accompanies by sensemaking support which in turn feeds into and out of human centred systems engineering re-design

process. A learning/explication support process is also needed which feeds into and out of technical and non-technical elements of the hybrid system, whilst treating it as a coherent whole (Stapleton (2003)).

But how does such a theoretical approach manifest itself in a practical systems engineering problem? The next section will set out a research study currently underway at the Waterford Institute of Technology and the Bulgarian Academy of Sciences. It shows how, by adopting a networking rationality, an entirely new application area emerges for assistive technologies.

3. ASSISTIVE TECHNOLOGIES FOR LEARNING DISABILITY

The American Technology-Related Assistance for Individuals with Disabilities Act of 1998, defines Assistive Technology as "any item, piece of equipment or product system...used to increase, maintain, or improve functional capabilities of individuals with disabilities" (P.L. 100-407 (1988)). The use of these application tools by people with learning disabilities generally falls under two methodologies; namely the *Compensatory* approach and the *Remedial* approach.

The compensatory approach applies when an assistive tool is used to circumvent the individual's deficit, thereby allowing them to avoid the implications of their disability. This is generally achieved by playing on their areas of established strengths rather than on their areas of weakness; for example, if an individual has poor or limited reading ability, then the use of taped texts or screen reading software allows them to avoid the necessity to read, rather than assisting them in the development of their own reading abilities. The remedial approach on the other hand does the exact opposite: AT is used to improve areas of deficiency, rather than simply compensate for them (Garner & Campbell (1987), Day & Edwards (1996), Raskind (1998)).

While both approaches can overlap, the compensatory approach is the preferred method when dealing with adults, and can be particularly appealing to those who have experienced 'burnout' from years of remedial solutions, that yielded little benefit (Raskind (1994), (Gray (1981), Vogel (1987), Mangrum & Strichart (1988)).

Individuals with learning disabilities are each unique in their profile, in terms of weaknesses, interests, strengths and experiences, therefore a tool that is of great benefit to one individual may be a hindrance to another. Similarly, what is suitable in one context or environment may be inappropriate in another (Raskind (1994)). In adopting an AT further consideration should include the level of learning disability involved, the individual's established strengths, abilities and skills, the environment in which the tool will be used, the context of interaction and the individual's "Technology Quotient", i.e. their ability and comfort with using technology (Bisango & Haven (2002), Raskind & Scott (1993)).

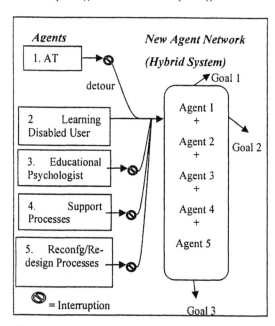

Figure 2. Interference & Goal/Function Transition diagram for the Learning Disabled Assistive System

For this paper the notion of the 'technology quotient' is very important. It specifically illustrates how the social context and social impact of the AT must be incorporated into the design process. We will see how this notion must be extended to ensure that the entire environment (social/technical context) must be co-designed according to human-centred (HC) principles which are proven to lead to increased self-esteem and other important individual and group benefits, in turn yielding far more effective technology-driven projects (Brown (1988), Reiff et.al (1992), Barton & Fuhrmann (1994)).

In the HC view, Assistive Technology usage must also be accompanied by the appropriate, on-going learning and psychological support processes. This will deliver sensemaking support into the socio-technical context (Stapleton (2003), Mills (2003)).

The considerations taken into account when choosing the method, by which training and support will be provided for the user, should be as stringent as those taken when choosing the tool. Individual deficits, previous experiences, preferred learning styles and personality traits should all be evaluated, addressed and catered for in terms of the approach taken in delivering the training material. This ensures that the technologically mediated environment makes sense as quickly as possible to the user.

It is important to note that, to a dyslexic individual, their learning disability can act as a brick wall or a locked door, between them and their understanding or communication capabilities. The way to unlocking that door is to find the correct key, but the key for each individual is as unique as they are themselves. By providing them with their own

distinctive 'Assistive System', enabling them to become system literate and removing the fears and negative implications of system usage, only then can they find the right key to their own locked door.

Current AT adopts a 'one-solution-fits-all' paradigm. The technology generally is developed with technological constraints in mind, with little appreciation for key aspects of learning disability. There are ethical concerns here. The technology itself does not recognize the individual's need to adapt to the AT. Indeed, the keyword in Assistive Technology is 'Technology' rather than 'Assist'. Assistance in this case means far more than simply providing the user with a series of complex tools or functionalities. Instead, there are key psychological factors, which will enable (or otherwise) the user to effectively utilize the tools. Indeed, for some dyslexics, the experience of marginalisation could potentially be heightened by the provision of a so-called assistive technology which they can neither utilize effectively nor understand, simply because the technology is not designed for THEM as individuals. In fact, the technology development process never incorporated their worldviews into the design process. This exposes certain darker aspects of power within the systems engineering discipline as we find the 'learning enabled' designing solutions for the 'learning disabled'. In engineering such a solution the difficulty then is two-fold (at least):

To provide a system which treats human and machine as a single system, centred upon the individual human's needs

If we adopt Latour's ANT view we require an Assistive System as shown in figure 2. It is most likely that this diagram could include bi-direction (or even cyclic) lines between the hybrid system and the goals. So, for simplicities sake, the model incorporates a specific interference transitions that the arrows are unidirectional. It is understood that in the theory the entire process is necessarily cyclic and is unlikely to be discrete.

The diagram replaces a traditional context diagram but tells us a lot more than the primary interfaces and scope of the system, as would be set out in a context diagram. Figure 2 indicates the primary agents whose interests must be managed in order to develop a successful system. The diagram also centres upon the disabled user rather than the technology i.e. the trajectory of the user remains unaffected, but the trajectories of all other agents are the ones with which the project interferes. We can also identify from the relatively simply diagram the primary components of a sensemaking support process (agents 1, 3 & 4) as well as the need to include re-configuration processes in the design of the overall system. Using this simple diagram, we can immediately recognize that the technological agent must comprise a highly flexible and adaptive technology in order to support the ongoing re-design/re-configuration process. Furthermore, an interest which emerges is the need in the support process to link to other similar users. This indicates the need for an e-community (human network). We begin to see that we need a new form of assistive technology if we are to manage the interests of the other agents (especially agents 2 and 3). The AT must provide both flexible, intelligent, reconfiguration capabilities AND the ability to operate in a distributed environment. Such a technology is the Soft Computing Agent (SCA).

4. SCAS AND ASSISTIVE SYSTEMS

The logistics and costs involved in creating a system that is as individual and adaptable as the user, make such a system potentially non-viable. Through the use of the SCA paradigm however, an intelligent distributive system is not only achievable but also affordable. By utilising the intelligent, adaptable and distributive capabilities of SCAs the development of a distributed hybrid system, encompassing all five actors becomes a reality. (This should be all five agents but that may cause confusion between network agents and computing agents)

The SCA paradigm is a combination of Intelligent Agents and Soft Computing driven expert systems, (Lakov & Kirov (2000)) which enables the provision of appropriate training and support materials and processes to be made available to the user according to their profile of abilities and learning styles, as opposed to their geographic location and local resources. Through the use of an Intelligent Fuzzy Agent, virtual groups of 'similar ability' users can be formed, with the membership of each group being determined according a specified Fuzzy Rule Base. Such virtual groups enable the user to not only interact with other users but also to avail of technical and psychological support provided specifically according to their individual needs, while enabling the intelligent distribution and provision of training to be delivered in such a manner so as to allow the user to obtain the maximum learning outcomes.

4.1. 'Success'-ful AT

In this context it is readily apparent that 'success' is an elusive term. The diagram chows how the new hybrid system will generate a new set of goals and functions which will be difficult to predict. The outcome of AT will hopefully be a successful system, but success can only be measured in terms of the effective support that the technology provides in the overall context of a supportive learning environment. We see that, in fact, the technology as a stand-alone system, cannot, of itself, be successful. Indeed, the nature of learning disability is such that, from individual to individual, success factors will vary, and even change as people become better able to address their own learning difficulties. Any AT to date which has been 'successful' therefore may be regarded as a 'lucky break', because the support needs for people with these conditions vary from person to person.

Theoretically, this poses a significant problems for systems engineers seeking to develop purely technical solutions according to the traditional methodologies (such as SSADM). What is required is an approach which recognizes the dynamic nature of success, from person-to-person and from day-to-day.

4.2. Assistive Systems: towards Human-Machine HYBRID Systems

The diagram indicates that further attention must also be given to any psychological support that may be required. As with any technology, the psychological impacts can often be considerable, and be even further compounded by the psychological effects of the disability.

On-going support must be provided, to ensure the user is gaining only the intended benefits without experiencing any of the potentially negative impacts. This is indicated in figure 2 by agent 4, which will focus upon processes for the effectiveness of the new agent network (an aspect of explication support during the sensemaking process (Stapleton (1999)). Only when the user is provided with the correct tool, the proper training in their usage of the tool and when the necessary psychological and physiological supports are put in place, will the user be able to gain the full compensatory and/or remediatory benefits of the new system. By ensuring that the user has the knowledge, ability and support to use the tool and as a result, develop their own 'hybrid system', the deficits and strengths of each user can be addressed.

However the logistics of this requires serious consideration. How can geographically dispersed users be provided with on-going psychological support and appropriate training, in a timely and economical manner? The interests of agents in this area must be addressed in the assistive system.

5. SUMMARY: CONSTRUCTING A HUMAN CENTRED ASSISTIVE ARCHITECTURE

In summary, a research project is needed which must bring together entirely new approaches to the problem of assistive technology for the learning disabled. In this approach it must place the power over the final system configuration in the hands of the individual. It must also treat the human/technology system as a single system, not isolating (in the design architecture) the human from the technology. The research team recognised that such a solution must, therefore, incorporate key elements in the design architecture:

1. Distributed Technology components

2. Psychological Processes

3. Humans (users, medical professionals etc.)

The solution must enable each of these elements to leverage each other, coexisting in a single, synergistic system.

5.1 Summarising the application environment

A project has been developed that strives to use the capabilities provided by the Soft Computing Agents, to identify the best means of developing the technology and the associated support processes, so that the overall hybrid system is as effective as possible, thereby allowing the user to gain the maximum benefits. The aim is to provide, not just Assistive Technology, but more specifically an 'Assistive Systems', i.e. tools, training and support; that are tailored to the specific requirements and experiences of the individual user. A further aim is to distribute these systems based on the user requirements, as opposed to the user's geographic location, in an efficient manner in terms of time and economics.

A project team has been established to develop solutions in this area. The international team comprises leading experts in intelligent systems, information systems methodologies and e-medicine. The local test site incorporate a team of researchers from the Information Systems, Organisations and Learning Research centre working with the Disability Support Unit. The project is to be targeted at students in higher education who experience learning difficulties. The project will ensue according to the following steps: -

1. Each student will be assessed in terms of their areas of weaknesses, strengths, personality, preferred learning styles and previous experiences. This information will then be used to build an individual profile of the students, according to a determined set of criteria.

2. This profile will then be fed into a central database, which will house the information relating to all students involved in the study. It is proposed that all participants will be under graduate students in their second/third year of study, and will be registered with the appropriate authorities within their academic institutes as being dyslexic.

3. A set of three groups will be defined using a Fuzzy Rule Base. These groups will be homogenous groups of students that have the same or similar levels of ability and disability. The membership class of each group will be determined using Fuzzy Logic, i.e. how strongly a students profile relates to the membership criteria of a particular group.

4. Systems of Intelligent Agents will be used to search through a central database and to assign each student to a group according to their level of membership class.

5. Data, such as training instructions and communication arenas will be dispatched throughout the system, with each student automatically being provided with only the information that is relevant to them according to their group membership.

6. SCA systems allows the formation, automated maintenance and interaction of groups according to specified criteria, as opposed to geographic location. This means that specialised training and support for the student is possible and can be designed, solely according to their individual needs and learning styles. This enables the provision of an 'Assistive System' consisting of the tool, support and training, as opposed to the traditional approach of an 'Assistive Technology' comprising of only tools & limited instruction.

6. CONCLUSION

Systems engineering methodologies research has not tackled directly the power relations associated with advanced systems. It is readily apparent that adapted agent network theory can be effective in helping us to understand key aspects of the complex relationships between humans and technology. By adopting the approach set out in this paper, the systems engineer can also identify human-centred aspects associated with the social impact of a new technology in a reasonably coherent way.

This paper sets out a project, currently underway at two European sites. The project shows how a revised, interdisciplinary and agent-network view of a well established problem (the provision of AT for learning disability) provides a fresh approach. It shows how the perspective applied here can illuminate potential points of failure, inappropriate underlying assumptions associated with the ways in which the human agents will respond to the new technology. More importantly, the approach denotes a paradigmatic shift which re-places humans into the centre of the technology research programme, with the associated ethical implications of that focus.

This paper focuses the attention of engineering research activities on the agents in both the centre of the process (laboratory equipment, the engineered technology) as well as the fringes of the society in which the research is conducted (for assistive technologies, the people with special needs participating as agents with a key stake in the project). Whilst this is a deliberate 'mis-reading' of Latour's instrumental realism, it is useful since it shows how a combination of Foucault's knowledge-power views and Latour's realism can provide a powerful basis for revising technological research. Whilst this paper specifically shows how such an approach can be utilised in a health informatics project, it has wider implications for a variety of automation engineering research positions.

ACKNOWLEDGEMENTS

The authors thank **Enterprise Ireland** for their support in funding this research.

REFERENCES

Barton, R.S. and Fuhrmann, B.S. (1994). Counselling and psychotherapy for adults with learning disabilities. In P.J. Berber and H.B. Reiff (Eds.), *Learning disabilities in adulthood: Persisting problems and evolving issues*, pp 82-92. Andover Medical: MA.

Bisango, J. & Haven, R. (2002). Customizing technology solution for college students with learning disabilities.*Perspectives, 1*, p 21-26.

Boland, R. (1985). 'Phenomenology: A Preferred Approach to Research on Information Systems', in Mumford, E., et. al. (eds), *Research Methods in Information Systems*, Elsevier: pp. 193-201.

Brown, C. (1988). *Computer access in higher education for students with disabilities* (2nd Ed.), U.S. D. of Ed: Washington

Clarke, S. & Lehaney, B. (2000). 'Information Systems as Constrained Variety – Issues and Scope', *Human-Centred Methods in Information Systems: Current Research and Practice*, Idea Group: London.

Day, S. and Edwards, B. (1996). A.T. for Postsecondary students with learning disabilities. *The Journal of Learning Disabilities*, vol. 29, pp 486-492.

Garner, J. & Campbell, P. (1987). Technology for persons with severe disabilities: practical and ethical considerations. *The Journal of Special Education*, vol. 21, pp 24-32.

Gray, R.A. (1981). Service for the adult LD: A working paper. *Journal of Learning Disabilities*, vol. 4, pp 426-431.

Ihde, D. (1998). '*Expanding Hermeneutics: Visualism in Science'*, Northwestern University Press: Ill.

Lakov D., G. Kirov, (2000). Information soft computing agents, *Proceedings of the International Conference "Automatics and Informatics'2000"*, Sofia, 2, , pp 135-40.

Mangrum, C.T. and Strichart, S.S. (1988). *College and the learning disabled student*. Grune & Stratton: Philadelphia.

P.L. 100-407 (1988). *The Technology-Related Assistance for Individuals with Disabilities..*

Latour, B. (1999). *Pandora's Hope: Essays on the Reality of Science Studies*, Harvard: MA.

Mills, J. (2003). Making Sense of Organisational Change, Routledge: NY.

Raskind, M.H. (1994). A. T. for adults with learning disabilities: Rationale for use, Gerber. P. & Reiff, H. (Eds.), *Learning disabilities in adulthood: persisting problems and evolving issues*. Andover:MA.

Raskind, M.H. (1998). Literacy for adults with learning disabilities through assistive technology. In S.A. Vogel and S. Reder (Eds.), *Learning Disabilities, Literacy and Adult Education*. Brookes: Baltimore, MD.

Raskind, M & Scott, N. (1993).Technology for postsecondary students with learning disabilities, in S.Vogel & P.Adelman (Eds.), *Success for college students with learning disabilities*,pp240-80. Springer-Verlag.

Reiff, H.B., Gerber, P.J. and Ginsberg, R. (1992). Learning to achieve: Suggestions from adults with learning disabilities. In M.L. Farrell (Ed.), *Support services for students with learning disabilities in higher education*, vol. 3, pp 135-144. Assoc. on Higher Ed. and Disability: Columbus, OH.

Siddiqi, J. (1994). 'Challenging Universal Truths of Requirements Engineering', *IEEE Software*, March, pp.18-20.

Sproull, L. and Kiesler,S. (1991). *Connections*, MIT Press: MA.

Stapleton, L., (1999). 'Information Systems Development as Interlocking Spirals of Sensemaking', in *Evolution and Challenge in Systems Development*, Zupancic, J., Wojtkowski, W., Wojtkowski, W.G., Wrycza, S., (eds.), Kluwer Academic/Plenum Publishers: New York, pp. 389-404.

Stapleton, L. (2001). *Information Systems Development: An Empirical Study of Irish Manufacturing Firms*, Ph.D. Thesis, Dept. of B.I.S., University College, Cork.

Stapleton, L. (2003). 'ISD As Folding together Humans and IT', *Information Systems Development: Advanced in Methodologies, Components and Management*, ed. Kirikova, M. et. al., Kluwer Academic, pp.13-25.

Stapleton, L. & C. Murphy (2002). 'Revisiting the Nature of Information Systems: The Urgent Need for a Crisis in IS Theoretical Discourse', *Transactions of International Information Systems*, vol. 1, no. 4.

Vogel, S.A. (1987). Issues and concerns in LD college programming. In D.J. Johnson & J.W. Blalock (Eds.), *Adults with learning disabilities: Clinical studies*, pp 239-275. Grune & Stratton: Orlando, FL.

Weick, K. (1982). 'Management of organisational change amongst loosely-coupled systems', in P. Goodman & Associates (eds), *Change In Organisations*, Jossey Bass: San Francisco, pp. 375-408.

SYSTEMS MODELLING IN TEAM BUILDING AND INNOVATION

Peter E Wellstead [*,1] Allan P Kjaer [**]

* The Hamilton Institute, NUI Maynooth, Co. Kildare,
Ireland
** Baslev Consulting, Copenhagen, Denmark

Abstract: This paper is based on the experiences of the authors, and seminars given to Research and Development teams in industry and scientific institutes. The seminars were developed over a ten-year period during which the authors have been particularly involved in science research management and research exploitation. Our particular aim has been to give a logical basis to the technological parts of innovation, and in a form that engineers can readily embrace. The seminars were based on the premise that the development of complex products and processes requires multi-disciplinary teams of disparate specialists. The assembling and management of these teams is a major issue in technology and innovation, with many opportunities for mis-understanding and loss of focus in research teams. In this paper we explore the role of systems mapping and modelling tools as a method of communicating across disciplines and enabling cross-disciplinary innovative steps. The paper concludes with concrete examples of where systems mapping and modelling created new opportunities to innovate and overcame technological roadblocks. Copyright © 2004 IFAC

Keywords: Innovation, Modelling, Management

1. PROLOGUE

This paper was drafted in consultation with the recruitment and personnel psychologist and master dry stone-wall builder, Edward Thurrell. Eddie continued discussing our work long after the initial draft was completed, and his insights and experience in team building have influenced our ideas greatly. Eddie died suddenly before we could properly bind his expertise into our methodology. The profession has lost an outstanding practitioner and we have lost a friend and potential collaborator. Any ideas that might be of merit in this paper are respectfully dedicated to the memory of Eddie and the human values that he embodied.

[1] Supported by the Science Foundation Ireland

2. INTRODUCTION

Continuous and rapid innovation is an economic imperative in the developed nations, and legions of books and journal articles are devoted to it. Distinguished authors discourse on the matter regularly, (Utterbeck, 1994),(Christensen, 1997), learned societies debate it and scientific journals are formed to study it. Management tools, e.g. (Kostoff and Schaller, 2001) are developed to map out needs for innovation and devise possible routes to success. These types of study have value in the 'top down' studies of innovation required in the planning of national development, and in large organisations and corporations. For example, road maps present a global view that is suitable for the purposes of overview and direction of technology. At a more granular level, requirement engineering and requirements management are deployed to focus innovation needs and targets. However, all the above methods are macroscopic methods that make the implicit assumption that the individual technical skills required exist in an innovation team, and that the team have a knowledge of their colleague's skills. However, it has been been our observation that while the individual scientific skills exist in large organisations, there are problems in the shaping, and operation of innovation teams. In small organisations team formation is crucial to success or business failure. At a basic level team building is hard because of human limitations, and difficulties in fusing the technical skills of the team in a focussed way. Throughout recruitment, team building and operation there are compromises that must be made. This paper aims to show one approach to reducing these compromises. There are two specific hurdles to successful technological innovation. First, increased specialisation in education and training is developing scientists and technologists who are not able to readily communicate with other specialities. Second, many projects, while loaded with GANTT diagrams, critical path charts and other organisational tools, do not have a mechanism with which to guide the scientific inquiries or technological research of a team. This is a major issue in small to medium sized enterprises, where nimbleness and speed to product is important. In previous articles we have written about the role of control in innovation (Wellstead, 2003). In this paper however the focus is upon on specific tools that meet the need, outlined above, for communication and facilitation tools for scientific investigation and technology development. This is a fundamental need in modern industry, and while methods exist, e.g. (Ekvall, 2002), we believe that systems oriented modelling methods, suitably linked to inspirational tools such as lateral thinking, mind-mapping and brainstorming, can perform a role that is appealing to research and development staff. As a sub-text to this is the theme that systems modelling should be a technology independent representation that bridges the gap between disciplines and between people (Wellstead, 1979).

3. INNOVATION MODELLING - THE MACRO VIEW

Innovation is a multifaceted process. According to the accepted wisdom it prospers on a wide scale when a combination of technical processes, social and economic conditions occur. These include:

- Social imperatives for economic advancement.
- The absence of constraints on innovation and its commercial exploitation.

- Easy access to the mechanisms for innovation.
- New ways in which to realise innovations.
- Access to funding and markets.

In this paper we will focus specifically on the fourth element of this and the use of mapping and modelling. It is however relevant to note how modelling and mapping are used when examining the other areas of innovation listed above. From a position of overview, modelling in innovation is widely studied by social scientists and management analysts. Even though this paper addresses the area of the technological processes of systematic innovation, it is useful to know that there are many excellent and thought provoking approaches to representing innovation at the general wide scale level. This is important because any systematic design of the technological innovation process presupposes that a prior study has been done to ensure that the innovation project - if successful - will have a market or receptive audience. There is a widespread belief that science and technology innovation operates in a cycle. Models based on a cyclical approach have their roots in the work of Kondratieff (Kondratieff, 1925), who developed the idea of waves of economic and social development. This idea was adopted by Schlumpeter (Schlumpeter, 1939) who further associated the wave theory with industrial and scientific activity. Actually the Industrial Revolution was proposed as the first of these waves.

The term wave is used specifically to distinguish between a periodic process (as implied by the term cycle) from the idea of an organic, self-sustaining movement that occurs from time to time but with no specific period. Many authors have given further clarification of the innovation process. Probably the most popular among these is Christenson and his idea of 'disruptive innovations' (Christensen, 1997)]. This idea that innovations are some how breaking a previously prevalent paradigm has echoes of the 'paradigm shift' model of scientific development (Kuhn, 1970), and for this reason is widely embraced by technology directors and science policy makers.

The ideas sketched out in the previous paragraphs exist on the grand scale of things. This paper is concerned with the more detailed, but vitally important, task of making innovation work when the project or area of invention has been identified, and a multi-disciplinary team is required. Needless to say, we will find that multi-disciplinary ideas drawn from control engineering play an important role.

4. INNOVATION AS A SYSTEMATIC PROCESS

The notion that technological development could be treated as a systematic process became important as the Industrial Revolution took hold. For example, Brunel's, (Clements, 1972), Whitworth's and Maudsley's (Rolt, 1986), (Rolt, 1970) work all hinged upon a systematic testing of ideas, a logical decomposition of problems into technically correct components and assembling structured teams to fulfil the overall innovative mission. In the USA the best know example of systematisation of innovation is Thomas Edison's formation in 1887 of a research and development laboratory in New Jersey. This was virtually an innovation factory, with intense focus on problem solving, driven forward by Edison's raw energy and passion (Josephson, 1992). Equivalent movements existed in most countries, usually associated with the passion and genius of one or two people. A particularly important figure in aircraft design was Sydney Camm (Fozard, 1991), an unremittingly intense engineer and leader, both in his dedication and the demands that he placed upon his research and development teams. His contributions span from early bi-planes in 1920's, through the Hawker Hurricane, jet fighters, the invention of vectored thrust, and finally the Multi-Role Combat Aircraft in the late 1960's. We dwell on the case of Edison and Camm deliberately, because each were disciplined organisers of the research and development process, but at the same time (from an historical perspective) act as a bridge between the heroes of the Industrial Revolution and the engineers and scientists of today.

Both Edison and Camm were men of genius who had the capacity to propagate the internalised models of their technical vision of the research and development process, and to instil it into their development teams. Such leaders structured the model of innovation, systematised it and made it available to the innovation team members in appropriate ways. This kind of behaviour is characteristic of many successful scientific and technological leaders. However, such leaders are not easy to find, and in a world of increasing complexity we look for ways of facilitating the search for leaders or automating their characteristics.

In the field of Personnel Development and Recruitment, the emphasis is upon the analysis of individual characteristics through personality profiling and the identification of specific leadership traits. In this paper, we take the complementary view that we should seek ways of replicating the internal scientific model that great people (such as Edison and Camm) used in ways that can be formulated logically and implemented in a computer. A question naturally arises here. How can a common scientific vision of a development process be held in a form that has the required features? Such as:

- Formative - the procedure allows a team to be built with a common vision of the task in hand, such that specialists from different backgrounds can develop a joint understanding of methodologies and requirements.
- Interdisciplinary - the representation is generalised in some way. It should not be focussed on only one aspect of the development e.g. a description of the electrical parts of a project are useless and incomprehensible to the non-electrical team members).
- Quantitative - the representation should, where appropriate, precisely characterise each aspect of an intended development.
- Analytical - the representations should interface clearly with principled analytical procedures.

An outcome of the above listing is that modelling in various forms is the key to structuring and operating an innovation team. There are many modelling tools that are used for this purpose, some are intellectual mapping tools on the way to solving a problem, others (particularly in the process planning area) are rigid models that are an integral part of the solution. In the next sections, some methods will be described that have been successfully used in team building and innovation management. We will then outline some particular cases studies that have been performed using such concepts.

5. MODELLING FOR INNOVATION: THE CONTEXT

As outlined above it is generally accepted that innovation today is radically different to the phases in history that saw Edison and Camm at the heights of their intellectual powers. It is no longer possible for an individual to understand all aspects of a modern product and to fit this into the needs and requirements of society and the economy. In its broadest sense, innovation in a product or service requires three domains of knowledge as indicated in figure (1).

In this paper is assumed that the high level modelling of innovation cycles has been done, so that DOMAIN 1 has been addressed. It is also assumed that techniques such as Technology Road Mapping (Kostoff and Schaller, 2001), have been performed and so DOMAIN 2 have been applied, at least in part. In this section we focus mainly on DOMAIN 3 - the scientific and technical components of a research and development programme. Let us assume that the analysis of

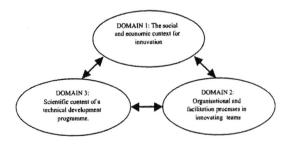

Fig. 1. Knowledge Domains Associated with Innovation

business and innovation cycles has indicated that a time window and market opportunity for success exists, and that a budget has been assigned to the project. We move then on to the technical issues of team building and performing the development programme. We deal first with the issues of team building.

6. MODELLING FOR INNOVATION: TEAM BUILDING

In building a team for technical and scientific purposes, personnel experts have become skilled in personality profiling the individual characteristics. Some of the procedures use graphical tools (we use mind-mapping (Buzan, 2003), but this is just one of many tools) as techniques to unlock or build upon ideas. Most procedures are based on free-association methods of recording and sharing ideas. These latter two points - recording and sharing - are at the core of things, and can be linked to specific project metrics as follows:

6.1 Technology metrics

In assessing the technical capabilities and scientific aptitude of potential innovation team members it is important that some level of objectivity and consistency be applied. Personal interviews are often subjective, so potential team members are also asked to record - using one of the many tools available - the scope of their speciality. They are then asked to map out their assessment of their particular strengths and weaknesses within this scope. The interviewer uses this process to build a picture of a candidate's knowledge base.

6.2 Organisation metrics

For external recruits this stage is important as new recruits to an organisation instinctively search for a role and identity within an organisation. Here the candidate is asked to generate a mind map of the organisation in which the innovation team is to be based. This clearly exposes the perceptions of the candidate and can be

used to orientate the candidate in the nature and aspirations of an organisation. The map for this stage will often reveal the candidate's motivations and drivers in a technically useful way - especially if used with the psychometric results.

6.3 Project metrics

This is used to orient and select people for a project or development process. Here each potential team member is asked to map out his/her view of a proposed project. Prompts are used to stimulate specific responses in (for example) a mind map and the candidates are usually given access to the decision processes that led to the investment in a particular innovative project. This has been found to be a most revealing stage of team building and the most useful. Specifically it is often possible to use ideas from the mind maps generated at this stage to enhance, or modify a project. Mainly however this is to determine the potential team member's vision of the project and look for a coherency and complementarity in the people selected.

6.4 Team metrics

Here the objective is to build a team style profile to complement the individual personality profile. The potential team member is asked to describe their role within a project or team. At this stage the team has probably been selected based on the previous stages and the task now is to determine the team style. This can be done objectively using, for example, Belbin's Team Style Inventory (Belbin, 1993) to construct a balanced team. This crucial stage can be further improved by adding a scientific dimension to team balance, because it is usually the case that a specialist will place their topic at the heart of the project, and associate key innovative steps with his or her speciality. To share this view with colleagues is most useful as it unveils scientific self-perceptions in an acceptable way and also helps broaden the understanding between the specialists. Here the fundamental aim is to show all the team members that they are part of a process - and none of them is important without the others.

6.5 Environmental Metrics

The physical and management environment within which work is performed is a further vital metric in innovation facilitation. Having build the team then building an appropriate climate to allow the individual and team contribution to be maximised is a priority. In this context the questions posed to the employer/managers include: What style of management is conducive to innovation? What physical environment can be provided to unlock a team's full potential. What operational atmosphere draws out contributions in the most effective manner?

7. MODELLING FOR INNOVATION: PROJECT AND PRODUCT MODELS

The previous section was essentially about modelling the team membership and their individual perception of an innovation team's objectives. Fundamental to this is creating a joint understanding of a project, isolating where the key innovative steps may be required and creating a common understanding of each person's role in the process. All of these should be carried forward into the active stages of a project. Specifically, to truly exploit the synergies between specialists and give a common platform for analysis it is important that generalised modelling tools be used across the team. Here we will talk about modelling tools that are representative of the objects used to give a logical and coherent framework to a project.

7.1 Generalised System Modelling

Central to the ability to integrate skills and systems components across all disciplines associated with a project is the ability to understand the physical basis of a range of subjects (in particular, electrical, mechanical, hydraulic and thermal systems). By understanding the performance characteristics of the components of a system - and more importantly how they interact within an overall system - team members can see where conflicts arise, where simplifications can be made and how changes in sub-systems may propagate through the system. In this spirit, system modelling is the single most important topic in the innovation team's tool kit. For example, using bond graph methods (Wellstead, 1979) the equations of motion of a system can be derived in a context free framework, simulated and the overall performance of the system to be assessed. This procedure is part of the control engineer's work and to describe it as an innovation-enabling tool may seem unusual. However within the engineering management community the benefits of modelling and simulation is not recognised. It is thus useful to emphasis their value in terms that management can best appreciate.

Allied to system model methods is the importance of analogies. These have a long history in engineering systems analysis and even in modern technical development they can be a powerful tool

in communication in a team with different technical backgrounds. Techniques such as bond graphs allow this since the resultant model of the system dynamics can be mapped into any domain. The power of generalised system modelling becomes clear when the resulting equations are embedded in corresponding simulation tools. These allow discipline independent analysis of scenarios and independent analytical determination of the behaviour of the planned product or process. In a very real sense, the systems tools of modelling, simulation, and mathematical analysis, act as the gatekeepers to system design and cohesion of the team contributions.

7.2 Reference Models

Systems modelling tools such as bond graphs are primarily intended for use in the development of physical products. In addition to this are the class of reference models, for example the Purdue Reference Model (Williams, 1991), which provide a systematic framework within which the role of each part of a process or system must be placed. The whole aim is to provide a logically consistent framework. In this context the ISA Standards S88 and S95, developed from the Purdue Reference Model, are examples of tools that allow all components of a system to be placed in its context, and that context to be clear to all team members. The dynamical modelling provided by the generalised modelling tool fits into this by describing the detail behaviour within the process components.

8. CASE STUDIES - PROCESS INNOVATION IN PROCESS REORGANISATION

This is an example of systems engineering for a tobacco preparation unit. The customer initiated the project because the Retailer Quality Index for important markets was unstable. The production chain was analysed and it was showed that the problem was in the primary production where the dried raw tobacco is treated in order to get a constant level of moisture and constant paper consistency. It was in this area that the batch variability was too great. A project redesign team was formed and asked to describe a method for improving this process stage, (Kjaer, 2003). It turned into a methodology consisting of six steps:

- Modelling of the functional demands in terms of brainstorming sessions.
- Modelling of material flow.
- Modelling of the physical layout of the plant in terms of the S88 physical model.
- Modelling of the plant network.
- Analysis of the information flow according to S95.

- Modelling of the recipes according to S88.
- Establishing the specification for the configuration of the software systems

Note that the methodology is based around models which give a common language of communication. The order of the steps in the model can change according to the particular problem, but provide a re-useable tool set, which allowed the team to innovate in the packing and preparation line. The methodology is transferable and has been successfully reused in other projects. One example which seems to have given good results concerned the production of print on plastic bags. There is no deep analysis in the method, but it systematizes the process of going from an idea to a complete specification such that software can be structured according to international standards. This gives the software an innovative ability because it is very easy to open up the software and implement new ideas. In this way the total cost of ownership is decreased and the life time value increased. Also, the fact that one follows international standards has the effect that the customer and the consultant/systems integrator are no longer the only ones pushing innovation, because innovation is pushed also by the supplier of standard software (which can be used because the specs are structured according to international standards) and the continuous development in that software. This fits well with the constant demand for productivity improvements.

9. CONCLUSION

Let us finish by quoting from an obituary for the Anglo-French engineer Isambard Kingdom Brunel. In this quotation, the emphasis is ours:

In all that constitutes an engineer in the highest and fullest and best sense, Brunel had no contemporary, no predecessor. If he has no successor let it be remembered that *the conditions which call such men into being no longer have any existence.*

The social and economic circumstances that allowed Brunel to flourish have indeed changed. In our complex and specialised world we can no longer equal them as individuals, but by systematic use of appropriate modelling tools we can unify our efforts within a team. In doing so we create a shared vision and can hope to emulate the greatness of our predecessors.

10. ACKNOWLEDGEMENTS

Peter Wellstead gratefully acknowledges the support of the Science Foundation Ireland and the E.T.S. Walton Visitor scheme. Both authors acknowledge the inputs of, and experiences with, colleagues at Control-Systems-Principles and our clients.

REFERENCES

Belbin, M. (1993). *Team Roles at Work.* Butterworth. Oxford, UK.

Buzan, T. (2003). *How to Mind Map.* Thorson Publications. New York, USA.

Christensen, C.C. (1997). *The Innovator's Dilemma.* Harvard Business School Press. Boston, Mass.

Clements, P. (1972). *Marc Isambard Brunel.* Longmans. London, UK.

Ekvall, G. (2002). Organizational conditions and levels of creativity. In: *Managing Innovation and Change* (K.M. Miller, Ed.). 2nd ed.. pp. 125–247. Sage Publications. London, UK.

Fozard, J.W. (1991). *Sydney Camm and the Hurricane.* Airlife. Schrewsbury, UK.

Josephson, M. (1992). *Edison: A Biograph.* J. Wiley. New York, USA.

Kjaer, A.P. (2003). The integration of business and production processes. *IEEE Control Systems Magazine* **23**, 50–58.

Kondratieff, J.J. (1925). The long wave in economic life. *Review of Economic Statistics* **17**, 105–155.

Kostoff, R.N. and R.R. Schaller (2001). Science and technology roadmaps. *IEEE Trans. on Engineering Management* **48**, 132–143.

Kuhn, T.S. (1970). *The Structure of Scientific Revolution.* University of Chicago Press. Chicago, USA.

Rolt, L.T.C. (1970). *Victorian Engineering.* Penguin Press. London, UK.

Rolt, L.T.C. (1986). *Tools for the Job.* HMSO. London, UK.

Schlumpeter, J.A. (1939). *Business Cycles, A Theoretical and Statistical Analysis of the Capitalist Process.* McGraw Hill. New York, USA.

Utterbeck, J.M. (1994). *Mastering the Dynamics of Innovation.* Harvard Business School Press. Boston, Mass.

Wellstead, P.E. (1979). *Introduction to Physical System Modelling.* Academic Press. London, UK.

Wellstead, P.E. (2003). Control and systems concepts in the innovation process. *IEEE Control Systems Magazine* **23**, 21–29.

Williams, T.J. (1991). *A Reference Model for Computer Integrated Manufacturing (CIM).* Instrument Society of America. North Carolina, USA.

ELSEVIER

IFAC

PUBLICATIONS
www.elsevier.com/locate/ifac

TACIT KNOWLEDGE AND HUMAN CENTRED SYSTEMS: THE KEY TO MANAGING THE SOCIAL IMPACT OF TECHNOLOGY

Fiona Murphy[1], Larry Stapleton[1] & David Smith[2]

[1]*ISOL Research Centre, Waterford Institute of Technology, Republic of Ireland*
E-mail: fmmurphy@wit.ie & lstapleton@wit.ie
[2]*University of Wales College Newport, P.O. Box 179, Newport, Wales*
E-mail: d.smith@newport.ac.uk

Abstract: Knowledge management is important in automation literature, in particular ICT (information and communications technology). To date, this literature focuses upon codifiable knowledge. However, this conceptualisation of knowledge for AMAT and ICT is extremely limited. In particular, in order to effectively design and manage complex technologies, we need to focus more on less-concrete forms of knowledge. These types of knowledge are often termed 'tacit' knowledge. This paper reviews the current literature on tacit knowledge and relates it to current research in AMAT and ICT. It then presents empirical evidence highlighting the importance of tacit knowledge in engineering design and development work. *Copyright © 2004 IFAC*

Keywords: Information systems, Information theory, Knowledge based systems, Social impact of automation, Systems design

1. INTRODUCTION

Knowledge is an extremely important concept in engineering research. Many modern systems incorporate concepts of knowledge management and capture into their designs, and purport to address these issues. However, many of these approaches fail to address human-centred-ness in the way they tackle systems engineering problems. So called 'hard' methods tend to ignore the relationship between knowledge and humans, preferring to emphasise codifiable data and information as 'knowledge'. This research trajectory is extremely limited, and fails to appreciate the enormous importance of tacit knowledge in the work of engineers, especially in the domain of the social-impact of the systems engineers create and deploy. This paper sets out a framework for addressing tacit knowledge, and indicates the current weaknesses in mainstream approaches to technology development. It then presents some empirical data to support the contention that tacit knowledge is extremely important to systems engineers in their work. It finally sets out some opportunities for research in this space, especially relating to human-centred systems.

2. TACIT KNOWLEDGE

Tacit knowledge is non-codifiable information that is acquired through the informal take-up of learned behaviours and procedures (Howells (1996)). Polanyi defines it as "knowing more than we can tell", meaning that we know how to perform a certain task, for example ride a bicycle, but we cannot explain to another person (s) how to perform that task successfully (Polanyi, (1961), p93). Tacit knowledge or "Intellectual Capability" is not easily catalogued. It is completely incorporated in the individual. It is ingrained in their practice and expertise, and can only be expressed and conveyed through proficient execution and through forms of learning that involve demonstrating and imitating (Fleck (1997)).

It is not possible to transmit tacit knowledge easily or directly. As task accomplishment and knowledge are specific to the individual involved and require the individual to make changes to their existing behaviour (Howells (1996)). Tacitness within the knowledge does vary and the more ambiguous this type of information is the harder it is for an organisation to assimilate it (Cohen & Levinthal (1990); Nelson & Winter (1982)).

Tacit knowledge is seen as an invaluable asset and a source of competitive advantage. Quinn (1992) observed that the competitive advantage of an organisation depends on 'knowledge-based intangibles', such as technological know-how and understandings. According to Baumard, "tacit knowledge is... a reservoir of wisdom that the firm strives either to articulate or to maintain if it is to avoid imitation" (Baumard, (1999), p23). If imitated the organisation would lose its competitive advantage. However, tacit knowledge can also cause problems for organisations, as it is difficult to formulate this type of unstructured knowledge. In addition this type of information "is often held in the minds of a handful of key persons and will be easily lost during any movement of staff" resulting in the firm losing its competitive advantage (Wong & Radcliffe (2000)).

Tacit knowledge is not easily distributed and can only be made known to other people through direct contact and socialisation. Wong and Radcliffe (2000) have stated that tacit knowledge consists of elements that can be successfully transferred via a demonstration process that is carried out by face-to-face contact between the user and the analyst.

According to Wong & Radcliffe (2000) there are six characteristics of tacit knowledge. They are:
- *Judgement facilitating*. This refers to the formation of an opinion about something: how the individual forms that opinion cannot be easily expressed.

- *Estimation and envisioning capability* involves understanding the current situation and actively evaluating what the possible outcomes may be, that is, the best guess.

- *Physical manoeuvring*, this includes physical body movement and co-ordination, for example sketching or using hand tools etc. These are often referred to as skills, but explaining and documenting them are impossible.

- *Efficiency enhancing* is generated through the possession of the knowledge learned in previous experience, and again this is difficult to vocalise and document.

- *Image formation and recognition*. When trying to complete a task the individual creates in his mind what the product of that task should be, for example creating a computer system. How the product is constructed is done using explicit knowledge, while the assembling and operation of the product is simulated using tacit knowledge already held by the individual.

- *Handling of human relationships*. This deals with the knowledge used in dealing with people in different circumstances, using the right criterion at the right place and time involves knowledge that is beyond articulation.

Wong & Radcliffe (2000) have stated that when a piece of information is used, it may display one or more of the tacit characteristics listed above and this highlights the tacit component of that piece of knowledge.

Grant & Gregory (1997) identified tacit knowledge as an accumulative process of learning. From this it can be deduced that tacit knowledge is continually being built upon and learnt (Howells (1996)). This dynamic know-how is developed through trial and error and from prior experiences of past successes or failures.

To summarise, tacit knowledge is accumulative knowledge that is embodied in the individual, escapes definition and quantitative analysis, is learned through trial and error analysis, and is transferred through socialisation, demonstration and imitation. Tacit knowledge is context specific. It is embodied within the social and organisational contexts of the individual (Roberts (2000)).

3. HUMAN CENTRED SYSTEMS

Kling & Star (1998) stated that Human Centred Systems refer to systems that are:

- Based on an analysis of the task being performed by a human that the system is aiding

- Performance monitoring in relation to human benefits

- Developed to take human skills into account and

- Easily adaptable to the changing needs of the human users

From this it can be deduced that Human Centred Systems are based on the social structures that surround the work and information being used by the individual. Human Centred Systems are developed to complement the skills of the user (Kling & Star (1998)). Tacit knowledge is an important factor in the way humans approach work, especially where they work with advanced technology. Consequently, an understanding and appreciation of tacit knowledge goes to the heart of human-centred systems approaches, and has been a central concern of the journal AI & Society from its very first issue (Cooley (1987)). A summary of recent European experience in one specific application domain is given by Brandt and Cernetic (1998).

The potential importance of the human-centred approach to tacit knowledge has been succinctly stated by Gill (1996), arguing that whereas knowledge is recognised as the new economic resource, divorcing it from its social and cultural roots effectively limits the potential of new technologies for the transfer of knowledge and models of experience between and across cultures.

The following section of the research briefly sets out the experiences of a manufacturing site involved in the design and development of heavy engineering products. It indicates the importance of tacit knowledge, and how, as part of the introduction of a new technology (CAD) the company managed the diffusion of tacit knowledge.

4. EMPIRICAL EVIDENCE FOR THE IMPORTANCE OF TACIT KNOWLEDGE

A research study was conducted into the organisational impact of advanced technologies, and how firms managed these projects. The research utilised semi-structured questionnaire designed to elicit detailed, rich stories of the experiences of interviewees in the firm. This was part of a larger study into the organisational impact of advanced, complex technological systems. The research findings are set out here in story form, as told by management and engineers in the firm. Due to the sensitive nature of the data, it was agreed wit respondents that all reported data would be published anonymously.

Company X is a large multi-national operating electrical engineering manufacturing sites in the Republic of Ireland. The manufacturing site studied here employed approximately one hundred and fifty people in the manufacture of electrical products for the European market. The production and engineering processes at the plant involved some of the most advanced automation systems around, and the business was run using enterprise resource planning systems, advanced data collection system and robotics. The production facility is heavily unionised and has been in operation in Ireland for forty years, experiencing industrial relations difficulties from time to time, and often associated with the introduction of new technology.

4.1 The Context

A study into the organisational impact of complex technological systems the research explored the experiences of an engineering group in this firm who introduced a new computer-aided design tool. Some of the engineers working in the facility have been there since the facility was opened in the 1950s. Consequently, there was a very large body of tacit knowledge within the group. This was recognised by the Personnel/Manufacturing Resources manager who was ultimately responsible for the change process, of which the introduction of the new CAD system was one component. This manager, D, had worked for years as an Engineer at this firm, and moved over to personnel. One interviewee had told me how

'His predecessor had spent years in and out of the labour court. ... The company's position was that you must pay in industrial relations problems now for new technology otherwise labour costs will increase forever, and you would stop development of the firm. It was brought to a head when D came in'.

The new manager had a complex problem to resolve. Firstly, he knew that the companies policy in this case

would not work. He recognised that forty years of engineering knowledge would be lost to the firm. Simple replacing an aging engineering group with new graduates would set the company back decades in terms of expertise. On the other hand, he knew that the new CAD system had to be introduced, and that there would be enormous resistance. Quoting D, 'we had to avoid extremes. We needed both the level of experience of the existing engineers and the energy of the youngsters. We had one guy, for example, who was a good designer, with forty years design experience. But, he couldn't use a PC.'

The feeling was that the firm couldn't lose him and all that knowledge and skill. Furthermore, engineers at his stage of career were close to retirement, and had little to gain in attending training courses and education programmes designed to get them up-to-speed with the new technology. But the company felt that they had to introduce the new technology, and that in itself would create a lot of resistance amongst people who had little to lose in not adopting it. This was the dilemma faced by manager D.

4.2 Developing a Knowledge Retention Strategy

Manager D knew that he could not bully people into using the new technology, and that he could not risk losing the support of the older engineers who they might need to bring in from time to time after retirement, and whose knowledge had to be inculcated into the organisation. D did a number of things:

1. He subsidised the older engineers to buy home PCs.

2. Introduced a logbook based system which was non-computerised

3. Worked hard to build the trust and good will of people for the new system, and from there built

4. Succeeded in establishing a 'chemistry' in the team

5. Introduced a new position - 'senior engineer'

This strategy was designed to retain knowledge within the organisation, whilst simultaneously successfully introducing the new technology into the group. This strategy worked as follows:

4.3 Subsidised Home PC

This was designed to address people's fear of the new technology. An essential factor here was the fear of using personal computer-based tools. The engineers had drafted designs on papers for decades and were recognised as a very successful and competent design group. The introduction of new technology threatened this. By providing the engineers with subsidised home computers they could play with them at home and become familiar with the technology. Furthermore, they could keep the system for their personal enjoyment and become familiar with a technology that was now quite ubiquitous in Ireland.

This also sent out another message. It demonstrated clearly that the company was willing to invest personally in the people involved. By doing this it showed to the engineers that, if they were willing to cooperate with the firm, their jobs were likely to be secured. Interviewees described this as a significant 'psychological and philosophical change'. The money wasn't as important as the willingness to invest in the people. Small bonuses were also provided in this respect to key people who were seen to be 'key players'.

4.4 Non-computerised Log-Book System

This provided a work around solution in case the system failed, again addressing fears associated with the new technology. Also, by providing a computerised and non-computerised approach, it addressed problems associated with the 'insensitivity of youngsters' i.e. the older engineers had access to equivalent technology at home, and had a non-computerised solution that was useful to the group and of which the 'youngsters' would have no knowledge. This neutralised potentially dysfunctional power imbalances in the group.

4.5 Trust, Good Will and Ownership

D recognised that in order for people to be committed to the project, and for them to have a stake in its success, goodwill and trust had to be inculcated. He saw this as a central plank of the strategy. As one interviewee put it 'if you treat an employee unfairly you can undo years of work'.

4.6 Chemistry

When recruiting personnel, and organising sub-groups, D tried to 'get the chemistry right' between people. This meant placing certain people together, and was described as something which required patience.

4.7 The Senior Engineer Role

This position was created in order to establish the seniority of the older engineers. It was the key to the retention of tacit knowledge. By creating the senior engineer position D could ensure senior managers acted as mentors for the new engineers. Furthermore, it meant that the new engineers, who found it easier to familiarise themselves with the new technology, could perform most of the computer-based design work, under the guidance of the senior engineer. This effectively reduced the amount of computer-based work the senior engineer had to perform, whilst ensuring that the design skills of the senior engineer were learnt by the new staff.

5. TACIT KNOWLEDGE AND HUMAN-CENTRED TECHNOLOGY DEPLOYMENT

By combining these elements with technical instruction and a solid technical deployment of the new system, the engineering group were able to train up new engineers in the CAD system, whilst simultaneously passing much of the older engineers tacit knowledge to the novices. By adopting a human-centred approach to the technology deployment problem, the organisation was able to utilise a new technology project to ensure that important tacit knowledge was diffused in the firm.

Although 'Knowledge Management' has enjoyed something of a vogue in corporate circles, it has also attracted some criticism. Scarborough (1999), for example, sees it as an essentially technocentric concept, based on groupware and intranets, rather than on an appreciation of the nature and dynamics of knowledge *per se*. He argues that this "...technology-driven view focuses on flows of information, as opposed to people, in the enterprise. The issue then becomes one of redesigning the people around the systems".

The critical managerial insight in this case study was to perceive the problem in terms of a human activity system, which used embedded technology, rather than as a technology system to which humans must in some way be accommodated. Such a shift of perception is at the heart of the Human Centred Systems approach – even if Manager D was not explicitly aware of the connection. It is perhaps telling that D was approach the issue from a Human Resource (personnel) standpoint rather from a conventional system engineering position.

It is readily apparent that, for this company at least, the management of tacit knowledge was central to the management of the social impact of the new technology. This suggests the needs for a research agenda, which addresses, comprehensively, tacit knowledge for technology deployment. The next section briefly reviews the tacit knowledge literature. This is designed to provide an overall research agenda for this domain.

6. DIMENSIONS OF TACIT KNOWLEDGE: A RESEARCH AGENDA FOR HUMAN CENTRED SYSTEMS

The previous section sets out some empirical evidence for the importance of tacit knowledge in company X. It then argues for a research agenda in this area, which might provide some direction for research in tacit knowledge and social impact. The literature indicates a number of important attributes for tacit knowledge which need to be taken into account in human-centred systems research. These traits can be divided into seven aspects of tacit knowledge for this research, namely:

1. *Implicitness.* This characteristic of tacit knowledge is extremely important. Throughout the literature tacit knowledge has been identified as being knowledge which one possess' but is unable to put into words (Polanyi (1966)). It has also been defined as a nebulous process, (Howells (1996)), intuitive

(Wong & Radcliffe (2000); Argyris (1987)), highly idiosyncratic (Roberts (2000)), inarticulable (Grant & Gregory (1997);), subjective (Baumard (1999); Nonaka & Konno (1998)), subsidiary awareness (Polanyi (1961)) and deeply rooted in ideals, values or emotions of individuals (Nonaka & Konno (1998)). Implicitness is knowledge that cannot be non-analytical (Wong & Radcliffe (2000)) and is typically learned and transferred through experience (Alic (1993)).

2. *Experiential.* This aspect of tacit knowledge is identified in the literature as accumulative knowledge (Grant & Gregory (1997); Howells (1996)), derived from experience (Wong & Radcliffe, (2000); Roberts (2000); Fleck (1997)). This tacit know-how is gained through experiences, and through trial and error (Howells (1996); Roberts (2000)). Polanyi (1962) identifies that this feature cannot be learned from books. It is only through experience that the user will be able to decide on the best course of action to pursue. Polanyi states "[experiential knowledge] guides integration of clues to discoveries" (Polanyi (1966), p2).

3. *Interactive-ness.* This feature of tacit knowledge is detected in the literature as 'in the corridor' style of learning that is codified in local practices and communities (Wong & Radcliffe (2000)). It is culture bound (Grant & Gregory (1997)); the knowledge is developed interactively (Roberts (2000)) through socialisation between co-workers (Fleck (1997); Baumard (1999); Nonaka, Takeuchi & Umemoto (1996)).

4. *Show-how.* An important attribute of tacit knowledge, show-how enables this knowledge type to be transferred among communities through on the spot learning (Fleck (1997)) and face-to-face contact between colleagues (Roberts (2000)). Show-how has been described as learning by watching, learning by doing and learning by using (Grant & Gregory (1997); Howells (1996); Fleck (1997)). Show-how is codified into the local practices and communities (Wong & Radcliffe, (2000)) and it is through demonstration (Roberts (2000)), imitation (Baumard, (1999); Polanyi, (1961)) and practice (Nonaka & Konno (1998)) that this feature of tacit knowledge is made available to others.

5. *Context.* The context of tacit knowledge is the knowledge that resides in individuals about how they perceive themselves in their society / organisational culture (Argyris & Schön (1974)). This is a form of tacit knowledge know-how as it allows us to make sense of the world (Polanyi (1962)). It transferred through informal local practices amongst co-workers (Howells (1996); Wong & Radcliffe (2000)).

This knowledge does not reside individually amongst workers but at an organisational level – it is specific only to that particular organisation (Cohen & Levinthal (1990)). The context of tacit knowledge can be further divided into:

a) *Social.* Informal way of learning through direct contact with co-workers (Howells, (1996) Fleck (1997)). This form of tacit knowledge is learnt on the job (Fleck (1997)) and is deeply rooted in ideals and values of the individual (Nonaka & Konno (1998)).

b) *Cultural.* Informal learning of behaviours within the organisation through socialisation with workmates (Roberts (2000); Howells (1996)). This tacit knowledge form has been described as being culture bound (Grant & Gregory (1997); Roberts (2000)) and critical knowledge that is firm specific (Cohen & Levinthal (1990)).

6. *Non-measurability.* In the literature this element of tacit knowledge is ascertained as being difficult to express (Fleck (1997)) and quantify (Howells (1996)), non-existent (Wong & Radcliffe (2000)), uncodifiable, (Wong & Radcliffe (2000); Roberts (2000); Grant & Gregory (1997)), and that it escapes observation and measurement (Baumard (1999)) as it is elusive and indeterminate (Polanyi (1966)).

7. *Personal.* In the literature tacit knowledge has been identified as having a personal trait. This has been defined as person-embodied (Howells (1996); Polanyi (1961)), second nature and highly proprietary (Wong & Radcliffe (2000)), subjective and intuitive (Baumard (1999); Nonaka & Konno (1998)), and mental processes (Polanyi (1966)).

Summarising, there are seven research issues that have been identified within the literature. These have been used to set out a research agenda for cross-cultural collaboration between Ireland and Wales. Waterford and Newport are geographically related regions of small countries, and both face the difficult transition towards a sustainable post-industrial economy. The problem of managing, developing and communicating the corporate tacit knowledge base in a state of rapid transition is a major concern for all companies. We are now planning to integrate research undertaken on both sides of the Celtic Sea in order to develop strategies which, whilst. Of immediate significance in our own localities, may also be of more general value – for example, throughout the enlarged European Union

7. CONCLUSION

It is readily apparent that tacit knowledge is central to any debate on the social impact of advanced technology in the workplace. It is also apparent that this form of knowledge is critical to engineers and technologists in very practical ways. At the same time, however, it is unusual to find an effective combination of domain tacit knowledge and human resource expertise in a modern company, and yet it is clear from our study that this is exactly what is needed.

Knowledge management is often greeted with suspicion by a skilled workforce, who may interpret it as the preface to deskilling or other forms of downgrading of their practice. The approach adopted in the project described above has demonstrated that this need not be the case. The Human Centred Systems Approach provides an effective set of tools for making tacit knowledge accessible throughout an organisation, whilst maintaining a sense of ownership and commitment on the pert of the skilled practitioners in whom the knowledge resides.

It is our contention that an effective organisation is one in which everybody both contributes to and has access to a culturally embedded corpus of tacit knowledge. Human Centred thinking offers a conceptual framework for the effective explication and transmission of aspects of the tacit knowledge components of skilled performances in a variety of domains, and, more importantly, for providing an understanding of the wider cultural contexts within which they are located. This makes it appropriate as a tool in the development of corporate knowledge management strategies for the twenty-first century.

8. REFERENCES

Alic, J.A. (1993). Technical Knowledge and Technology Diffusion: New Issues for US Government Policy. *Technology Analysis and Strategic Management*, **5**, 4, 369-384.

Argyris, C. and D.A. Schön (1974). *Theory in Practice: Increasing Professional Effectiveness*. San Francisco, Jossey-Bass.

Argyris, C. (1987). Reasoning, Action Strategies, & Defence Routines. *Research in Organisational Change & Development*, **1**, 89-128.

Baumard, P. (1999). *Tacit Knowledge in Organisations*. London, Sage Publications.

Brandt, D. & J. Cenetic (1998). Human Centred Approaches to Control and Information Technology: European Experiences. *AI & Society*, **12**, 2-20.

Cohen, W.M. and D.A. Levinthal (1990). Absorptive Capacity: A New Perspective on Learning and Innovation. *Administrative Science Quarterly*, **35**, 128-152.

Cooley, M. (1987). Human Centred Systems: An Urgent Problem for Systems Designers. *AI & Society*, **1**, 37-46.

Fleck, J. (1997). Contingent Knowledge and Technology Development. *Technology Analysis and Strategic Management*, **9**, 4, 383-397.

Gill, K.S. (1996). *Knowledge and the Post-Industrial Society*. London, Springer Verlag. In: *Information Society* (K.S. Gill, Ed.). 3-29.

Grant, E.B. and M.J. Gregory (1997). Tacit Knowledge, the Life Cycle & International Manufacturing Transfer. *Technology Analysis and Strategic Management*, **9**, 2, 49-160.

Howells, J. (1996). Tacit Knowledge, Innovation &Technology Transfer. *Technology Analysis and Strategic Management*, **8**, 2, 91-106.

Kling, R. and L. Star (1998). Human Centred System in the Perspective of Organisational and Social Informatics. *Computers & Society*, **28**, 1, 22-29.

Nelson, R.R. and S.G. Winter (1982). *An Evolutionary Theory of Economic Change*. Cambridge, MA, Harvard University Press. In: Howells, J. (1996). Tacit Knowledge, Innovation &Technology Transfer. *Technology Analysis and Strategic Management*, **8**, 2, 91-106.

Nonaka, I. and N. Konno (1998). The Concept of "ba": Building a Foundation for Knowledge Creation. *California Management Review*, **40**, 3, 40-55.

Nonaka, I., H. Takeuchi and K. Umemoto (1996). A Theory of Organisational Knowledge Creation. *IJTM; special Publication on Unlearning and Learning*, **11**, 7/8, 833-845.

Polanyi, M. (1961). Knowing and Being. *Mind N. S.*, **70**, 458-470.

Polanyi, M. (1962). Tacit Knowing: It's Bearing on some Problems of Philosophy. *Review of Modern Physics*, **34**, 601-616.

Polanyi, M. (1966). The Logic of Tacit Inference. *Philosophy*, **41**, 155 1-18.

Quinn, J.B. (1992). *Intelligent Enterprise: A Knowledge and Service Based Paradigm or Industry*. New York, The Free Press. Nonaka, I., H. Takeuchi and K. Umemoto (1996). A Theory of Organisational Knowledge Creation. *IJTM; special Publication on Unlearning and Learning*, **11**, 7/8, 833-845.

Roberts, J. (2000). From Know-how to Show-How? Questioning the Role of Information and Communication Technologies in Knowledge Transfer. *Technology Analysis and Strategic Management*, **12**, 4, 429-443.

Scarborough, H. (1999). System Error. *People Management*. 8/4/99, 68-73.

Schön, D.A. (1983). *The Reflective Practitioner*. New York, Basic Books.

Wong, W.L.P. and D.F. Radcliffe (2000). The Tacit Nature of Design Knowledge. *Technology Analysis and Strategic Management*, **10**, 2, 247-265

PUBLICATIONS
www.elsevier.com/locate/ifac

THE DHOKRA ARTISANS OF WEST BENGAL: A CASE STUDY OF REGIONAL INTERVENTION

David Smith[1] and Rajesh Kochhar[2]

[1]Newport School of Art, Media and Design, University of Wales, Newport,
PO Box 179 Newport, Wales, NP18 3YG UK david.smith@newport.ac.uk
[2]National Institute of Science, technology and Development Studies (NISTADS)
KS Krishnan Marg, New Delhi 110012, India

Abstract: The case study presented in this paper is based on a multimedia-based ethnographic study of the traditional cire perdue brass industry of Bikna, West Bengal, India. It demonstrates a 'Human-Centred' application of Checkland's Soft Systems Methodology to a problem of technology transfer in the developing world. It describes the success of this approach in the context of a traditional community of practice which had apparently resisted a range of apparently well-founded attempts to improve its products and processes. Analysis of the Bikna dhokra community as a human activity system permitted the construction of a holistic overview ('rich picture') of this system in relation to a complex network of other contiguous activity systems. In particular, it was possible to identify the vital role of community identity as a factor in enabling or inhibiting technology transfer. We believe that there are important lessons here which go far beyond the specific case, and which may be widely applicable to technology transfer within and between communities of practice Copyright © 2004 IFAC

Keywords: Human centred design; Knowledge Transfer; Knowledge Engineering; Rural Development; Developing countries; Soft Systems Methodology

1. INTRODUCTION

This paper deals with a process of technological change in the traditional *cire perdue* (dhokra) brass-making craft as it is practised by one group of families in Bikna Village, near Bankura in West Bengal, India. This change was initiated and coordinated by the Indian CSIR (Council for Scientific and Industrial Research) agency NISTADS (National Institute for Science, Technology and Development Studies). It involved replacing an ancient traditional but inefficient metal-foundry technique with another which is almost as ancient but more efficient. The impact of this apparently simple change on the dhokra practice has been both profound and rapid.

The paper is based on an ethnographic study using multimedia technology (Smith & Hall, 2001; Smith, 2003). This research documents a period during which a group of traditional craft workers began to adapt their ancient way of working to the demands and possibilities of both a new technology and a new commercial environment. It therefore provides a unique contemporary record of a historic living tradition undergoing rapid and fundamental change. At the same time, it provides a clear demonstration of the need for rigorous application of what Checkland (1981) called "systems thinking" to issues of technology transfer in regional economic development.

1.1 The Dhokra Artisans of West Bengal

The name 'Dhokra' or 'Dokra' was formerly used to indicate a group of nomadic craftsmen, scattered over Bengal, Orissa and Madhya Pradesh in India, and is now generically applied to a variety of beautifully shaped and decorated brassware products created by the *cire perdue* or 'lost wax' process. The craft of lost-wax casting is an ancient one in India, and appears to have existed in an unbroken tradition from the earliest days of settled civilisation in the sub-continent.

The situation of the West Bengal dhokra trade is interesting because it is in many ways typical of similar artisan craft traditions throughout the world. The craft was developed and sustained in a particular social and economic context which has been swept away by a complex mixture of economic and social changes. The loss of the 'natural' market has forced the artisans to attempt to diversify their products and to find new markets.

According to Sen (1994), the dhokra makers were originally nomads, who moved from village to village in the south-western districts of Bengal, repairing old and broken brass utensils and selling small images made in a very strong and primitive folk style. These images were traditionally installed in the household shrines of newly married Hindu couples to bring prosperity and happiness. They also made and sold measuring bowls in different sizes. These were highly prized by those villagers who could afford them. A variety of other products were made, some of them small masterpieces of casting. However, despite its antiquity, it appears that the work of the dhokra makers was always marginal to the domestic economy of India, and this may be a factor in the vulnerability of the craft to the socio-economic changes of the second half of the 20th century.

As the traditional markets declined, a small demand arose for dhokra work from urban Indian families, as well as in the tourist trade. However, this was never enough to generate a living wage. By the last years of the 20th century, most of the remaining dhokra communities were extremely poor, and their economic condition had caused many families to leave the craft to find wage employment in local manufacturing centres or in metropolitan centres such as Kolkata (Calcutta). The craft was threatened with extinction

Applying an essentially Marxist political analysis, Sen (1994) attributed the roots of this failure to

> "...Greedy dealers in handicrafts [who] took advantage of the failure of the government and the voluntary organisations to provide adequate price protection for the producers".

However, as we shall show, the situation is far more complex than simply being a matter of economic exploitation.

1.2 The Bankura Dhokra Community

The casting of finely detailed metal artefacts by means of the *cire perdue*, or lost wax, technique is almost as old as settled civilisation. The earliest known examples of *cire perdue* work include the famous bronze 'dancing girl' found in Mohenjo-Daro in the Indus Valley (Agrawal, 1971). Even at such an early stage, this finely observed bronze figure already shows the highly developed creativity and mastery of the production technique typical of *cire perdue* at its finest. The technique is simple to describe (but difficult to perfect). A detailed account can be found in Smith and Kocchar (2002; 2003). There are many refinements and variations, but most of the traditional styles of *cire perdue* work still extant throughout the world follow a basic technique which might have been recognised millennia ago.

One of the major remaining foci for the dhokra craft in India is some kilometres to the north of Bankura in West Bengal. Thirty six related families live in a close-knit clan community in Bikna village. According to Dhiren Karmakar, interviewed in September 2001, their forefathers were nomads who came from Chhota Nagpur (see also Sen, 1994, op. cit.). It has not been possible to confirm this from extant records (perhaps indicating that the migratory way of life had ended some time before these groups attracted the attention of the great and good). But the traditional metal founding technology used by the people of Bikna village was certainly more appropriate to a migratory than a settled way of life (Smith & Kochhar, 2002 op.cit.).

Although the craft had been essentially stable for hundreds of years, the past twenty years did see some significant changes. In the 1980s, personal and state patronage both played important roles in bringing about interaction between the artisan communities and creative sculptors like Meera Mukherjee and Pradosh Das Gupta. These artists successfully incorporated techniques and motifs of the Dhokra art into their own work and, once accepted as insiders, introduced the Dhokra artisans to new forms. It was during this phase that the 'Bankura Horse', a stylised, decorated horse with long upright neck and pointed ears, was successfully adapted for casting in metal from prototypes from the work of Khumbkars (clay artisans).

Despite the aesthetic impact of these interventions, they were not successful in raising the incomes of the artisans and their families. They were, like Sen's analysis, based on an incomplete understanding of the factors underlying the commercial decline of the dhokra craft. Whilst agencies such as the West Bengal Arts Commission were correct in identifying (and attempting to remedy) the inappropriateness of the traditional repertoire to modern market conditions, they saw the production process itself as a core feature of the craft tradition, and failed to address underlying technological problems which effectively undermined the potential market advantages of the new repertoire. Demand continued to decline, and the craftspeople were reduced to making items of very low quality for a weak tourist market. Poverty and low financial returns reduced many dhokra makers to using road pitch as a modelling medium, rather than beeswax or the preferred 'mom' resin, and this further reduced the quality, and hence the appeal of their products.

During the mid-1990s, the Indian development agency NISTADS (National Institute for Science, Technology and Development Studies) attempted to improve the lot of the Bankura Dhokra community by offering them access to improved technology. It was quickly appreciated that there was a particular problem with the temporary furnaces used by the Bikna dhokra makers. Whilst the traditional furnace was seen by the earlier agencies as an essential component of the craft, NISTADS field staff saw that it was very inefficient, wasteful of fuel and difficult to control.

NISTADS funded Bengal Engineering College to design and develop a fuel-efficient permanent furnace. The initiative appeared at first to be successful. The new furnace technology was enthusiastically adopted by Netai, from Petrasayer in Bankura District. In 1997, NISTADS helped Netai to modernise his facilities. He was subsequently able to obtain substantial production orders for dhokra items: a fact which was well known to the Bikna artisans. However, despite Netai's obvious prosperity, the 36 Bikna families, the bulk of the craft community, made no move to adopt the new technology.

It was all too easy - and indeed rather conventional - at this point to apply a conventional 'technology tranfer model' and to attribute the Bikna villagers' apparent lack of enthusiasm for the new technology to a kind of laggard conservatism. After all, furnaces such as that developed for them by NISTADS are almost as old as civilisation itself. However, a field visit to Bikna and Petrasayer carried out by Professor Rajesh Kochhar of NISTADS and David Smith of NSAMD in November 2000 revealed a different, more interesting and more complex picture. The specific focus for this study arose in the context of an EU-India Cross-Cultural Communications Project (Gill, 2003). Smith and Kocchar were concerned with issues of technology-based craft innovation, and the visit to West Bengal

was intended to provide illustrative case studies. (See Smith & Kochhar, 2003a,b)

Smith and Kocchar used direct observation of craft processes in action together with individual and group interviews with craft people in Bikna. Interviews were conducted in Bangla or Hindi. Digital video and still images were used, following Collier and Collier (1992), to elicit comments and support group discussions. The immediate availability of digital images was a major advantage, both from the point of view of direct data gathering and as a focussing tool for discussion. It took little time to discover that the reluctance of the Bikna artisans to adopt the new furnace technology (with which Netai had been so successful) was far more complex than simple conservatism or entrenched caste tradition.

In comparison with the primitive working conditions at Bikna village, Netai had set up what was, in effect, a "micro-factory" of a kind which might have been witnessed anywhere throughout Europe during the early days of the industrial revolution. At the time of the first field visit, the workshop was given over to batch production of brass drinking beakers, using a modern oil-sand investment moulding technique and using scrap brass water pots as the source of metal. A small electric grinder had been installed for finishing the products, and the business appeared to be flourishing, supporting several families in Petrasayer.

2. ART, CRAFT OR INDUSTRY?: A HUMAN-CENTRED ANALYSIS

It was hypothesised that the previous interventions had failed because they were focussed onto incomplete analyses of the problem. We therefore adopted A "Human-Centred Systems" approach, giving attention to the social structures that surround the craft practice (Gill, 1996). Using Checkland's (ibid.) Soft Systems Methodology (SSM) in the context of Wenger's (1998) analysis of communities of practice, it was possible to arrive at a "rich picture" of the human activity system engaged in the dhokra trade as practiced at Bankura. The reasons for the reluctance of the Bikna artisans to adopt the furnace technology with which Netai Karmakar had been so successful proved be far more complex than simple conservatism or entrenched caste tradition.

Firstly there was the issue of poverty. No detailed study has been carried out of the micro-economics of dhokra production at Bikna, but such evidence as there is, together with anecdotal field interview evidence point to the fact that the nett money earnings of the artisans are very low indeed. They could not raise the finance to pay for a permanent furnace except by borrowing from a local moneylender at interest rates of around 2% per day.

Secondly there was the question of the sociodynamics of the craft. An attempt was made to trace routes and sources of influence within the community (see Rogers & Kincaid, 1981). This revealed a major weakness of the previous attempt to introduce the improved furnace. Despite his evident prosperity, Netai was not regarded by the bulk of the dhokra craft community as a good role-model. His craftmanship was not admired in any case, but his location in Petrasayer, several hours' journey from Bikna, put him outside a tight-knit circle of closely related families. In fact, the Bikna people regard him not as a 'true' Karmakar like themselves, but as an inferior outsider.

A third factor concerns the technological competence of the Bikna artisans. They expressed concern that they might not be able to adapt their known and established moulding and casting methods to a new way of working. They thought that pouring molten brass into open moulds might be particularly dangerous. This opinion had been previously as evidence of conservative resistance to change. However, it now seems possible that there is a real problem with the suitability of locally available clay for the construction of moulds for open casting of dhokra articles.

This was clearly apparent when Netai arranged a demonstration of dhokra casting, using an open mould rather than the traditional closed mould used in Bikna, the clay mould was clearly too weak and 'leaked' molten brass as the metal was poured. It is worth noting that artisans in South India reinforce their moulds with iron wire, as well as firing ('biscuiting') them before moulding (Krishnan, 1976).

A final, critical factor concerns the extent to which the Bikna artisans' sense of identity is connected with the integrity and status of their craft. Netai's success ultimately rests on the abandonment of the dhokra craft as such. Although he still makes dhokra items to order, the bulk of his income comes from the mass-production of low craft-content industrial items. But the identity and self-esteem of the artisans of Bikna is deeply invested in their craft. The core group of families at Bikna remained committed to dhokra making. Any change which effectively meant the death of the craft was unthinkable.

Additionally, and perhaps decisively, there was an ingrained suspicion of "initiatives". It emerged that the Bikna artisans were owed the (for them) huge sum of 175000 Rupees (Rs 50/- = approximately US$1) for goods previously supplied to official Crafts Emporia.

3. THE INTRODUCTION OF A NEW TECHNOLOGY INTO BIKNA

After the field visit in November 2000, Rajesh Kochhar of NISTADS initiated a project to develop an efficient furnace for Bikna village. A NISTADS Field Scientist, met with Juddha and Mahdav Karmakar, two of the most senior and highly respected artisans in the village, and also arranged meetings with Netai. The object was to collaborate with the craftsmen in achieving a design which would not only be technically appropriate, but where there would be a sense of ownership. A prototype furnace, based on Netai's, was built in Bikna during December 2000. NISTADS agreed to finance the development, but it was made a condition that the furnace should be a community resource, rather than the property of a single artisan or his immediate family.

A permanent furnace needs to be protected from the weather. Fortunately, protection was available in the form of three large shelters build some years previously under a West Bengal regional development. The new furnace was built in one of these shelters. Experience soon showed that the first prototype was too large and would be too expensive to operate in the long run. The design was therefore modified to create a smaller furnace. This proved to be a complete success, and over the next three months, a further five were built, so that there were two in each of the three village shelters. All of them were used as communal resources.

3.1 How the Craft has Changed

It was expected that the introduction of a new furnace technology would catalyse major changes in the dhokra craft at Bikna. What was not anticipated, however, was the speed and extent of this change. The advantages of the new furnace were so apparent to the Bikna artisans that the old traditional way of doing things was changed within the space of a few months.

Whereas it had been anticipated that take-up of the new furnace would follow a classic technology transfer profile, the new furnace was adopted almost immediately by all of the families. The inefficient 'nomad' furnace was relegated to the secondary role of pre-firing charcoal for charging the new furnace heating scrap brass (this makes it brittle and easier to break up), and, interestingly, for baking the moulds.

As the location of the furnace has moved from the open air to the cover of the shelters, the production process has followed suit. Most of the work is clustered around the furnaces. This allows for fuel efficiency, since as soon as the furnace is finished with one batch of moulds, it is cleared and re-charged, making use of the heat stored in the body of the furnace and reducing fuel requirements. The furnace can also be used for secondary purposes, such as baking moulds or pre-heating scrap brass. The artisans have also developed a range of new tools appropriate to the improved processes. However, they have not followed the example of Netai and changed over to open crucible casting. They acknowledge that this would probably be more efficient, but they feel cautious about the safety aspects of handling molten brass. Also, as Raneswan Karmakar pointed out, they are not sure whether the moulds would be suitable. Our observation in Netai's workshop (see above) suggests that this caution is probably well justified.

3.2 A New Creative Confidence

The introduction of the new furnaces has had an immediate beneficial impact on the output of the better artisans. It is now possible to maintain effective control over the casting of artefacts containing relatively large amounts of brass. New products have been created, such as the "polybonga" (based on a popular terracotta form). This has encouraged a renewal of creative confidence, and craftsmen like Dhiren have begun to develop quite stunning works of original artistry.

Equally importantly, however, they are able to concentrate once more on the quality of their products. They see this as more important than developing new products. Dhiren Karmakar is happy making the traditional dhokra repertoire, and believes there is a market for it if high quality can be maintained.

In parallel with the development of the new furnace technologies, NISTADS actively catalysed a range of developments intended to move the artisans' business methods in line with their new commercial opportunities. For example, Kochhar was able to persuade the senior craftsmen to reactivate a defunct village Cooperative Society. This would give them access to 'soft' loans through the formal banking system, rather than high-interest 'hard' loans from local moneylenders.

3.3 The Future of the Dhokra Craft in Bikna Village

Despite the new sense of confidence among the older artisans, the future of dhokra making lies in the balance. Dhiren Karmakar and his relative Raneswan Karmakar were ambivalent. They felt that they themselves were better off than their fathers. The market for their products is good, and they are able to have two square meals a day, so there is no hunger any more. They do not save to accumulate capital - they do not think in that way at all. Any money that is accumulated is spent on social events or medical treatment. Both Dhiren and Raneswan saw a reasonably secure future for themselves in the dhokra trade. They feel able to cope with changing market conditions, but acknowledge that rising costs may eventually cause problems.

All the same, they feel that they are better placed than those who have left the craft to take up wage labour in cities and towns such as Bankura. Although wage labourers have more secure incomes, they do not have the prestige of independent craft artisans, and this is crucially important to the dhokra makers' sense of personal and community identity. In the end, however, they are quite clear that there is no long-term future in the dhokra trade, either, and they would prefer it if the young people of the village had alternatives - other than becoming wage-labourers.

4. SUMMARY AND CONCLUSIONS

The ancient craft of the dhokra artisans of Bankura is in the balance. The new furnace developed under the auspices of NISTADS has eliminated a major source of inefficiency from their work, which should therefore become more profitable. In addition, a new professionalism is beginning to be apparent in the artisans' trading practices, thanks largely to the advice, support and guidance of NISTADS. A major factor in this success has been the engagement of the artisans in such a way that ownership and control have been perceived as internal to the community rather than with external agencies.

All this, coupled with the creative confidence and attention to quality documented here, means that the immediate future for the dhokra craft is reasonably assured. In the long term, however, the artisans face serious decisions about the craft. On one hand, they may choose to follow the route to industrialisation, illustrated here by the case of Netai Karmakar. On the other hand, and this is what they appear to prefer, they can develop towards a consumer market based on high quality high aesthetic value artefacts. This could possibly be found supplying high craft content artefacts to a growing tourist and indigenous middle class market.

The continuation and development of the dhokra industry depends on the artisans finding a stable market niche for themselves and their products. Whatever it proves to be, this market needs to be developed and supply chains established. It is easy to demonise the middle-men, but if the economic conditions of the Karmakars become less marginal and their terms of trade can be improved, then there is no reason at all why existing middle-men may not have a major role to play in this market development.

The case study presented here demonstrates an application of Checkland's Soft Systems Methodology to a problem of technology transfer in the developing world. It relates to a traditional community of practice which had apparently resisted a range of apparently well-founded attempts to modernise both products and processes. Analysis of the Bikna dhokra community as a human activity system permitted the construction of a holistic overview ('rich picture') of this system in relation to a complex network of other contiguous activity systems. In particular, it was possible to identify the vital role of community identity as a factor in enabling or inhibiting technology transfer. Previous attempts at intervention had been correct in their partial evaluations of the problem, but failure to think "out of the box" had led to 'solutions' which were entirely pre-defined by the current paradigm, whether aesthetic, political or technological. A soft systems summary of the relative focal points of the different regional interventions is shown in table 1.

Table 1: Comparison of intervention perspectives using Checkland's 'CATWOE' Inventory

Checkland SSM Category	Market Aesthetic	Political	Techno-centric	Human Centred
Client	Patrons	Artisans	Regional economy	Artisans
Agency	Patrons	Politicians	Techno-logists	Artisans & their state mentors
Transform-ations	Aesthetic improvement of product marketability	Marxist supply chain economics	Transfer of 'new' technology	Adapt the human activity system
Worldpicture	Market & social patronage	Marxist economics	Technical patronage	Human centred
Ownership	Patrons	Politicians	Techno-logists	Artisans
Environmental constraints	Artisans' creative deficiency	Political economic hegemony	Laggard conser-vatism	Social values & poverty

We believe that there are important lessons here which go far beyond the specific case, and which may be more widely applicable to technology transfer within and between communities of practice (Stapleton, Smith & Murphy, 2005, in press).

5. ACKNOWLEDGEMENT

The research described above was funded in part by a grant to David Smith by the Arts and Humanities Research Board, UK.

6. BIBLIOGRAPHY

Agrawal DP (1971) The Copper Bronze Age of India. Mushiram Manoharlal, New Delhi.

Checkland P. 1981. *Systems Thinking, Systems Practice*. John Wiley. Chichester.

Collier J & Collier M 1992. *Visual Anthropology*. University of New Mexico Press, Albuquerque.

Gill KS (1996) *Knowledge and the Post-Industrial Society*. London, Springer Verlag. In Gill KS (ed) Information Society pp 3-29.

Krishnan MV (1976). *Cire Perdue Casting in India*. Kanak Publications, New Delhi

Rogers EM and Kincaid DL (1981) *Communication Networks: towards a new paradigm for research*. Free Press, New York

Sen P. (1994) *Crafts of West Bengal*. Mapin Publishing, Ahmedabad,

Smith DJ (2003) *Communicating Tacit Knowledge: case studies of the use of multimedia archiving in modern and traditional craft practices*. Paper given at International Visual Sociology Association Conference, Southampton,

Smith DJ & Hall J (2001) Multimedia know-how archiving in aviation industry training. In *Proceedings of the 7th IFAC Symposium on Automation Based on Human Skill: Joint Design of Technology and Organisation*, ed. Brandt, D. and Cerenetic, J., Elsevier, Oxford.

Smith D.J. & Kochhar R.J. (2002) The Dhokra Artisans of Bankura. *AI & Society* **16, 4,** pp 350-365

Smith DJ & Kochhar RJ (2003). The Dhokra Artisans of Bankura. In Brandt D. (ed) *Navigating Innovations: Indo-European Cross-Cultural Experiences. 1 Entrepreneurial Case Studies*. India Research Press, Delhi.

Stapleton S, Smith DJ and Murphy F (in press). Engineering methodologies, tacit knowledge and communities of practice. *AI & Society* **20.**

Wenger E (1998) *Communities of Practice: Learning, meaning and identity*. CUP, Cambridge.

METHODOLOGY OF RESEARCH OF THE EFFICIENCY OF INFORMATION CONTROL IN SOCIAL SYSTEMS

Kononov D.A., Kul'ba V.V, Shubin A.N.

Russian Academy of Sciences
Institute of Control Sciences
Tel.: (007)(095)334-90-09
Fax: (007)(095)334-89-59
E-mail: kulba@ipu.rssi.ru

Abstract Challenges of today urgently demand an accessible means to start a sanction of the problem: to develop a modelling scenario methodology (Kononov et al 1999, 2001, 2004) for a new research object — information control in social and economic systems (SES). This work contains a description of new results for the researches started in (Kononov et al. 2003). The original conceptual base is suggested. Based on «the information action» concept, the following definitions are given: information field, information control, information action, etc. A detailed analysis of information processes in social and economic systems is performed. The basic scheme of studying objects and subjects of information control in SES is worked out. This scheme can be applied to collect qualitative and quantitative estimations of efficiency for information control by various social objects on selected classification groups.
Copyright © 2004 IFAC

Keywords Information control, efficiency methodology.

1. INTRODUCTION

Verbally, control in a society is understood as a way of an influence inducing people to ordered behaviour, performance of required actions. An information control is understood as a mechanism of control when managing influence has implicit, indirect information character. At the same time the controlled object is granted with a certain information picture being guided on which it as though independently chooses a line of its behaviour. Information processes are an important branch of a human activity. At the same time the modern state of the researches is those, that neither methodology nor conceptual base necessary for it exists. Moreover, models of information control have sketchy character.

The following questions require verbal and formal answers to develop initial definitions and classifications:

- What is the social system as an object of control?
- What is an influence and control in SES?
- What does an information influence and control look like in social systems?
- What classification of information influences do we need to consider?
- What is an information control in social systems?
- What objects of an information influence do we need to consider?
- What are the ways of an information influence and affects?
- How to estimate an efficiency of applying these ways?
- What is information as an object and means of influence?

We have to define a number of concepts to answer these questions. Initial terms for these concepts are:

an object, a condition, an influence, a subject, a control, data, information, uncertainty, etc. They are defined in (Kononov et al. 2003).

The modern theory of the information basically is technocratic: it was created within the framework of cybernetics for the description of control processes first of all by technical systems. At the same time distribution of its basic positions on a social and economic field is represented by rather important and perspective direction of researching. An effective description should be somewhat universal, i.e. covering the common laws both poles the Nature - Society. The concept «object of the Nature» $O \in NAT$ is an initial one, allocating an initial component of the Nature (element) which is exposed to studying and is precisely outlined by a researcher. The basic property of an object is its objectivity, i.e. existence irrespective of learning subject. The set NAT from the point of view of mathematics is an universal set, and «Society», being a part of the Nature $Soc \in NAT$, it is possible to examine as object of the Nature. At the same time Soc can be considered as a subset NAT.

An universal characteristic of an object is its **state**. This category characterizes a process of changing and developing of objects which, finally, is reduced to change of their properties and relations. A set of such properties and relations determines an object state. Representation of an object state is performed with the help of an object state **data collection**. Depending on circumstances (including inwardness and a condition of an environment) the data are aggregated in **information elements**.

Another universal attribute of an object is its opportunity of **information affect** on other objects. *Information affecting represents a process that results in the fact that information elements of one object are perceived by other object as the data on its own state.* That is the point of the world unity. Thus, information affecting, in particular can be, that according to Laws of the Nature on another (object of affecting) can result information influence of one object in change of a state of the last. Realization of an opportunity of affecting demands the certain conditions (circumstances). The occurrence of such conditions caused by life of the given object, we shall name its **influence**. Thus, affecting is directly influence.

In the formalized kind influence as object of modelling we shall present as a cortege

(1) $inl = (O, P, \mathbf{Re}\,s\,(O, P), I)$.

Here O is a source of an influence, P is a recipient, $\mathbf{Re}\,s\,(O, P)$ are results of the influence, I are circumstances of the influence.

An object that is able to carry out the will i.e. to formulate and realize its purposes, we shall name a subject.

To formalize the problem we introduced in [4] a *data collection about an object of the Nature* O

(2) $SD(O, \mathbf{SI}) = \bigcup_{I \in \mathbf{SI}} B(O, I)$,

as a collection of images (phenomena) $B(O, I)$, *received in aggregate of circumstances* $\mathbf{SI}$ *at the description of an object of the Nature* O.

It allowed creating information elements as relations on the data collection (2) received by a certain way $\mathbf{Mn}(O, I)$ in concrete circumstances I. *Information collection* (**information**)

(3) $Inf(O, \mathbf{Mn}, I, \Lambda) =$

$\{(Inf(O, \mathbf{Mn}_\lambda(O, I)), \lambda \in \Lambda\}$

represents collection of information elements about object O, received by ways $\mathbf{Mn}$.

If information collection of object O is as circumstances I of reception of information elements of object P than information collection of object P represents result (1) of process of information influence of object O on object P, i.e.

(4) $\mathbf{Re}\,s\,(O, P) = Inf(P, \mathbf{Mn}, I, \Lambda_P) =$

$Inf(P, \mathbf{Mn}, Inf(O, \mathbf{Mn}, \mathbf{I}, \Lambda_O), \Lambda_P)$.

Process of information influence which results in change of data collections both source of influence, and the recipient, refers to as **interference**.

Based on the specified terminology, concepts «information», «adequacy», «knowledge», «uncertainty», «the conflict of the data» are defined [4].

These definitions allow introducing a concept of **information connection**. Specifically, objects O and P are **connected on information** (*infoconnected*) in the given information collection if there is a sequence of circumstances by means of which it is possible to receive an image of object P from an image of object O.

Presence of connection between objects O and P can be considered as a potential opportunity of their information interaction. Thus, the mechanism of realization of this opportunity essentially depends on objects O and P, and also circumstances of their representation.

Taking into account the following collection for circumstances $\mathbf{SI}$ and objects $\mathbf{SO}$

(5) $SD(\mathbf{SO}, \mathbf{SI}) = \bigcup_{O \in \mathbf{SO}} \bigcup_{I \in \mathbf{SI}} B(O, I)$,

it is possible to define «an information field» and to introduce a number of its characteristics.

2. INFORMATION CONTROL IN SOCIAL AND ECONOMIC SYSTEMS

The suggested scheme allows to carry out classification of kinds of information influences, *classifying circumstances of realization of process of information influence*. Thus, classification of circumstances can be examined on the basis of the essential attributes describing subject domains of research, each of which is characterized by original ways of the description; the purposes and conditions of realization of investigated processes in the specified subject domain, and also used methods of research which determine essential conclusions about character of investigated processes.

We represent a process of information influence of one object on another on the following stages.
1. Collecting the data (image of a state of a source of influence).
2. Creating of information elements and information sets.
3. Transferring of the information of influence by a source.
4. Receiving of the information by a recipient.
5. Generation of a data set of influence by an object.
6. Generation of information elements and creation of new information sets of an object of an influence.

Each of these stages has their features and should be described by original characteristics. Meanwhile the unity of their realization consists that they are shipped in the common information field.

Structuring and studying of functioning of social system, and process of information control result in necessity of their detailed description.

The elementary decomposition of managerial process consists in allocation of two stages: acceptance of decisions - performance of decisions.

Process of decision making consists of stages of manufacture of the decision and is finished, when the decision-maker precisely formulates the purposes of control and ways of their achievement. Realization of the accepted decision means process of influence on controlled object for maintenance of achievement of the given purposes of management and is carried out with the help of the intermediary - control systems. At a verbal level the controlled object is meant as a «passive» element of examined system, and a control system is its «active» element. Thus controlled object is the passive element of organizational system which are not having the own purposes, the distinct from purposes of management. As decision-maker carries out manufacture and realization of the accepted decision through the intermediary - the appropriate part of a control system which represents the subject of

action in some cases (for example, diversion, corruption etc.), There is a serious mismatch between the purposes of management and executed decisions of the subject. It can result in distortion of the administrative decision accepted to execution.

Basic difference of information influence consists in social and economic systems that information processes are carried out under the relation and by means of the social objects, representing subjects of action. As against objects of the natural Nature, submitting to its Laws, subjects of action have an opportunity to carry out a formulation and realization of the own purposes, i.e. the realized independent intervention in information process.

Classification of circumstances of information influence in social and economic systems assumes allocation of its essential system elements: social objects and social institutes.

The social subject of action is a subject of action or social institute of social system. During information influence of model of measurement of state M_{MO} of the social subject and conditions of its environment M_{ME} it is necessary to examine as converters of the information. On their input the information on external or internal parameters acts, and on an output the appropriate opinion about true, in its opinion, values of these parameters is formed. Thus, converter M_{MO} forms an internal components of information field of the subject of action, and M_{ME} - its external components. Efficiency of influence depends, in particular from condition M_{MO} and M_{ME}. For social objects such changes are formed depending on subjective representation of the subject of action.
Managerial process by social object is carried out by means of the mechanism of control - classification group of ways of the control used for achievement of the given purposes of control.

Let the mechanism of control $\mathbf{Mu}(O^{(u)}, I)$ is given. Here $O^{(u)}$ is controlled object, I - circumstances of application of mechanism $\mathbf{Mu}$.
Information field of the mechanism of control we shall name information collection

(6) $\quad Inf(AdO^{(u)}, \mathbf{Mu}(O^{(u)}, I))$.

It forms for realization of the managerial process applied to area of description $AdO^{(u)}$ of controlled object $O^{(u)}$. Experts represent such data as information model of control.

In view of the entered concepts information control is set of ways of influence on an information field of the mechanism of control with the given purposes, i.e. information control - a way of influence at which the information can be object of influence and means of influence.

The careful analysis of an existing information field that is the basic condition for construction and realization of purposeful information influence is necessary for increase of efficiency of realization of operations within the framework of information control.

Concrete information influences on elements of social and economic system can be classified to the following attributes: on a source of occurrence (inside SES, outside of SES); on duration of influence (single, periodic, constant); on a nature of occurrence (caused by the current structure SES, external influences on elements SES) (Kulba 1999)[5]. For definition and classification of information influences it is necessary to carry out classification of system elements SES and structuring the object of influence and the subject of influence. For an estimation of efficiency of concrete ways of information control, it is necessary to determine the typical problems and tasks solved with the help of information influence to carry out the analysis of process of formation of information operations and to develop criteria of their estimation.

The offered definitions enable development of axiomatic construction of the concept of information control in social and economic systems.

The submitted analysis of information elements, ways and mechanisms of information control allows starting modeling system of information activity SES. Thus, it is necessary to pay attention to the decision of the following problems:

— to determine qualitatively various conditions of social objects of management as objects of information influence;

— to define characteristics of an information field, including force of information influence as «physical variable», describing change of a condition of object of purposeful information influence in comparison with its initial condition;

— to specify characteristics and properties of the entered characteristics in relation to various social objects;

— to develop practical ways of measurement of efficiency of information influence of objects in an investigated information field.

In a basis of the decision of these questions the concept of the **information action** lays.

3. CONCLUSIONS

On the basis of suggested principles of modeling a number of information actions, including actions of direct information influence, actions of information counteraction, advertising actions as element of information economy can be formalized. It is possible to hope, that by more detailed development of methods of the analysis and synthesis of information control on the basis of the suggested approach can be received not only rational decisions at control of social systems, but also processes of information economy, and also new appendices are investigated, including some problems of linguistics, philosophy and other sciences traditionally named humanitarian are investigated.

REFERENCES

D.A. Kononov, V.V. Kul'ba, S.A. Kosyachenko and A.N. Shubin(2004). Methods of formation of scenarios of development of social and economic systems. [in Russian], V.A. Trapeznikov Institute of Control Sciences, Russian Academy of Sciences, Moscow.

D.A. Kononov, S.A. Kosyachenko, V.V. Kul'ba. (1999) Analysis of scenarios of development of socioeconomic systems in emergency control systems: models and methods, *Avtom. and Rem. Cont.*, **Vol. 60. Part 2. No. 9**, 1303-1320

D.A. Kononov, V.V. Kul'ba and A.N. Shubin (2001). Stability of socioeconomic systems: scenario investigation methodology. *8th IFAC Conference on Social Stability: The Challenge of Technological Development; SWIIS'01. Preprints Volume.* 27-29 Sept. 2001; Vienna, Austria, pp. 91-96.

D.A. Kononov, V.V. Kul'ba, and A.N. Shubin. 2003; Scenario analysis of information conflicts. *10th IFAC Conference on Technology and International Stability; SWIIS'03*, 3-4 July, Waterford, Republic Ireland, Preprints Volume pp. 36-40.

V.V. Kul'ba, V.D. Malugin, A.N. Shubin and M.A. Vus. (1999) Introduction in information control. [in Russian], S.-Petersburg.

ELSEVIER

IFAC
PUBLICATIONS
www.elsevier.com/locate/ifac

GOALS IN NEW COMPLEX DEVELOPMENTS

Robert Genser

IFAC-Beirat Austria and
Austrian Society for Automation and Robotics
Malborghetg. 27-29,6/6, A-1100 Vienna, Austria
rgenser@aon.at

Abstract: The present development of complex systems with new technologies is scrutinized. Goal systems and risk management are analysed. Stabilizing factors for getting human-oriented design are pointed out. *Copyright© 2004 IFAC*

Keywords: Complex systems, human-centred design, goals, risk management

1. INTRODUCTION

The establishment of new advanced systems causes much dissatisfaction not only to users rather also to industry. Many flaws can be recognized.

The usability of mobile phones is not available as it was the case with the classical telephone in the past. People with no great flexibility or having the visual acuity reduced had no great problem to handle it, may it be in Africa, America, Asia, or in Europe.

The complexity of possible functions and wide area of application of new technologies increases the risk for all groups afflicted. The problem becomes aggravated because strong and experienced actors for smooth development on a global level are now missing.

2. PRESENT SITUATION

Air transportation shows that new technology can be transferred according the state of art to safe application on a global level for the advantage for the users. The applicability is developed in close co-operation of research, industry and the experienced user, the military system. This structure stimulates research, makes it possible to get strong requirements for development by getting feedback of real application and to build the market for economical solutions. Civil air transportation can use these results on a global level in an harmonized way by strong actors like FAA (Federal Aviation Administration) and by the the experience of using TV-equipment at home. But

experience gained in industry as well as by transfer of know-how between military and civil aviation.

In the case of the telephone, intelligent investigations have been done for designing the human-machine interface in the past, see for example (Oden, 1968). But most of the present design may be even a hazard in case of emergency, especially for elder people. The important keys for calling have a layout, which is specific to the ideas of some artist of a company. The knowledge available about ergonomics is neglected.

But also projects seeming well designed and supported by the Commission of European Union for example do not come in operation as it was foreseen before.

Besides the requirements of application also the goals in a system and not only the own targets have to be recognized.

3. SYSTEM OF GOALS

Inadequate development is in many cases caused by not considering goal systems of reality.

3.1 Dynamics of Goals

Missing feedback of application causes unsuitable requirements and causes neglecting the need of users. A director of a company for television sets may have Even conflicting objectives are weighted and the targets of the company force him to neglect this knowledge. It is similar to a car driver who is approaching a zebra crossing not wanting to reduce the speed. The same person as pedestrian becomes angry with the reckless drivers. Goals are changing with situation or context.

But objectives are changing also with time. This may be by accustoming apart situation may change because of dynamics of system. **Figure 1** show schematically the influence of planning process and reality on objectives.

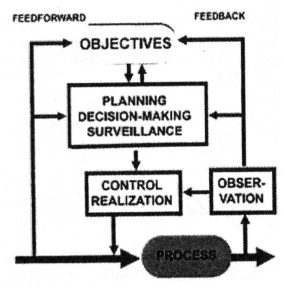

Fig. 1. Influence of planning process and reality on objectives

Figure 2 shows the change of importance of objectives depending on experience and skill of users in case of equipment for CAD (computer aided design) (CAD, 1986).

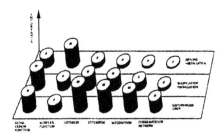

Fig. 2. Importance of objectives depending on experience of users

Most of the optimization methods for decision making used at present are not fitting for dynamical objectives. be become not adequate for hyper-system.

combined in one expression for getting single criteria function. The ability of powerful and efficient information processing is not considered for receiving flexibility and transparency (Genser, 2001; Wierzbicki and Makowski, 1992).

Moreover, the solution would be also to apply dynamical planning. Modern office buildings consider the change of demand and allow very flexible to adapt the outlay of rooms according to given situation.

3.2 Goals given but wrong decisions

The missing of one strong actor for setting the requirements and granting a market place for economical success causes for manufacturers in competition a situation similar to prisoner's problem of game theory (Lloyd, 1995). The results of missing trustable information about requirements of future developments can be recognized in the area of digital video recording. The industry is not able to get an agreement for a harmonized standard for the advantage for customers and for themselves in a holistic point of view. A stable framework for long range strategic planning is a precondition for optimal economic success.

Also the conflict of goals can be resolved in best way if the goals in the upper level of hyper-system are taken in account.

4. ORGANIZATIONAL STRUCTURE

Market forces can improve quality and fitness of a system by competition given and, most important, by severe consequences for wrong actions or behaviour. But free market can endanger complex long living systems or hyper-system itself if the feedback about dangerous impacts for the system needs longer than the time for reaching a disastrous state of no return. The history of infrastructure has enough examples about the problems of both ideologies, either free market with privately owned enterprises or strong regulated market with state-run one. The kind of operation changed for some infrastructure several times already. As it is proven for complex conditions by many examples in reality, too much and too few are objectives of fools.

Auditing offices etc. control generally state-owned enterprises. Inefficiency is a result of no or not recognizable consequences for wrong behaviour. Also the selecting process for actors may be hampered for example by political reasons. It is a human characteristic that by the time an organization may become inefficient. Research shows that also non-profit organizations like Labour Unions have the tendency to become mediocre (NZZ, 1996) and may for shaping the social and legal framework fitting to new technologies. This obligation cannot be left to

A close fruitful co-operation between real **application**, **research** and **industry** can be found in the military area, as already mentioned. Many standards used in civil sector have been enforced by military or space research background. When new technology entered civil application like safety related systems for railways, it was even difficult to adapt the industry on such requirements like life- cycle till scrapping, skill besides knowledge, user instead of customer etc., which are no topics for the delivery of rockets.

But now also experienced users of civil areas, like railways and EDF (Electricité de France), have only very limited resources for investment in long range developments. This is caused by the governmental policy and market situation at present. Road transportation, especially the automotive industry, becomes very engaged also with much support by EU. But by historical background, some lack of systems engineering skill has to be overcome and the organizational structure is very complex. The numerous users are very diverse and the access to them is very weak.

Of course somebody could believe that consumer protection groups would take over this duty in their own interest and for advantage of the clientele. It is needed to install a pool of experienced users because the application area is broad and it consists the similar organizational complexity as it is seen for road transportation. But it seems, not this difficulty is the reason of no engagement in the development process for human-oriented systems. Many groups still stick to the strategy to wait till a solution is installed and then to say no because it should be otherwise. They have not recognizing the change of product development to system development by new technologies. But the change of complex solutions is difficult and needs extreme effort. Just Prof. Fred Margulies, well know not only in the IFAC community, showed that all groups concerned by new technology gain much if Labour Union, science and industry co-operate in an early stage of design.

Best fitting structures for development in complex systems have **matrix organizations** (Genser, 1987), see **figure 3**. They allow access to different areas of knowledge and skill, are flexible, efficient and effective. They enable participation, foster also learning of problems of other fields, and improve acceptance. The power of matrix organizations was shown in the global and complex development of EDIFACT (Electronic Data Interchange for Administration, Commerce, and Transport) (DIN, 1994). EDIFACT was enforced by UN/ECE (United Nations/ Economic Commission for Europe) and showed also the importance of commitment of top actors on international and national political levels.

Active commitment of government is a first essential

new technologies. This obligation cannot be left to market forces and diverse groups split up.

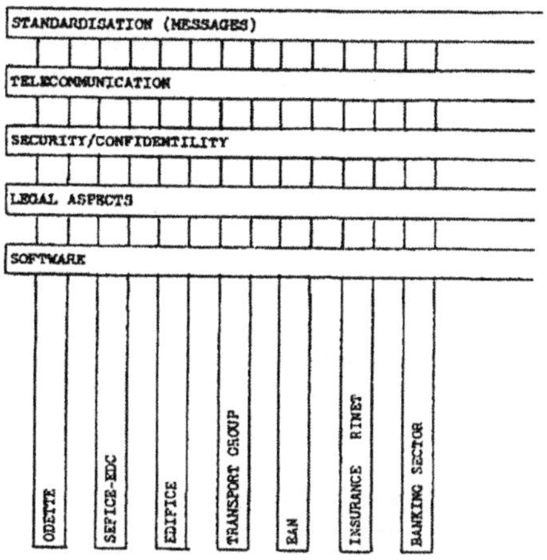

Fig. 3 TEDIS (Trade Electronic Data Interchange System) Project of CEC for EDIFACT

5. RISK MANAGEMENT

Requirements are unsure and unreliable if people are asked before they had already the possibility to use new equipment really. Moreover human limitations do not allow grasping complex long living system and its dynamics. The risk at decision-making cannot be full prevented. The discussion on gen technology shows the uncomfortable situation. Prof. Ortwin Renn distinguishes following types of risk management according the problem given (CORDIS, 2004b).

Risk-based management is applied for complex risks like industrial plants with hazardous materials. A high degree of modelling will be used (Wierzbicki, et al., 1999). Because of extension of new technologies, like embedded systems, to safety related applications in medicine or in air transportation, an intensive strive for new approaches can be found. CORAS (Fredriksen, et al., 2002)) is one of such examples.

Precaution-based management is considered in case of high level of uncertainty like new epidemics or green biotechnology. Also the railway had in the early stage severe restrictions by law for use on public area. The time of precaution-phase is not only needed for learning the use of new technologies and getting the skill rather for adapting social and legislative systems to new demand if stability and human solutions are strived for.

Discourse-based management is needed if the risk is seen very controversial like it is the case for genetic engineering. But transparency, participation and two-way communication are a precondition for reasonable results. Many times not recognized, but time is not only needed for learning by test or pilot projects rather time is required for learning to understand and to accept each other (Genser, 1986). In some cases, a better solution has been developed by this approach than the so-called experts group has considered before.

Prevention will be the measure when it is dealt with imminent danger.

The discussion on acceptance of risk is not finished rather new aspects are taken in consideration being due to the new technology-based safety related systems in medical area (EWICS, 2004). Also the term risk is not settled; see (Ladkin, 1998).

The difference of possibility and probability is not taken in account in many cases. Mathematical approach prefers probabilities. The dynamics of objectives and the dependency of data on context concerning information make pure mathematical approaches difficult. It is a mistake to assume probabilities distributions for handling possible events with a simple algorithm, as it is done with the fuzzy set approach. If something is not known the missing knowledge should not be covered by sophisticated assumptions.

6. HUMAN-ORIENTED DEVELOPMENT

Developing new complex systems requires being aware of human peculiarities, to be conscious of the fuzziness of information on future and about the limitations to grasp present as well as also past facts.

To tackle the human problem, it is a prerequisite to involve all groups concerned in the developing process from very beginning. This is not only for getting acceptance and trust, but also for improving knowledge for all. Missing information and limited trust may cause irrational behaviour.

Prof. Bryan Wynne stated at the EuroNanoforum in Trieste (CORDIS, 2004a) that public ignorance is not the cause of mistrust and scepticism; this has been proved by Eurobarometer surveys. The cause is what as seen as a denial by scientists of scientific ignorance. The novel nature of nanotechnology means that there are many knowledge gaps, and the well-meaning but mistaken behaviour of institutions involved in nanotechnology leads to doubts.

Participation presupposes actors who bring the groups together. At the present situation some kind of support is needed also after start up. The DG XIII (Directorate General for Information Technology) of CEC (Commission of European Communities) made a break through in Europe with installing the Purdue Workshop European Branch 1973. Even at present

without further outside support, the European Workshop on Industrial Computer Systems TC Reliability, Safety & Security (http://www.ewics.org) is active with members from industry, research, authorities, and users.

The Japanese are very active to bring the population in contact with new technology in an early stage. The Japan Rail not only invites seniors to the first part under test of the Magnetic Levitation Line Tokyo to Osaka, autumn 2003 a country wide prize competition was running for riding.

DG XIII of the CEC recognized by effort of French experts the importance of a **top down** approach for systems development. The automotive industry has had only experience in developing products when they started a **bottom up** development of telematics in the framework of EUREKA.

The decision-making process should allow **two-way communications** and foster **transparency** according the possibilities of state of art of knowledge presentation (Genser, 2001).

Bang-bang control is used in classical control theory for reaching a target in short time. The ability of self-organization in human systems should be not ignored. Ideology prevents usually to counteract rightly for keeping the equilibrium.

Human-oriented development is considering human limitations. Uncertainty and fuzziness of information needs adequate procedures. This may be:

- **Multi-step decision-making process**
  + from rough model to context oriented model for details
  + step by step reduction of solution's space
- **Object oriented approaches** (Rumbaugh, et al., 1991)
- **Liberty in scope for adaptation**
  + Robust solutions
  + Dynamical planning
  + Limitation of obligations
  + Bearing in mind possible occurrences
- **Evolutionary optimisation** (Bäck, et al., 1997)
  + Co-operative competition

Fuzziness caused by human limitations of grasping can be reduced by:

- **Abstraction**
  + Mathematics, which concentrate and reduce the amount of information and proofs by investigations off-line without time racing problems
  + Formal object oriented methods like Vienna Definition Method (Jones, 1990)
  + Standardisation
- **Human oriented presentation**
  + Graphs like Petri nets (Voss and Genrich, 1987)

+ Pictures, like modern simulation tools are using.

Important is the visualisation of information. Visualised data are not only easier to grasp by decision makers (usually more than one is needed for complex problems) rather it fits better, like pattern classes, than number crunching for handling fuzzy information.

Human oriented presentation requires recognised global harmonised standards as it is given with matrices in the field of mathematics. Standardisation not only reduces the learning effort rather avoids misunderstanding.

In the nature, units with diverse behaviour or strategies tackle unknown environment or future. These evolutionary processes, based also on genetic selection, evaluate the fittest solution.

Co-operative competition is a prerequisite for the stability of the hyper-system. This means, the objectives that should be gained and the rules for competitors have to be given. It is accepted that for sport events, like at Olympics, the goal is known and the rules are stated and controlled.

Objectives have to be stated, which support achieving the targets given in spite of human shortcomings. **Objectives for reaching targets** are also for smoothing adaptability, and to improve early perceptibility of changes:

- Control of results
- Responsibility **and** consequences
- Facilities for comparison with other developments
- Observation of and access to global development
- Co-operative competition
- Securing freedom for correction and control actions
- Robust and praxis oriented solutions
- Early correction of shortcomings
- Early conflict solving
- Motivation.

7. CONCLUSION

Development of complex systems needs time for getting experience and acceptance. Usually the legal system is a stabilizing factor because of its inertia. This was true for EDIFACT and for using electronics in safety related areas of railways for example. But Justitia is blind and tends to judge yes or no. It cannot be decided if the sun is good or bad. The sun is needed for life, but even at the same time the sun can be deadly.

The Commission of EU tries again to improve the co-operation of diverse groups by installing EU platforms (http://europa.eu.int/yourvoice/). These platforms should evaluate what and how the developments for the future should be. It has to be strived for that the real user will be involved in the early stage of design.

But new complex systems are a global task as it was given with EDIFACT and as it is with disaster management, which is under the auspices of the UN Space Office.

REFERENCES

Bäck, T., D. B. Fogel and Z. Michalewicz (Ed) (1997). *Handbook of Evolutionary Computation.* Oxford, New York.

CAD (1986). Rechnergestützte mechanische Konstruktion. *Elektrisches Nachrichtenwesen,* **60**, 315.

CORDIS (2004a). Experts examine the barriers to public acceptance of nanotechnologies. *CORDIS FOCUS,* **236**, 4.

CORDIS (2004b). The precautionary principle. *CORDIS FOCUS,* **236**, 5-6.

DIN (1994). UN/EDIFACT – *Organisation und Ansprechpartner.* Normenausschuß Bürowesen im DIN, Berlin.

EWICS (2004). Medical Group, http://www.ewics.org

Fredriksen, R., M. Kristiansen, B. A. Gran, K. Stolen, T. A. Opperud and Th. Dimitrakos (2002). The CORAS Framework for a Model-Based Risk management Process. In: *SAFECOMP 2002* (Anderson S., et al., (Ed)), 94-105. Springer-Verlag, Berlin.

Genser, R. (1986). Learning in Decision Making. In: *Large-Scale Modelling and Interactive Decision* Analysis (Fandel, G. (Ed)), 138-147. Springer-Verlag, Berlin.

Genser, R. (1987). Austrian General Conception of Transportation. In: *Control of Transportation Systems (1986)* (Genser, R., et.al. Ed), 47-54. Pergamon Press, Oxford.

Genser, R. (2001). Aspects of Decision Making. In: *Preprints IFAC SWIIS 01* (Kopacek, P. (Ed)), 127-130. TU Wien IHRT, Vienna.

Jones C. B. (1990). *Systematic software Development using VDM.* Prentice-Hall International, Englewood Cliffs, New Jersey.

Ladkin,P. B. (1998). www.rvs.uni-bielefeld.de/publications/Reports, Article RVS-Occ-98-01.

Lloyd, A. L. (1995). Computing Bouts of the Prisoner's Dilemma. *Scientific American,* **281**, 80-83.

NZZ (1996). Der strukturelle Zwang zur Mittelmässigkeit. *Neue Züricher Zeitung.***262.**15.

Oden, H. (1968). Der Mensch und das Telefon. *ETZ-A,* **89**, 688-694.

Rumbaugh, J., M. Blaha, W. Premerlani, F. Eddy and W. Lorensen (1991). *Object Oriented Modelling and Decision.* Prentice-Hall

International, Englewood Cliffs, New Jersey.

Voss, K. and H. J. Genrich (Ed) (1987). *Concurrency and Nets - Advances in Petri nets.* Springer-Verlag, Berlin.

Wierzbicki A. P. and M Makowski (1992). Multi-Objective Optimization in Negotiation Support. *WP-92-007.* IIASA, Laxenburg.

Wierzbicki, A., M. Makowski and J. Wessels (Ed) (1999). *Model-Based Decision Support Methodology with Environmental Applications.* Kluver Academia, Dordrecht.

ELSEVIER

IFAC

PUBLICATIONS
www.elsevier.com/locate/ifac

NEURO-FUZZY PHILLIPS-TYPE POLICY MODELS
FOR
FISCAL AND MONETARY DECISION-MAKERS

Yukio Ito

Department of Information Management, Osaka University of Economics
Higashiyodogawa, Osaka 553-8533, Japan
E-mail: itou@osaka-ue.ac.jp

Abstract: The purpose of this paper is to consider applicability and theoretical framework of the Neuro-fuzzy Phillips-type PID stabilization policy with genetic algorithm for decentralized fiscal and monetary decision makers under hierarchical policy structure with the upper level policy-coordinator. This policy is formulated by the Phillips-type stabilization policy problem consisting of feedback rule policies with varying the feedback gain parameters leading to the desired targets subject to the econometric models characterized by the respective fiscal and monetary decision-makers. In order to compute the feedback gain coefficients, the neural networks algorithm after selecting some policy rules by genetic algorithm are used. *Copyright©2004IFAC*

Keywords: Fuzzy control, neural networks, genetic algorithms, PID control, economic systems

1. INTRODUCTION

In recent year Japanese economy has been depressed more than 10 years since we had bubble-burst of land's and securities' prices in 1992. The government has taken the stringent fiscal constrained policy targeting to banish the bad loans among all over Japanese banks to reorganize the hierarchical structure of Japanese industries and public spendthrift policies despite that we have to pump up the aggregate demand of Japan. There has been issued whether activist or monetarist (so-called market economist) policy would be more effective to the long-stagnated Japanese economy. However, there has been no clear empirical evidence which such policies have the strong efficacy. In the declining process of the US economy power, the advanced industrial and countries have been aiming at economic policy necessitating reorganization of industrial structure due to development of information communications among the hybrid industrial transactions and international finance cooperation such as shift foreign-exchange rate intervention policy through the advanced IT. The world economy expects Japanese economy would steer up its economic growth and lower the

unemployment rate without bad loans. In this sense Japan should confront these difficulties by cooperating fiscal and monetary policymakers represented by Japanese government and Bank of Japan. However, these policymakers' policies have not always been consistent in their objectives and effectiveness in the history. In order to avoid the unnecessary policy conflicts between them, their behaviors have to coordinate in means and results.

Recently, there have been the increasing number of applications of neural networks to stock markets (Omatsu, 1996), fuzzy control and, exchange-rate markets (Franses, 2000) in finance and economic policy (Ito, 1998). Despite of this efficacy of combined use of neural networks and fuzzy control algorithm to economic stabilization policy, there is few research whichever optimal control or the Phillips-type feedback stabilization policies should be applied. Since Phillips first initiated PID feedback economic stabilization policy approach more than fifty years ago, there have been still some more unsolved problems he assigned. One of these problems is nonlinearity due to irreversible or unknown relations among economic variables and their interconnected varying parameters.

Progresses have made progressed in structuring and estimating for nonlinear economic relations by the development of econometric methods and Systems Dynamics approach at present. However, It has been faced that the uncertain and dynamic world economy from the viewpoint of Japanese stagnated economy as the adaptive mechanism is modeled with the ever-changing economic environment, the complex and time-varying parameters. Thus, it have been naturally to deal with the economic stabilization policy suitable to the this changing economy, in particular with stock prices and the foreign exchange-rates indicating individual economic health influenced by the various factors happening at the current time transition. Thanks to the rapid development of the down-sizing computer and statistical software, economic modeling and estimating can be dealt with ease in recent years.

Although the traditional Phillips-type stabilization policy has been replaced by optimal stabilization (control) policy incorporating the state space models in the modern control system theory, the Phillips-type stabilization policy approach solved as the stated above has not well-developed in theoretical and empirical studies. In the most recent years, the PID control laws invented in engineering has revived from the practical and familiar tools as the important idea of automatic controls in engineering science due to the rapid computer technology. The practical use of fuzzy control, neural networks and genetic algorithms have also owed to this development of information technology. Particularly, these computer-oriented algorithm approaches for nonlinear systems cannot be the powerful solution of modeling and controlling for the complex systems like economy and corporate management without combining their merits. This is why we focus on the new technology of fuzzy concepts and neural networks together with genetic algorithm for selecting policy rules. Therefore, the purpose of this paper is to apply these integrated advanced technology and the idea of this hybrid algorithm on modernized Phillips-type PID feedback control policies for economic stabilization of the simplified macroeconomy with fiscal and monetary variables as a typical example of the intelligent policy models. By these methods it could be understood how causes to results are related and how much intensity of connectedness between among these vivid economic variables exist. These casual relations can be verified by fuzzy-linguistic rules and computed by ANN with GA selection. Thus, we could find how influential these causal relations between fiscal and monetary policy on GDP, prices and unemployment are. This paper considers efficacy of this neuro-fuzzy stabilization modeling for macroeconomy. The theoretical frameworks of traditional economic policy due to Tinbergen-Theil approach including the fixed target economic policy and optimal decision rules derived by optimizing theory have been not enough to propose the dynamic optimal stabilization policy with the

conflicting decision-makers adjusted by an imaginary policy-coordinator in the upper level has not existed and there has been no example applied on the estimated econometric models. The approach proposed in this paper there is the policy coordinator as the decision-making organization determines policy adjusted and integrated in the upper level by exchanging information on the subsystem decision-maker in the lower level. Thus, two types of policy structure can dealt with, i.e. (i) coordinating and (ii) non-coordinating due to information relation between two decision-making in the lower level and that in the upper level. Therefore, this paper shows the policy structure of hierarchical Phillips-type stabilization policy-maker in the upper level with fiscal policymaker (Ministry of Finance) and monetary one (central bank, i.e. Bank of Japan) and simulates the dynamic performance of a coordinator and decentralized policymakers by using fiscal (Keynesian) and monetary econometric models.

The organization of this paper is as follows. In section 2, the Phillips-type stabilization policy in terms of various policy information between a policy-coordinator with fiscal and monetary policy-maker under hierarchical policy structure. In section 3, neuro-fuzzy networks PID policy modeling with genetic algorithm is presented for policy expressions Section 4 shows the small-size econometric model of Japanese economy to ascertain applicability of policy theory. In section 5 policy simulation results using the econometric model will be shown and their effectiveness will be considered. Finally, some remarks on this policy approach and applicability will be made.

2. STABILIZATION POLICY UNDER HIERARCHICAL POLICY STRUCTURE

In this section it is to formulate, and derive solutions of Phillips-type PID stabilization policies in policy structure of the policy coordinator in the upper level adjusting two decentralized fiscal and monetary policymakers in the lower level under the hierarchical system and consider its implication. The coordinator has the Phillips-type PID feedback stabilization policy to be implemented into the original structural econometric model. The decentralized policymakers have their own reaction functions with their counterparts' control variables and coordinator's target variables. The respective decentralized policymaker exchange information through coordinator's target variables. There are two policy structures due to how to transmit information as (i) coordinating and (ii) non-coordinating types. The latter, the case of one taking the optimal behavior without considering the counterparts', is divided into three policy structures as (a) Fiscal policymaker leading, (b) Monetary policymaker leading, and (c) Isolationist policies. Let us formulate these policies.

The estimated reduced form econometric model to be controlled can be written by

$$y_t = Ay_{t-1} + B_1u_{1t} + B_2u_{2t} + Dd_t + \eta_t$$

where
y : m-dimensional target vector
u_1: n1-dmensional fiscal policymaker control vector.
u_2 : n2-dimensional monetary policymaker control vector.
d : l-dimensional data vector
η : m-dimensional residual vector assuming that: $E(\eta)=0$, $E(\eta_t\eta_t`)= Q_\eta$, $E(\eta_t\eta_s`)=0$ for $t\neq s$
A, B, C and D are the estimated reduced-form parameters.

The Phillips-type stabilization problem for hierarchical policy structural system has the center problem the policy-coordinator in the upper level adjusts the interaction between fiscal and monetary policymakers and the local problem in the lower level. Coordinating type can be considered to be the centralized stabilization problem with a single government. In this case there is no policy-conflicting between fiscal and monetary policymakers can control both control instruments simultaneously. Therefore, two decision-makers are respectively simply the subset of the single centralized policymaker. On the other hand, (c) in (ii) non-coordinating case each policymaker has only its own targets and controls but no other targets and controls. (a) and (b) have their reaction functions incorporating counterpart's control and a whole econometric model.

Thus, the Phillips-type stabilization policy problem under hierarchical policy structural system is defined as the Phillips-type stabilization policy from the viewpoint of macroeconomic policy from available information on targets transmitted by two policymakers in the lower level, aiming its own objectives subject to its own reaction function and whole macroeconomic structure. The objectives at each decision-makers aim the constant desirable values of specified target variables, for example, 2 percent annual growth rate of GDP for the policy-coordinator, 2 percent unemployment rate for the fiscal policymaker and 2 percent inflation rate for the monetary to be controlled by each policymakers' control variable such as government expenditures and interest rate. The above formulations are summarized as follows:

Center Problem:

$$\text{Min} (\Delta Ln(GDP) - \alpha)^2$$
subject to
$$y_t = Ay_{t-1} + B_1u_{1t} + B_2u_{2t} + Dd_t + \eta_t$$

Local problem:

(a) Fiscal Leading Type: $\text{Min} (\Delta Ln(U) - \beta)^2$
subject to $u_{2t} = F_1y_{t-1} + F_2u_{1t} + F_3d_t$
$y_t = Ay_{t-1} + B_1u_{1t} + B_2u_{2t} + Dd_t + \eta_t$

(b)Monetary Leading Type: $\text{Min} (\Delta Ln(CPI) - \gamma)^2$
subject to $u_{1t} = M_1y_{t-1} + M_2u_{2t} + M_3d_t$
$y_t = Ay_{t-1} + B_1u_{1t} + B_2u_{2t} + Dd_t + \eta_t$

(c) Non-interventional Isolationist Policy:

Fiscal policymaker $\text{Min} (\Delta Ln(U) - \beta)^2$
Monetary policymaker $\text{Min} (\Delta Ln(CPI) - \gamma)^2$

In (c) both policymakers are subject to

$$y_t = Ay_{t-1} + B_1u_{1t} + B_2u_{2t} + Dd_t + \eta_t$$

These policy structures are diagrammed in Fig. 1.

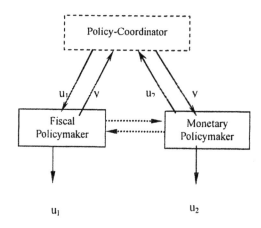

Fig. 1 Hierarchical policy structure

where real and dash boxes denote existence of policymaker, the former expresses existence and the latter non-existence, respectively, and real and dash lines denote direct effect and indirect effect, respectively.

3. NEURO–FUZZY PID POLICY MODELING

This section is to deal with the basic framework and empirical study of the Phillips-type PID stabilization policies for economic models, which was first initiated by well-known T.W. Phillips as a economist more than four decades ago, with the gain parameter estimation of the controlled equation by neural network approach. Let us consider the following estimated nonlinear econometric models in terms of ARMAX vector-type :

$$y_t = f\{y_{t-1}, y_{t-2}, ..., x_t, x_{t-1}, x_{t-2}, ..., z_t, z_{t-1}, z_{t-2}, ..., e_t\} \quad (1)$$

where y_t : target vector, x_t: control vector, z_t : data vector, and e_t : residual vector.

Defining the first difference of each variables vectors eq.(1) above represented by taking natural logarithm, we can describe the difference type equations as:

$$\Delta y_t = f\{\Delta y_{t-1}, \Delta y_{t-2}, ..., \Delta x_t, \Delta x_{t-1}, \Delta x_{t-2}, ..., \Delta z_t, \Delta z_{t-1}, \Delta z_{t-2}, ...\} \quad (2)$$

where Δ implies the first difference for each variables: $\Delta y_t = y_t - y_{t-1}$, $\Delta x_t = x_t - x_{t-1}$ and $\Delta z_t = z_t - z_{t-1}$

In eq. (1), the policy-maker aims to minimize the deviation between the actual targets and the desired so that for the first difference of natural logarithm-type equation (2) the aim can be replaced by minimizing the deviation between the first difference variables and certain constant values. We call this the policy deviation as:

$$d_t = \Delta y_t - \alpha \quad (3)$$

where α is the desired constant values vector which the policy-makers wish to attain for example, the desired GDP growth rate. We can show PID feedback stabilization policy as follows:

$$\Delta x_t = f_P(d_t - d_{t-1}) + f_I d_t + f_D(d_t - 2d_{t-1} + d_{t-2}) \quad (4)$$

where f_P, f_I and f_D are proportional, integral and derivative coefficients row vectors, respectively. They are neural outputs denoted by F_P, F_I and F_D.

The section is to deal with the basic framework and rewrite as

$$y_t = f(Y_t) + e_t \quad (5)$$

Assume the linguistic representation of model (1) is given by a set of fuzzy rules, the ij-th fuzzy rule is

$$r_{ij}: \text{IF } (Y_t \text{ is } A^i) \text{ THEN } (y_t \text{ is } B^j) \quad (6)$$

where A^i is the multivariate fuzzy set obtained form the intersection of k univariate fuzzy sets of each input, and B^j is the univariate fuzzy set of the output. The antecedent (Y_t is A^i) related to every possible consequent of (y_t is B^j) (j = 1, 2, ...). This can be shown by the following examples:

Rule 1: IF <Government Expenditure = POSITIVE> THEN <GDP =POSITIVE>

Rule 2: IF < Interest Rate =POSITIVE> THEN <Exchange Rate VERY POSITIVE >

Antecedent Part:

Government Expenditure   Interest Rate
Very Negative Negative Zero   Positive Very Positive

Consequent Part:
GDP   Exchange-rate
Very Negative Negative Zero   Positive Very Positive

Fig. 2. Examples of Simple Fuzzy Economic Rules and Its Corresponding Membership Function for Antecedent and Consequent Clauses

The fuzzy output set is obtained from fuzzy rules using union operator. Assuming the real-valued inputs are represented by fuzzy singletons B^j, being symmetric and bounded for all j with addition and product operators. Using the center of gravity defuzzification algorithm, the defuzzied output is

$$\mu_j = \prod_{i \in I} \mu_{ij} \quad (7)$$

$$y_t = \frac{\sum_{j=1}^{J} \mu_j w_{kj}}{\sum_{j=1}^{J} \mu_j} \quad (8)$$

where I is input number, j is fuzzy rule number and k is the data number. The shape of membership function is the radial basis function with insensitive range c that is useful for reducing the membership functions. The membership function of an input value and the j-th fuzzy rule is expressed by

$$f(I_i) = \begin{cases} -b_{ij}(|I_i - a_{ij}| - c_{ij})^2 & \text{if } |I_i - a_{ij}| \geq c_{ij} \quad (9) \\ 0 & \text{if } |I_i - a_{ij}| \leq c_{ij} \quad (10) \end{cases}$$

$$\mu_{ij} = \exp\{f(I_i)\} \quad (11)$$

The gradient decent method to tune antecedent and consequent parts of parameter variation rules are presented as follows:

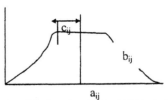

Fig. 3. Membership function by Radial Basis Function

Consequent Part: $w_{ij} = w^*_{ij} - k_w \cdot \partial E_p / \partial w_{pj}$

where w, p, and k_w are the consequent value, the rule number, and the learning coefficient, respectively.

Antecedent Part:

$$a_{ij} = a^*_{ij} - k_a \cdot \partial E_p / \partial a_{ij}, \qquad b_{ij} = b^*_{ij} - k_b \cdot \partial E_p / \partial b_{ij}$$
$$c_{ij} = c^*_{ij} - k_c \cdot \partial E_p / \partial c_{ij} \qquad (12)$$

where k_a, k_b, k_c are the learning coefficients, and E_p is the deviation between actual target value and the desired, where a, b, c are the coefficients forming the shape of membership functions.

The neural networks is given by minimizing the quadratic criterion as:

$$E_p = \frac{1}{2} d_t^2 \qquad (13)$$

The combined weighting coefficients are obtained by the steep-decent method as follows:

$$\Delta w_{t;1}^{kj} = -\eta \frac{\partial E}{\partial w^{kj}} + \beta \Delta w_t^{kj} \qquad (14)$$

Defining

$$\delta_k = -\frac{\partial E}{\partial net_k} \qquad (15)$$

Then the following equations can be obtained

$$-\frac{\partial E}{\partial w^{kj}} = -\frac{\partial E}{\partial net_k} \cdot \frac{\partial net_k}{\partial wkj} = \delta_k F_j \qquad (16)$$

$$-\frac{\partial E}{\partial \Delta y_t} = -\frac{\partial E}{\partial d_t} \bullet \frac{\partial d_t}{\Delta y_t} = d_t$$

$$\frac{\partial F_k}{\partial net_k} = f'(net_k) = F_k(1 - F_k)$$

$$\frac{\partial \Delta x_t}{\partial F_k} = \begin{cases} d_t - d_{t-1}; k = 1 \\ d_t; k = 2 \\ d_t - 2d_{t-1} + d_{t-2}; k = 3 \end{cases}$$

Applying the chain rule of differentiation gives

$$\delta_k = -\frac{\partial E}{\partial net_k} = -\frac{\partial E}{\partial \Delta y_t} \cdot \frac{\partial \Delta y_t}{\Delta x_t} \cdot \frac{\partial \Delta x_t}{\partial F_k} \cdot \frac{\partial F_k}{\partial net_k} \qquad (17)$$

Finally, the following equation is obtained

$$\delta_k = d_t \frac{\partial \Delta y_t}{\partial \Delta x_t} F_k (1 - F_k) \frac{\partial x_t}{\partial F_k} \qquad (18)$$

Therefore, the alternation formula is represented by

$$w_{t+1}^{kj} = \eta \delta_k F_j + \beta \Delta w_t^{kj} \qquad (19)$$

$$\delta_k = d_t \frac{\partial \Delta y_t}{\partial \Delta x_t} F_k (1 - F_k) \frac{\partial x_t}{\partial F_k} \qquad (20)$$

In the middle layer, we can represent as :

$$\Delta w_{t+1}^{kj} = -\eta \frac{\partial E}{\partial w^{jt}} + \beta \Delta w_t^{ji}$$

$$= \delta_k w^{kj} f'(net_j) = \sum_k \delta_k w^{kj} F_j (1 - F_j)$$

$$-\frac{\partial E}{\partial w^{ji}} = -\frac{\partial E}{\partial net_j} \cdot \frac{\partial net_j}{\partial w^{ji}}$$

Taking account of the relation as:

$$net_j = \sum w^{ji} F_i + \theta_j$$

$$\delta j = -\frac{\partial E}{\partial net_j} = \sum_k \frac{\partial E}{\partial net_k} \cdot \frac{\partial net_k}{\partial F_j} \cdot \frac{\partial F}{\partial net_j}$$

$$\Delta w_{t+1}^{ji} = \eta \delta_j F_i + \beta \Delta w_t^{ji}$$

Therefore, the alternation rule of the middle layer can be shown as:

$$\delta_j = \sum_k \delta_k w^{kj} F_j (1 - F_j)$$

Summarizing the above expressions, we have the following PID stabilization policy by neural networks Fig.2.

GA Parameter Selection

Selection is executed by ranking the tune strings based on fitness value F in the following equation (14),

$$F = \alpha MSE + \beta MAXSE + \gamma R + \delta N \qquad (21)$$

where MSE, MAXSE, R, N, and α, β, γ, δ are the mean square error between tutorial and output values, values, maximum square error between tutorial and output the number of rules, the number of membership function, and maximum square error between tutorial and output the number of rules, the number of membership function, and the coefficients, respectively. GA selection constitutes cross over, selection and mutation process by coding each policy rule by different gain parameter.

4. NONLINEAR MACROECONOMIC MODEL

A simplified nonlinear with logarithm-type annual macroeconometric model for the sample periods from 1968 to 2000 of Japanese macro data in this section is shown. This model is estimated models for the economic relations among three aggregate target variables as GDP, consumer price index, unemployment rate, and control variables as Government Final Consumption,, Money Supply. The Unemployment Rate is output layer as a target variable, GDP and Consumer Price Index input for the middle layer, and Government Final Consumption and Money Supply for the input layer as controls. These variables were chosen by investigating the correlation between them. They have the hierarchical structure for two inputs, Government Final Expenditure and Money Supply, and one output, unemployment rate through the intermediary variables such as GDP and CPI. The three equations are expressed in terms of the first difference equations by natural logarithm type as follows:

$$\Delta \ln(GDP) = 0.0456 - 0.0092U + 0.1455 \, \Delta \ln(G)$$
$$(2.977) \quad (-1.984) \quad (1.048)$$
$$- 0.0514 \, \Delta \ln(M)$$
$$(-1.912)$$

$$R^2 = 0.2371, \quad SE = 0.0193, \quad DW = 2.177$$

$$U = 3.23 - 12.637 \, \Delta \ln(GDP) - 11.58 \, \Delta \ln(CPI)$$
$$(19.686) \quad (-4.243) \quad (-5.364)$$

$$R^2 = 0.6137, \quad SE = 0.5346, \quad DW = 0.7176$$

$$\Delta \ln(CPI) = -0.0371 + 1.0999 \, \Delta \ln(GDP) - 0.0145U$$
$$(3.828) \quad (-4.195) \quad (-3.931)$$

$$+ 0.01842 \, \Delta \ln(G)$$
$$(2.999)$$

$$R^2 = 0.5902, \quad SE = 0.0282, \quad DW = 0.9943$$

where
G = real government final government expenditure, billion Yen
GDP = real gross domestic products, billion Yen
M = M2+CD, billion Yen
U = unemployment rate (%)
CPI = consumer price index 1995 =100
Source: National Economic Accounts (Statistical Bureau), Research and Statistic Department, Bank of Japan.
Δ implies the difference between the current time and the previous time for the natural logarithm transformed variable Z. Symbols: R^2 is the adjusted coefficient of determination, SE, the standard error of residual, and DW, Durbin-Watson statistic.

5. POLICY SIMULATIONS

Some simulation can be executed to investigate how the gains, f_P, f_I, f_D can obtained by neural networks with GA selection process by giving a desired target vector, α, β and γ. It implies that the growth rate of GNP could be predicted once the gains are automatically scheduled by giving the growth rates of the government expenditure and money supply. It will be shown that this method is applicable to any nonlinear econometric models as well as the linear once the appropriate neural networks with any PID stabilization policies. can be constructed.

6. CONCLUDING REMARKS

The Phillips-type PID Economic Stabilization Policy is more sensitive to parameter variations of the coefficients in the econometric equations. The more higher GDP growth rate is, the more distinctive each policy is. The most closest policy to actual values of GDP growth rate is the derivative policy as the least powerful to depress the growth rate of GNP, what the most powerful policy can lower the growth rate is the integral policy, and the proportional policy take the middle position to lower the growth rate among three policies. The policy simulations ascertain how effective the variations and the following calculation of target trajectories, and design the computer programming are.

REFERENCES

Franses, P. H. and van Dijk, D. (2000). *Nonlinear Time Series Models in Empirical Finance*, Cambridge University Press,. Cambridge.

Ito, Y. (1998). Optimal Stabilization Policies by Neural Networks In *Advances in Artificial Intelligence and Engineering Cybernetics, Volume IV: Systems Logic & Neural Networks* (G. E. Lasker (Ed.)) pp. 81 - 86. The International Institute for Advanced Studies in Systems Research and Cybernetics, Windsor.

Omatsu, S. (1996). Self-Tuning Controls using Neural Networks (in Japanese) In *Self-Tuning Controls* (Omatsu, S and T. Yamamoto, *et al.* (Ed.)), pp. 67-98, Society of Instruments and Control Engineers, Tokyo.

ELSEVIER

IFAC

PUBLICATIONS
www.elsevier.com/locate/ifac

OPTIMAL MACROECONOMIC POLICIES FOR SLOVENIA: AN OPTIMAL CONTROL ANALYSIS

Reinhard Neck[1], Klaus Weyerstrass[2] and Gottfried Haber[3]

[1]*Department of Economics, University of Klagenfurt, Klagenfurt, Austria*
[2]*Institute for Advanced Studies, Klagenfurt, Austria*
[3]*Ludwig Boltzmann Institute for Economic Policy Analyses, Vienna, Austria*

Abstract: This paper presents an application of optimum control theory to derive economic policy recommendations for Slovenia. In particular, it analyzes the optimal design of macroeconomic policies for Slovenia after its integration in the EU. For this purpose, the model SLOPOL4, a macroeconometric model for Slovenia, is built and simulated over the next few years. Moreover, optimal macroeconomic policies for Slovenia are determined as solutions of optimum control problems with a quadratic objective function and the model SLOPOL4 as dynamic constraint. A scenario approximating Slovenia's entry into the European Economic and Monetary Union (EMU) is constructed. Simulation and optimization experiments under different assumptions about the set of available economic policy instruments are carried out. It is shown that the best policy results are obtained when the average tax rate on labor is available as an active policy instrument. *Copyright © 2004 IFAC*

Keywords: optimization, optimal control, economics, nonlinear system, application.

1. INTRODUCTION

Optimum control theory has been applied successfully to problems of macroeconomic policy for several years; see, for instance, Chow (1975; 1981), Kendrick (1981). Especially for short-term policy problems, such as the adequate response to transitory shocks in the course of the business cycle, this theory has been shown to be a helpful tool for supporting policy-makers' decisions. So far, only a few studies of this kind were undertaken for transition countries. This is mainly due to a lack of consistent data over a longer time period, which is typical for these countries. Slovenia is a case in point: not only has this country undergone structural changes after having got rid of the command economy of the Communist era; the state of Slovenia as an independent unit did not even exist before its departure from former Yugoslavia. In the meantime, however, Slovenia has caught up with the industrialized Western European countries in an impressive way and is now a member of the European Union (EU) since May 1, 2004. It is likely that Slovenia will enter the exchange rate mechanism of the European Monetary System EMS-II and eventually become a member of the Euro System in the European Economic and Monetary Union (EMU) very soon. It is obviously of interest how macroeconomic policy goals such as high GDP growth, low inflation and unemployment as well as external equilibrium and a balanced budget can be achieved under these conditions.

In this paper, it will be shown that optimum control techniques can be used to answer some of the questions relevant for macroeconomic policy design in Slovenia over the next few years. Simulation and optimization experiments are conducted that differ with respect to the set of economic policy instruments assumed to be available to Slovenian policy-makers. The simulation and optimization experiments are carried out using the optimum control algorithm OPTCON and SLOPOL4, a macroeconometric model of the Slovenian economy.

The paper is organized as follows: In Section 2, the econometric model SLOPOL4 is briefly described, and results of simulation experiments are sketched. Section 3 gives an overview of the optimum control algorithm OPTCON and details the optimization design used. Section 4 addresses the optimization results obtained with the average "tax" rate available as an active policy instrument on the one hand and under the assumption that this tax rate is exogenous. "Taxes" in this context include both income taxes and social security contributions. Section 5 concludes the paper. Details about the model and the simulation and optimization results can be found in another paper (Neck et al., 2004).

2. THE SLOPOL4 MODEL

SLOPOL4 (SLOvenian economic POLicy model, version no. 4) is a medium-sized macroeconometric model of the small open economy of Slovenia. It consists of 45 equations: 15 behavioral equations and 30 identities. The former were estimated by ordinary least squares (OLS), using quarterly data for the period 1992:1 (where available; 1994:1 otherwise) until 2001:4. The model combines Keynesian and neoclassical elements. In this section, the behavioral equations are sketched very briefly. A more detailed description of an earlier version of the model can be found in Weyerstrass et al. (2001).

Consumption of private households is explained by a simple Keynesian consumption function, depending on current disposable income and on lagged consumption. Real gross fixed investment is influenced by the change of total domestic demand, the user cost of capital (approximated by the real interest rate), and by the capacity utilization rate, i.e. the ratio of actual to potential real GDP. Real exports of goods and services are a function of the real exchange rate and of foreign demand for Slovenian goods and services. As the aggregate euro area is by far Slovenia's largest trading partner, the rest of the world is approximated by the euro area. Therefore, foreign demand is measured by euro zone real GDP, and we consider the exchange rate between the Slovenian tolar and the euro only. Slovenian real imports of goods and services depend on domestic final demand and on the real exchange rate.

Money demand depends on real GDP and the short-term interest rate. The long-term interest rate is linked to the short-term rate in a term structure equation. An exchange rate equation determines the nominal exchange rate between the Slovenian tolar and the euro as depending on the interest differential and the price ratio between Slovenia and the euro area.

Labor demand (actual employment) is influenced by final demand for goods and services and by the real gross wage, while labor supply depends on the real net wage and on the size of the population. The wage rate is determined by the price level, by the difference between the actual unemployment rate and a proxy for the natural (or not-accelerating-inflation) rate of unemployment (the NAIRU), by labor productivity, and by the average labor tax rate, which is defined as the difference between the average gross and net wages as a percentage of the gross wage (hence "labor taxes" include income taxes and social security contributions). Consumer prices depend on unit labor costs, the capacity utilization rate, and the nominal money stock; in addition, Slovenian prices depend on the oil price and on the nominal exchange rate.

Total government expenditures are linked to government consumption and to transfer payments to households, total government revenues are linked to labor tax revenues. The budget deficit is given by the difference between total government expenditures and revenues. Potential output, which is determined by a Cobb-Douglas production function with constant returns to scale, depends on trend employment, the capital stock, and autonomous technical progress. Trend employment is defined as the labor force minus natural unemployment. The NAIRU is approximated by a four-quarter moving average of the actual unemployment rate.

As a first step, the model SLOPOL4 is simulated over the period of interest, which is 2003 to 2008. For this purpose, forecasts of the exogenous and the control variables over this period are determined and used as inputs into the model to obtain predicted values of the endogenous variables. Slovenian and international estimates for these variables are used; for example, a slowly decreasing short-term rate of interest, constant growth rates of euro zone variables and of Slovenian government expenditures, a constant labor tax rate, etc., are assumed. The purpose of this exercise consists in simulating "business as usual", i.e., continuation of previous trends, in particular for the policy instruments. In order to mimic Slovenia's membership in the EU

Exchange Rate Mechanism ERM from 2007 on, a scenario of a crawling peg of the Slovenian tolar in the EMS-II until 2006 and its membership in the ERM thereafter is constructed, which finally results in a fixed exchange rate to (in reality: the adoption of) the euro. The exchange rate is assumed to be 232 SIT/EUR in 2003, 236 in 2004, 238 in 2005, 239 in 2006, and 240 from 2007 on.

The results of the simulation are shown for the most relevant variables of the model in Figures 1 to 8. Although quarterly data are obtained as output from the quarterly model SLOPOL4, only annual data are shown here to ease comparisons and interpretations. The simulation experiment results in fairly smooth time paths of the endogenous variables, with a decreasing rate of unemployment and an increasing current account surplus and government budget deficit. Government consumption and transfers grow parallel to each other. This simulation experiment is to be regarded as benchmark for the optimization experiments to be described next.

3. THE OPTIMUM CONTROL APPROACH

The aim of this study is to calculate time paths of macroeconomic policy instruments that are "optimal" according to an objective function of a hypothetical policy-maker for Slovenia. To obtain optimal economic policies, the OPTCON algorithm, which was developed by Matulka and Neck (1992), is applied. OPTCON determines approximate solutions of optimum control problems with a quadratic objective function and a nonlinear multivariable model. The objective function has to be quadratic in the deviations of the state and control variables from their desired values.

The objective function has the following form:

$$L = \frac{1}{2}\sum_{t=1}^{T}\begin{bmatrix} x_t - a_t \\ u_t - b_t \end{bmatrix}' W_t \begin{bmatrix} x_t - a_t \\ u_t - b_t \end{bmatrix},$$

$$W_t = \alpha^{t-1} W, \quad t = 1,\dots,T,$$

where x_t denotes the vector of state variables, u_t denotes the vector of control variables, a_t and b_t are the desired values of the state and control variables, W_t is the matrix containing the weights given to the deviations of the state and control variables from their desired values, respectively, and α denotes the discount factor. The dynamic system has to be given in a state space representation. Although OPTCON can solve deterministic and stochastic optimum control problems, here only deterministic optimizations are considered.

Optimizations are carried out for the period 2003 to 2009, but the final year is neglected to avoid terminal point effects, hence the period of interest is 2003 to 2008. In the present paper, five "main" and several "minor" objectives are considered. The "main" objective variables correspond to the most important macroeconomic goals Slovenian policy-makers will want to achieve in the medium-term future: adequate real GDP growth, low unemployment, low inflation, a balanced government budget and external equilibrium. Specifically, for the optimization experiments a desired real GDP growth rate of 4.5 percent p.a. is assumed. The desired rate of unemployment is assumed to be reduced by one percentage point per year from 9 percent in 2003 to 4 percent in 2008. The desired rate of inflation declines gradually from 6 percent in 2003 to 2 percent in 2008. The government budget and the current account (both in percent of nominal GDP) are assumed to be aimed at being balanced.

As "minor" objective variables, real GDP and its components (consumption of households, investment, government consumption, exports) and imports are considered. For these variables, ideal values consistent with the desired 4.5 percent growth rate of real GDP are specified. The introduction of "minor" objective variables shall reflect policy-makers' aim of obtaining smooth paths of the main macroeconomic aggregates, but serves also as substitute for introducing inequality constraints on state variables, which is not feasible in OPTCON. In addition, the policy instrument (control) variables are regarded as minor objective variables; this serves to reflect costs to the policy-makers of changing instruments, but is also due to formal requirements of the OPTCON algorithm and in order to prevent erratic fluctuations.

In the weight matrix of the objective function, all off-diagonal elements are set to zero. In addition, all endogenous variables of the model not mentioned get the weight zero, implying that they are not of direct relevance to policy-makers. The "main" variables are assigned the weight 10,000, whereas the "minor" objective variables are given a weight of 1, except for the control variables, which get weights of 10. These weights reflect both the relative importance of the "main" and "minor" objective variables and their different orders of magnitude.

In the optimization experiments, the question is addressed whether unemployment can be reduced significantly by cutting non-wage labor costs. For this purpose, in the first optimization experiment the average labor tax rate, i.e. the difference between the average gross and net wages as a percentage of the gross wage, is introduced as a policy instrument in addition to government consumption and transfers.

The results for this case are compared to those where this labor tax rate is fixed at a constant level over the entire optimization horizon.

4. RESULTS OF OPTIMIZATION EXPERIMENTS

First, optimal macroeconomic policies with the average labor tax rate as a policy instrument are described. The set of control variables consists of government consumption expenditures and government transfer payments to private households, both in nominal terms, and the average tax rate as defined in the previous section. The results for the "main" objective variables and some other variables of interest are displayed in Figures 1 to 8.

The optimal paths of the "main" objective variables are considerably better (closer to their desired values) than those in the simulation. Both unemployment and inflation are reduced more over the six-year period, the budget deficit remains below 2 percent of GDP, thus fulfilling the Maastricht fiscal deficit criterion fixed in the Stability and Growth Pact of the EU, the current account improves over the optimization horizon, and the average growth rate of real GDP is 2.1 on average. This favorable result is brought about mainly by low values of the labor tax rate, combined with only moderate increases of government consumption and transfers. The optimization experiments show that the loss of monetary policy as an instrument of stabilization policies resulting from pegging the exchange rate to the euro can be compensated for by cutting labor taxes once and (more or less) for all at the beginning of the optimization period. This policy, which has both favorable supply-side and demand-side effects, reduces the wedge between gross and net wages, thus lowering upward pressure on gross wages, thereby stimulating employment and reducing inflationary pressure at the same time.

Consider next optimal macroeconomic policies that are obtained when the average tax rate, i.e. the difference between the average gross and net wages as a percentage of the gross wage, is not available as an active policy instrument. Only two fiscal policy instruments remain at the policy-maker's disposal, namely government consumption and transfer payments to households. Again, results can be found in Figures 1 to 8.

It can be seen that the performance of most objective variables is considerably worse than under the low-tax rate regime, showing the importance of this instrument in the Slovenian econometric model under consideration. Except for the first period (2003), real GDP growth is lower, and the rate of unemployment is higher (and even higher than in the simulation

without optimization) in every year. The rate of inflation, the current account and the government budget deficit show developments similar to those with an "active" tax rate. There is a strong shift from transfers (which remain approximately constant in nominal terms) to government consumption. These results show that, in particular, the rate of unemployment can be very effectively and favorably influenced by a lower labor tax rate without putting unduly pressure upon the government budget. Optimal policies improve on the "business-as-usual" scenario of the simulations, but require not only counter-cyclical demand-side reactions (either through automatic stabilizers or through discretionary policy) but also structural (supply-side) reforms, such as a shift of government expenditures from transfers to purchases of goods and services and a reduction of the level of labor income taxation.

5. CONCLUDING REMARKS

In this paper it was shown how an optimum control approach can be used to obtain insights into the design of economic policy decisions for a country on the way into EMU membership. Assuming that over the optimization horizon 2003 to 2008, Slovenian policy-makers aim at high GDP growth rates, low rates of inflation and unemployment, balanced government budgets and balanced current accounts, effects of variations in the set of available policy instruments on optimal economic policies and on the objective variables were investigated. It turns out that when the average labor tax rate is available as an additional policy instrument, the loss of the monetary policy instrument resulting from adopting a pegged or even fixed exchange rate regime can be compensated by a different mix of the fiscal policy instruments, which can secure high GDP growth and a reduction of unemployment without endangering the aim of an almost balanced budget. If, on the other hand, the labor tax rate is held constant over the optimization horizon, higher unemployment cannot be avoided.

It has to be stressed that factors like structural imbalances between labor supply and demand, which may be very important determinants of unemployment, cannot be captured with an aggregated model like SLOPOL4. In addition, only the variables contained in the objective function are taken into account in the present analysis, while other measures of economic welfare, which are linked to the corporate ownership structure or the income distribution, for example, cannot be adequately considered within the framework of a highly aggregated macroeconomic model, which is used in this paper. Moreover, we have not touched upon the problem of the so-called Lucas critique, which is a

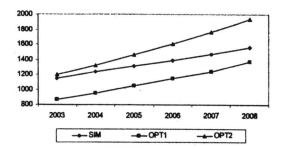

Fig. 1. Government consumption (nominal)

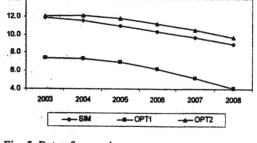

Fig. 5. Rate of unemployment

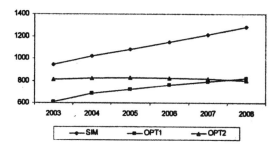

Fig. 2. Transfer payments to households

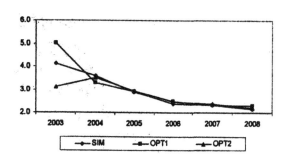

Fig. 6. Rate of inflation

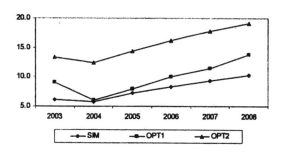

Fig. 3. Short-term interest rate

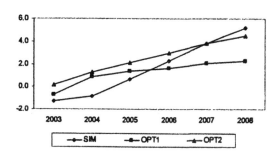

Fig. 7. Current account

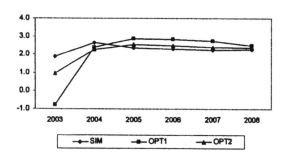

Fig. 4. Rate of real GDP growth

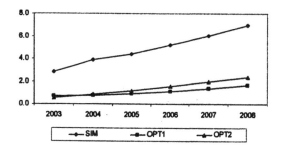

Fig. 8. Government budget deficit

SIM: simulation
OPT1: optimization with labor tax rate as policy
instrument
OPT2: optimization with fixed labor tax rate

fundamental objection against the approach followed here. Incorporating changes in the public's expectation with the recognition of a new policy regime into the model, as required by the Lucas critique, will certainly be a major improvement, although the short time series available for Slovenian data makes an attempt at executing it still more difficult than for countries with a longer history without structural breaks.

7. ACKNOWLEDGEMENTS

The present research was supported by the Jubiläumsfonds of the Austrian Central Bank (project no. 9506) and by the Ludwig Boltzmann Institute for Economic Analyses, Vienna. Comments by an anonymous referee are gratefully acknowledged. The views expressed here need not coincide with those of the Austrian Central Bank.

8. REFERENCES

Chow, G.C. (1975). *Analysis and Control of Dynamic Economic Systems.* Wiley, New York.

Chow, G.C. (1981). *Econometric Analysis by Control Methods.* Wiley, New York.

Matulka, J. and R. Neck (1992). OPTCON: An Algorithm for the Optimal Control of Nonlinear Stochastic Models. *Annals of Operations Research,* **37**, 375–401.

Neck, R., K. Weyerstrass and G. Haber (2004). Policy Recommendations for Slovenia: A Quantitative Economic Policy Approach. In: *Slovenia and Austria: Bilateral Economic Effects of Slovenian EU Accession* (B. Böhm, H. Frisch and M. Steiner (Eds.)), 249–271. Leykam, Graz,

Weyerstrass, K., G. Haber and R. Neck (2001). SLOPOL1: A Macroeconomic Model for Slovenia. *International Advances in Economic Research* **7** (1), 20–37.

A LOGISTIC REGRESSION ANALYSIS IN THE ISTANBUL STOCK EXCHANGE

Hakan Aksoy

Senior Portfolio Manager, Mutual Funds Division, Koc Portfolio

Abstract: By using Altman's famous bankruptcy prediction method, a model for predicting next period's stock return is developed with the discriminant analysis and the logistic regression by using 81 financial ratios before the factor analysis. The results are statistically significant that "good" companies perform better than "bad" companies in the ANOVA analysis. Then, the weighted average of the z score or the logit score of the stocks forecast the next period return of the ISE100 Index. In case of the restrictive properties of the discriminant analysis for the distributions of the variables, the result of the logistic regression analysis has better estimation power. *Copyright © 2004 IFAC*

Keywords: Stock Market, Efficient Market Hypothesis, Logistic Regression, Discriminant Analysis, Bankruptcy Prediction.

1. INTRODUCTION

With respect to the rules regarding the technical analysis, the fundamental analysis should be a more realistic approach to estimate the index or price levels. To construct some rules from the balance sheets and income statements of the companies, Altman z-score discriminant function can be used as a fundamental modelling. The goal of the discriminant function analysis is to predict group membership from a set of predictors. The major purpose of discriminant analysis is to predict membership in two or more mutually exclusive groups from a set of predictors, when there is no natural ordering on the groups.

Furthermore, a discriminant function of the Turkish companies can be calculated and the average z-score of the companies will be a very good indicator for

estimating the index level. This study attempts to predict equity returns with a statistical method and provides an improved predictive accuracy using data from the Istanbul Stock Exchange. A comparative analysis of discriminant and logit methods reveals that the methods end up in similar findings for the data used between 1996 and 2002.

Altman z score discriminant function for bankruptcy prediction has been one of the most challenging tasks in accounting after Fitzpatrick in 1930s and the early researches (Ramser and Foster 1931; Fitzpatrick 1932; and Winakor and Smith 1935) focused on the comparison of the values of financial ratios in failed and non-failed firms and concluded that the ratios of the failed firms were poorer. Later, Altman's theoretical and empirical research has been successfully completed (Altman 1984). Two main studies were conducted for the fundamental analysis of healthy and poor companies. The first study was on the empirical search of financial ratios for lowest misclassification rates. The second one dealt with the search for statistical methods for improved prediction accuracy. In the 1930s, there were no advanced statistical methods or computers available for researchers. Thus, the values of financial ratios in good and bad firms were compared with each other, with bad firms observed to be poorer (e.g. Fitzpatrick, 1932). In 1966, Beaver showed the method of discriminant analysis and in 1968, Altman expanded this study to multivariate analysis. In Altman (1968), multiple discriminant analysis (MDA) is used to classify financial ratios and to develop a bankruptcy prediction model. The discriminant analysis model for bankruptcy prediction results in accurate predictions up to two years prior to actual failure. However, accuracy rapidly diminishes after the second year. Other researches for bankruptcy prediction by using multiple discriminant analysis will come up after this study (Altman, Haldeman and Narayanan 1977; and Blum 1974). In Deakin (1972), there are some modifications in the methodology of the discriminant analysis by calculating the probability of the bankruptcy. Taffler (1983) calculates performance score for the companies to modify the discriminant analysis. Performance score is used to grade the bankruptcy risk of the companies. After a detailed research on which financial ratios would better serve to differentiate healthy firms from ailing firms, selected market ratios were employed in the academic studies to analyze the financial strength of a company. In Levy (1978), beta plays no role in price determination and variance provides a better explanation for price behaviour in an imperfect market. Besides, in 1980 Aharony, Jones and Swary found that variance of the mean returns for the failed firm is significantly larger than those for the control group for every week prior to bankruptcy week. In Joy and Tollefson (1975) and Altman and Eisenbeis (1976), there are different aspects of the discriminant analysis in size and financial profile of business failures and financial reporting techniques. In 1977, Altman, Haldeman and Narayanan developed the prediction model, dubbed the ZETA credit risk model, to be applicable to non-manufacturing firms and private companies whose stocks are not publicly traded. The ZETA model is more valid in bankruptcy classification in five years compared to Altman's original model with the modification of the previous model in emerging market scoring.

However, in the method of discriminant analysis, the assumption of normality of the financial ratio distribution is almost impossible to realize. The discriminant analysis was the sole method used until 1980s. After this period, the logistic regression analysis was developed. In addition to the distributional normality problem in discriminant analysis, there is also the problem of matching the procedures used in this analysis. Logistic regression

allows one to predict a discrete outcome such as group membership from a set of variables that may be continuous or discrete. For both methods of the discriminant analysis and logistic regression, the categories in the outcome must be mutually exclusive. One of the ways to determine whether to use logistic regression or discriminant analysis is to analyze the assumptions. The logistic regression is much more relaxed and flexible in its assumptions than the discriminant analysis. Unlike the discriminant analysis, the logistic regression does not have the requirements of the independent variables to be normally distributed, linearly related, and equal variance within each group. The logistic regression analysis requires larger sample size. At least 50 cases per independent variable might be required for an accurate hypothesis testing. However, when the assumption of the distribution is met, the discriminant function analysis is more powerful than the logistic regression. As a result, Ohlson (1980) firstly constructs the logit model by analyzing the bankruptcy of the companies. Zavgren (1985) develops the classification success of the logit model.

In both the discriminant analysis and the logistic regression, financial ratios have been selected based on their ability to increase the prediction accuracy. There are some efforts to create theoretical constructions in failure prediction concept, but there are no significant developments to be construed as a basis for the theoretical ratio selection. A stepwise selection procedure is used to identify the empirical characteristics of the ratios on statistical grounds.

In the first step, factor analysis is used to search for predictors that make higher classification rates utilizing 81 financial ratios documented by the previous studies. In the second step, discriminant analysis and logistic regression are used to improve the prediction accuracy through a comparative analysis of each method using data from the Istanbul Stock Exchange. The data span a period of deeply effective crises, which had severe ramifications on the corporations, leading several firms to bankruptcy. Fortunately(?), the data serve to pinpoint the critical variables in turbulent environments. As a result, the quarterly balance sheets and income statements of the industrial companies traded in the four major sectors (textile, non-metallic mineral production, food and beverage, and metal sectors) of the Istanbul Stock Exchange during the 1996-2003 period are used as input data for the analysis.

2. FACTOR ANALYSIS

Discriminant analysis tries to derive the linear combination of independent variables that will discriminate best between a priori defined groups, which in our case are failing and non-failing companies. This is achieved by the statistical

94

decision rule of maximizing the between group variance relative to the within-group variance. However, in logistic regression, maximum likelihood function is calculated and maximized by using derivation. For the database, we have 81 ratios to define the companies. Since such a broad number of ratios is excessive to calculate a general discriminant function or to run the logistic regression, the factor analysis is firstly used to eliminate the unnecessary ratios. In other words, the size of huge data is reduced without loss of information (or with a negligible loss of information) in factor analysis. These 81 financial variables are grouped into twenty-two categories after the factor analysis including profitability, liquidity, solvency, degree of economic distress, leverage, efficiency, variability and size. Factor analysis is conducted to identify the variables that seem to be doing the best job in producing the necessary database. Consequently, 22 variables are identified by stepwise selection and are used in subsequent analyses to reveal those that are significant predictors of failure. The results indicate that in case of total variance explanation, ratios related to the financial expense item, liquidity item and turnover item seem to play an important role.

Factor analysis rests on the quarterly financial statements of the firms. The number of observations per variable is approximately seventy, which is above the minimum ratio of five observations per variable. The variables are selected so as to have an absolute value of minimum of 0.5 loading. The observations that are +/- 3 standard deviations from the mean are regarded as outliers and are eliminated. The factors are labelled and the variable that has the highest loading on each factor is selected. The analysis reveals that 22 variables that are selected explain 76.76% of the variance and the addition of one more variable contributes to the explained variance by less than 1.25%. Ten factors in the analysis have a 57.18% total variance.

By using these 22 variables after the factor analysis of 81 financial ratios, the data will be ready for the discriminant analysis and the logistic regression.

3. DISCRIMINANT ANALYSIS

To start the discriminant analysis, the stepwise selection begins with the variables, which are chosen according to their highest factor scores in each of the 22 factors. Since multicollinearity is a serious problem for the discriminant analysis, some of the unnecessary variables should be eliminated by using factor analysis from the data set with 81 ratios. In each of the factors, only one variable is selected and this variable has the highest factor score in this factor. Finally, before starting with the discriminant

analysis, factor analysis has eliminated 59 of the ratios and one factor is represented by only one variable.

Empirical experiments have shown that especially failing firms violate the normality condition. In addition, the equal variances condition is also violated. Moreover, multicollinearity among independent variables is often a serious problem, especially when stepwise procedures are employed (Hair et al., 1998). However, empirical studies have proved that the problems connected with normality assumptions were not weakening its classification capability, but its prediction ability.

In both the discriminant analysis and the logistic regression, the independent variable should be selected before running the program. The next period returns of the industrial companies in the ISE are calculated. Since the balance sheets and the income statements are announced quarterly, the testing period (next period) for the returns are calculated in the next 3 months to follow the announcements of the companies. However, the pure returns are not enough. In order to compare them, the return of the ISE100 Index in the same period should be excluded from the pure returns by using the following formula:

$$\hat{\mu}_{Company\_i}^{Next\_Period} = [(1 + \mu_{Company\_i}^{Next\_Period})/(1 + \mu_{ISE100}^{Next\_Period})] - 1 \quad (1)$$

The calculated real returns for each company are used as independent variable for the analysis. If the real returns are positive, the categorical independent variable is recorded 1. Otherwise, it is recorded 0.

The discriminant analysis rests on the data generated from the quarterly financial statements of the firms and is carried out to identify the most important predictors in return prediction using the variables identified by the factor analysis. At the final step there are five variables:

A) Cash Ratio = (Cash + Marketable Securities)/ Current Liabilities. Liquidity is seen as the most important financial ratio because of the frequently erupting crises in Turkey. Since the funding of debt increases with soaring interest rates in the crisis environment, the liquidity problem is aggravated and firms write losses.

B) Annual Volatility in Quarterly Period. Before upward or downward movements are noted in any given stock, a rise in volatility is observed in the stock.

C) Sales/ Tangible Fixed Assets. Since the turnover ratio decreases during crisis, firms increase the profitability margin in order to survive. In recession times, prices move upwards and government's inflation target was abandoned. Turnover factor is strongly related to the efficiency of the economy. In the crisis period, firms' goods cannot be marketed and sold. By using the increasing returns to scale assumption in the economy, firms increase the profit margins to fund the fixed costs. Thus, efficiency and turnover of firms decline in crisis periods.

D) Market Value of Equity/ Book Value of Total Liabilities. This variable is a measure of long-term liquidity as well as solvency. It shows by how much the firm's assets can decline in value before liabilities can exceed the assets and the firm becomes insolvent. Consequently, the lower mean of MVETL ratio for the firms reflects unfavourable expectations of the capital market investors and/or the relatively higher amounts of liabilities in the capital structures of these firms.

E) Total Assets/ WPI. This variable is a proxy of size and reflects the total asset values divided by the wholesale price index (1994=100). This adjustment in asset values eliminates the adverse effects of two-digit inflation in Turkey on assets that are recorded at historical costs. Interestingly, the financially distressed firms are significantly larger than the firms in the other group supporting Altman et al. (1977). This variable has been selected as a significant predictor of failure in discriminant and logit analyses but its coefficient is small relative to other predictors in both models.

The discriminant analysis yields a correct classification rate of 54.4%. With 55.7% among 2124 samples, bad performed companies and with 52.8% among 1819 samples, good performed companies are correctly classified. 54.2% of cross validated grouped cases are also correctly classified for validation. The coefficient of variable TOAS is seen as very low value because the variable of the data is significantly large, but it does not affect the analysis negatively. Consequently, the final discriminant function is:

$$Z = -1.668 + (-0.0000059 \text{TOAS}) + (0.019 \text{STFA}) + (0.597 \text{MVETL}) + (2.038 \text{ VOL}) + (-0.956 \text{CASH}).$$

Finally, the comparison of the means for the failed and nonfailed companies can be analyzed to validate the results of discriminant function's accuracy. The value discriminant function is higher for the unfailed companies (Priori=0) and lower for the failed companies (Priori=1). With a total of 2124 data, the unfailed companies have an average value of 0.1 with 1.12 standard deviation. Besides, with a total of 1819 data, the failed companies have an average value of –0.1 with 0.95 standard deviation. Consequently, with

0.1 and –0.1 means of discriminant functions around 1 standard deviation for the failed and the nonfailed companies, respectively, the model is criticized to be not adequately descriptive. In this kind of model, we should not expect a high significance level because of the noisy financial environments especially in emerging markets, which entail frequent episodes of crises. In the next stage, a logistic regression is used to develop another model for the return prediction in the ISE.

4. LOGISTIC REGRESSION

The logistic regression model reveals four variables as predictors of the next period real returns. Tests of statistical significance and overall model fit for the logistic regression model reveals that it is a better method to be used when normality assumption is violated. The correct classification rate of this model is 54.4%, which is equal to the correct classification power of the discriminant analysis.

Consequently, logistic regression model that does not have the restrictions of the discriminant method seems more appropriate as a method of failure prediction in this study. The first three variables are as same as the variables in the discriminant case. The forth variable in the logistic regression is EBIT/Paid in Capital. The last variable of the logistic regression is similar to the fourth variable in the discriminant analysis since the profitability and the sales are directly correlated to each other. The last variable in the logistic regression is:

- EBIT/ Paid in Capital. Profit before interest payments and taxes over paid in capital gives generally the financial return of the equity for the given period. While the dependent variable in the logistic regression is the expected returns of the given equity with respect to the index, this variable gives the real life return of the company in the same period.

The logistic regression results are analyzed by using the balance sheets and the income statements of approximately 192 industry firms traded on the ISE in the period between the last quarter of 2002 and the first three quarters of 2003. If the logit-value of the company is higher for the given periods, the expected return for that company will be higher, too. To test this hypothesis, the company returns of the four quarters -- which are not used in the analyses -- for approximately 192 industrial companies of the ISE are calculated and after dividing with the return of the ISE, we end up with the real returns.

$$\hat{\mu}_{Company\_i}^{ith\_Period} = [(1 + \mu_{Company\_i}^{ith\_Period})/(1 + \mu_{ISE100}^{ith\_Period})] - 1 \quad (2)$$

When we analyze the total data set for the real returns, we see that the average real return of these companies for the last quarter of 2002 and the first three quarters of 2003 is –4.05% with 28.06% standard deviation. Thus, the ratio of the mean to the standard deviation is –14.43%. Finally, for each logit value, there is a real return. The total data set can be divided into two pieces with respect to the average of this logit value (the average of the logit values of the total data set is 0.89). As a result, according to our data set for the industrial firms, if the logit value is greater than 0.89, we will have 368 companies. Similarly, if the logit value is smaller than 0.89, we will have the other 385 companies as in the following table.

Table 1 Anova Table of the Logistic Regression

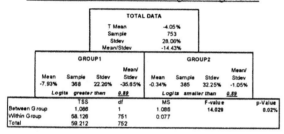

Consequently, the mean of the nonfailed companies is -0.34% with 32.25% standard deviation. However, the mean of the failed companies is –7.93% with 22.26% standard deviation. The standard deviation of the nonfailed companies is higher because of the higher the return higher the volatility assumption. Besides, the means are significantly different between the nonfailed and the failed companies. The difference is also seen at the ratio of the mean to the standard deviation. It is –1.05% for the nonfailed companies, while being –35.65% for failed companies. As a result, the risk in the nonfailed companies is incomparably higher than the risk in the failed companies. The Anova test of these two groups with a cut-off logit value of 0.89 has enough evidence to discriminate the good return expected companies to the bad return expected companies with 14.0 F value and 0.02% significant level.

As in the discriminant analysis case, by using the book value of the stocks, weighted average of the logit score can be calculated. If there is a correlation between the Istanbul Stock Exchange Index and the weighted logit score, it will be useful for the forecast of the ISE100 Index. As seen in Table 2, between the periods 1996/1 and 2002/3, the calculation period, the correlation between the 3- month period returns of the ISE100 Index and the weighted logit score is 26.4%. However, the correlation jumps to 70.6% between the index and the weighted logit score in the testing period. Because of the Russia Crisis in August 1998 and Turkish Liquidity Crisis in February 2001, there are huge fluctuations and these fluctuations distort the correlation of the calculation period. Since between 2002/4 and 2003/3, there have been no

major crises in Turkey, the economy is in a recovery process. Thus, the correlation in this period is significantly better than the correlation in the calculation period.

Table 2 Correlation of the ISE100 Index and the Weighted Score

Year	Quarter	Index	Weighted Score	ISE100 Return(t+1)
1996	1	1	0.889	5.14%
1996	2	2	0.948	2.78%
1996	3	3	0.995	34.69%
1996	4	4	0.955	65.29%
1997	1	5	0.897	15.13%
1997	2	6	0.952	39.63%
1997	3	7	0.994	33.09%
1997	4	8	0.922	-5.56%
1998	1	9	0.897	25.81%
1998	2	10	0.910	-44.73%
1998	3	11	0.846	14.65%
1998	4	12	0.913	75.29%
1999	1	13	0.918	8.70%
1999	2	14	0.913	22.65%
1999	3	15	0.933	150.52%
1999	4	16	0.900	4.67%
2000	1	17	0.848	-9.13%
2000	2	18	0.946	-21.54%
2000	3	19	0.931	-16.85%
2000	4	20	0.765	-14.98%
2001	1	21	0.773	39.65%
2001	2	22	0.842	-31.94%
2001	3	23	0.884	80.74%
2001	4	24	0.833	-15.27%
2002	1	25	0.868	-19.68%
2002	2	26	0.899	-5.74%
2002	3	27	0.909	17.28%
2002	4	28	0.845	-8.63%
2003	1	29	0.821	14.87%
2003	2	30	0.934	19.95%
2003	3	31	0.966	29.22%
Correlation between 1996/1 and 2002/3				**26.4%**
Correlation between 2002/4 and 2003/3				**70.6%**

We have one dependent variable and one independent variable for the correlation analysis. If we regress the independent variable, the result will be as seen in the following picture:

$$Return\_ISE100(t+1) = -1.6 + (1.9 * We\_Logit\_Sco) \quad (3)$$

with 7.39% coefficient of determination value, which is quite considerable for our analysis.

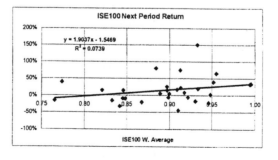

Fig. 1. Analysis of the ISE100 Index According to Logistic Regression Results

5. CONCLUSION

For the fundamental analysis, a factor analysis is done with 81 financial ratios of the stocks and then by using Altman's famous bankruptcy prediction method, a model for predicting next period's stock return is developed with the discriminant analysis and the logistic regression. The results are statistically significant so that "good" companies perform better than "bad" companies in the ANOVA analysis. Then, the weighted average of the z score or the logit score of the stock exchange companies forecast the next period return of the index for the stock market. In the testing period of the forecastibility power of both the discriminant analysis and the logistic regression analysis, it is seen that the failed companies are more than thirty times riskier than the nonfailed companies. In case of restrictive properties of the discriminant analysis for the distributions of the variables, the result of the logistic regression analysis of the model is more useful.

Furthermore, cash ratio is the most important factor in the analyses. EBIT and the sales factors of the companies are also considerably important. However, the stock market reacts to the crises promptly but the financial data follow with a lag. Because of the delay, volatility of the prices and the market value of the equity are adjusted on time and can be important variables. The delay makes the correlation between the return of the stock market index and the results of both analyses to decline sharply. In the long term, the correlation can be around 30%, which can attract the attention of the stock market investors, but in the short term, it fluctuates between 15% and 80%.

REFERENCES

Aksoy, H., İ. Sağlam (2001). Sınıflayıcı (Classifier) Sistem ile İMKB'de Yeni Bir Anomali Gözlemi. *Working Paper, Department of Economics, Boğaziçi University*, **15**.

Aksoy, H., M. Ugurlu (2003). Prediction of Corporate Financial Distress: Evidence From An Emerging Market. *Unpublished Working Paper*.

Altman, E.I. (1968). Financial Ratios, Discriminant Analysis and the Prediction of Corporate Bankruptcy. *The Journal of Finance*, **23**, September, pp: 589-609.

Altman, E.I., R. Eisenbeis (1976). Financial Applications of Discriminant Analysis: A Clarification. *Working Paper, New York University*, **79**.

Altman, E.I., R. Haldeman and P. Narayanan (1977). ZETA Analysis: A New Model to Identify Bankruptcy Risk of Corporations. *Journal of Banking and Finance*, **1**, June, pp: 29-54.

Altman, E.I. (1984). A Further Empirical Investigation of the Bankruptcy Cost Question. *Journal of Finance*, **39**, September, pp: 1067-89.

Blum, M. (1974). Failing Company Discriminant Analysis. *Journal of Accounting Research*, Spring, pp: 1-21.

Deakin, E. (1972). A Discriminant Analysis of Predictors of Business Failure. *Journal of Accounting Research*, Spring, pp: 167-179.

Fitzpatrick, P. (1932). A Comparison of the Ratios of Successful Industrial Enterprises with Those of Failed Companies. The Accountants Publishing Company.

Hair, J., R. Anderson, R. Tatham, and W. Black (1998). Multivariate Data Analysis. Macmillan Publishing Company.

Joy, O.M., J.O. Tollefson (1975). On the Financial Applications of Discriminant Analysis. *Journal of Financial and Quantitative Analysis*, **10**, December, pp: 723-739.

Levy, H. (1978). Equilibrium in an Imperfect Market: A Constraint on the Number of Securities in the Portfolio. *American Economic Review*, **68**, September.

Ohlson, J.S. (1980). Financial Ratios and the Probabilistic Prediction of Bankruptcy. *Journal of Accounting Research*, **19**, pp: 109-131.

Ramser, J., L. Foster (1931). A Demonstration of Ratio Analysis. *Bureau of Business Research*, **40**, University of Illinois.

Taffler, R.J. (1983). The Z-Score Approach to Measuring Company Solvency. *The Accountant's Magazine*, **87**, No: 921, pp: 91-96.

Winakor, A., R. Smith (1935). Changes in the Financial Structure of Unsuccessful Industrial Corporations. *Bureau of Business Research*, **51**, University of Illinois.

Zavgren, C.V. (1985). Assessing the Vulnerability to Failure of American Industrial Firms: A Logistic Analysis. *Journal of Business Finance and Accounting*, **12**, pp: 19-45.

ELSEVIER

IFAC
PUBLICATIONS
www.elsevier.com/locate/ifac

THE WORLD COMMUNITY CONTROL PROBLEMS
DURING THE GLOBALIZATION PROCESSES

Shubin A.N., Kulba V.V. Malugin V.D

*The Institute of Control Sciences, Russian Academy of Sciences,
Profsoyuznaya st., 65, Moscow, 117997, Russia
e-mail: shoubine@ipu.rssi.ru*

Abstract: The globalization is treated as a social development phase, which provides advanced capabilities for international collaboration due to consolidation of interrelationship and interpenetration of different kind human activities, including economical, political, social and intellectual ones. A drastic deepening of both between nations and between different States is the main distinguishing feature of this globalization process as a new stage of a global evolution in a qualitative sense. This interrelationship spreads to all areas of social life, including the governmental control of economy. *Copyright © 2004 IFAC*

Keywords: global control, social stability, world community.

1. INTRODUCTION

The collapse of bipolar model of the world community did not increase the world safety and stability. Relatively stable structure of the world community has been changed for a permanent instability evoked by the tendency of some countries called victors in the cold war to establish a new world order with an evident unipolarity. All abovementioned factors only intensify old contradictions and stimulate occurrence of new ones. International organizations (U.N.O, OSCE) have been substituted for military blocks.

Uncontrolled globalization processes may stir up such negative phenomena as nationalism, racism, regional opposition enforcement, ethnic conflicts and result in a social tension and splitting the world community.

2. GLOBALIZATION PROCESSES

The world civilization as a uniform system evolves non-uniformly under impact of ongoing integration and differentiation processes. Any epoch and each model of a social community have its own specific economic, legal and social relations. New loop of integration associated with using information flows, improving comprehensive data communications, creating worldwide economic nets as well as interrelated business interaction works for a large-scale unity of different social systems.

The idea of globalization testifies that problems of the survival of the humanity become worsened and that joint efforts are required for their overcoming. By the time the world economy started its development towards implementation of a new-type global economy it turned out that some countries were not ready for changes, and as a result, their resource potentiality lost its former significance and their major development activities failed to match with requirements of a "novel economy."

Changing role of underdeveloped countries ("world outskirts", inhabited by the majority of humankind) becomes visible in the world economy, international security and global informational processes. There are indications of a gradual breaking down of the world outskirts into a part actively involved in globalization processes and one outside the major global trends of development. The last part includes a considerable part of Africa where stagnation and marginalizing processes gather momentum.

The growing impoverishment of African population and the ongoing continent marginalizing are not exclusively African problems but a real threat for a global stability also.

The beginning of new century is a critical moment in the humankind history. A population explosion is anticipated in the year 2020, when the Earth population will run up 10 billion, and most of people will be dwellers of the Asian region. Further aggravation of the humankind survival problem will occur as well as resource depletion problems. The anti-global activity phenomenon will gain further acceptance, including its Latin-American constituent. Departure from universalism ("progress for all") in favor of implementing strategies of social and Darwinian selection ("progress for elite") is responsible for a global social crisis.

The globalization attacks the traditional world frequently in the most aggressive manner. We are witnesses of an aggressive behavior of some countries against other ones in order to override them to own economic and political interests. The USA is the most striking instance. Those countries, which do not support in the full measure the American model of globalization, are subjected to different punishment measures – from soft propaganda tricks to using force. The USA and allies aggression against Yugoslavia and Iraq gave a powerful incentive to reappraisal of the world relations and policy. The world community saw one of aggressive forms of globalization, which resulted in death of thousands of innocent people and in seizure of national riches of the independent country.

It is necessary to understand that the entire system of the world capitalism resides in a state of dynamic equilibrium. Any incentive (a disturbance) may turn it into the non-equilibrium state, resulting then in a crisis.

3. THE WORLD COMMUNITY AS A CONTROLLED PLANT

Against a background of globalization, people's striving to search ways of a harmonic humankind amalgamation is observed based on consolidation of their culture and originality. However, dismantling of national cultures passes ahead of implementing global interests.

Partly paradoxical situation arises in this point: the former controlled plant breaks apart, while a new one yet grows ripe. Besides that, it is unknown whether the new plant will be controllable. To control it under such conditions means contribute to maturing of new organizational and functional structure of the national

economy, introducing into it elements and systems time-proved by the worldwide experience.

As a preliminary it is necessary to become aware of existing qualitatively new geo-economic framework of the world economical system, which defines to a great extent organizational and management system of a single state national economy.

The main distinguishing feature of modern international relations lies in forming and development of powerful regional coalitions allowing protect own interests on the world scene.

The world trade exchange turns increasingly into the geo-economic operation of national economies within the existing reproductive internationalized kernels. Market grouping proceeds through the development of regional economic unions: NAFTA (North-American Free Trade Agreement), EC (European Community), APEC (Asia Pacific Economic Cooperation), etc. The contribution of these unions to the world GNP adds up to 80% approx.; they share about 82% of all national budgets of the world and 85% of the world export. The central place in the international economic system belongs to the "Great Seven" and WTO (World Trade Organization); the last was established by the USA initiative. So many countries become global entrepreneurs, being at the head of new entities "nations – systems". Other national economies tend to occupy certain units of worldwide reproduction structures to provide themselves with a part of the world property. Globalization results in changing social policy of the European Community as the largest and the most integrated region of Europe. Variations in EC social strategy are in evidence as well as in a social integration process and developing a common social space within it. All these changes arise from EC expansion, activity of transnational corporations at the international labor market, establishing areas of social security and eliminating regional imbalances. Taken together these facts point to the possibility of developing self-dependent European pole of force.

Rejection of global game rules is pregnant with destruction of national infrastructures and irrevocable transformation of national economies.

The economic integration of underdeveloped countries into globalizing world economy is a major task for these countries and their partners in the frame of development process. Should they fail to adapt for new international competitive environment then their isolation from the global economy may be aggravated.

What actions are required from the world community to give all countries (including underdeveloped ones) a chance to overcome intricate contradictions of the global process of development?

There is a need to take specific actions directed to strengthening capability of governmental institutes and institutes of civil society for efficient operation covering its three components: Institutional Control, Society-assisted Control and Economic Control.

Institutional Control must provide a triumph of democracy and law through democratic reforms along with reinforcing independent executive, judicial and legislative powers. This form of control includes reforms of judicial system and state administration, programs of education and implication of citizens as well as measures intended to improvement of accountability and transparency at all levels of institutional structure.

Society-assisted Control must provide participation of different social strata in determination of the development concept and its implementation through decentralization of the state administrative system and strengthening organizational capabilities of local units of the civil society.

Economic Control is directed toward improving management activity, tightening governmental control over the reformation programs, stricter parliamentary supervision over the government, reducing a scope of poverty, assurance of democracy principles and respecting human rights in programs of economic development.

Partners on the program of development of underdeveloped countries must increase their technical help with counteraction to excess expenditures on defense technologies, provided that international legal conditions matching to goals and tasks of effective management would be assured.

Countries that can't keep up pace and tempo of reproduction processes "take a back seat" in the world economy playing a part of auxiliary manufacturing within the frame of ongoing processes. Through the foreign commerce, these countries provide their national wealth for supporting a worldwide reproduction process, being removed from redistribution of the world profit.

Corporate structures and primarily financial and industrial groups can become one of key control units in terms of assimilation of the global geo-economic space. To pave the way for corporate structures the appropriate foreign policy is essential. As these take place, a responsibility for developing the foreign policy is very high, because the cases is a survival and secure development in the modern globalizing world.

4. PROBLEMS OF SECURITY WITHIN THE FRAME OF GLOBALIZATION

There is much speculation among politicians and philosophers that the world community stands on a crossroads. Where will it go? If the world community splits up and a part of it follows one path while the other its part selects the different path, this might result in a disaster in the form of the war of civilizations.

To select the prospective path envisaging further development for peoples, active forms of cooperation between countries based on mutually beneficial economic, political and cultural communications are essential. Therefore, the globalization process must proceed, uniting nations with different levels of development, keeping cautiously the originality of their cultural heritage without any aggressive manifestations. The global and regional security depends on contacts between the civilizations.

The history of Islamite civilization evolution points to the existence of different branches and tendencies of Islam and the sphere of its impact expands due to rapid increasing Islam's followers all over the world against the background of aggravating crisis that seizes now Moslem countries.

A closer look at the origins and reasons of this crisis is called for, to consider relations between Western and Moslem civilizations in the historic retrospective and today, when Moslem countries become actually pariahs of globalization.

Problems of interrelations between states and ethnical groups of different cultural fields appear to be the most pressing global challenge of the present.

In this connection, the problem of global security arises as related to preventing armed conflicts and control over them together with a further restriction of offensive and defensive strategic armaments, and collaboration in fighting against terrorism.

There is a need to make predicted assessments of the most pressing foreign-policy problems to be solved by the world community.

It is essential to examine also the historic evolution of international relations as well as to reveal relationships between force, violence, security in the modern world and defining a role of international conflict as the instrument of stability protection.

It is well to analyze fundamental problems of power organization in transition-type communities and fist of all make a socio-political assessment of corruption and negative economy, which give rise to destructive and socially dangerous forms of political management.

Account must be taken of global and regional consequences of new political forces coming to power in Brazil – the largest country of the South-

American region. It is necessary to analyze also global and regional consequences of the unexampled economic and political crisis in Argentina and Columbia. The problem of sovereignty, internal and external security of Latin American countries is noteworthy within the frame of globalization as well as an outlook of transmutation of these countries to the real entity of the world policy.

CONCLUSION

The latest decades have demonstrated a violent evolution of socio-economic processes within the world community both in the sense of democratization of social relations and toward the totalitarianism.

Today the globalization processes have come under the control of the world financial oligarchy that may result in the global network totalitarianism.

Possession of information and management of its flows become the major instruments of a total control and global domination. In solving the above said problems, there is no escape from making a note of the USA leading role in the system of international multilevel and multivariate socio-economic interrelationship.
One of geo-economic consequences of globalization is the reduction of economic and socio-cultural breach in spite of comprehensive interrelationships between national and global processes of development.

Moving toward the new form of existence requires different relations within the world community as well as a different philosophy of interaction and different style of management within the society.

It's the scientists' opinion that a competitiveness of present-day situation may be resolved by implementing the multi-polar and many-sided community of counties, nations and cultures.

REFERENCES

Chesnut H. and P. Kopacek (1989), Supplemental Ways for Improving International Stability. In: *Report on the IFAC/EPCOM Working Group (WG 7.2) "Control Engineering and International Conflict Resolution"*, Vienna.

Grishutkin A.N. and Tsyganov V.V. (2001). Progressive Adaptive Mechanisms of Globalization. In: *Proceedings of the International Conference on Systems Cognition.* **Vol. 2**, *pp. 101-107. IPU RAN*, Moscow.

Konohov D.A., Kulba V.V., Shubin A.N. (2001). Stability of Socio-Economic Systems: Scenario investigation Methodology. In: *Proceedings of the 8th IFAC Conference on Social Stability "The Challenge of Technology Development"*, Vienna.

Kulba V.V., Malugin V.D., Shubin A.N. (2001). Techniques of Planning the Set of Measures for Preventing and overcoming Reasons and Consequences of Emergency Situations. In: *Proceedings of the 8th IFAC Conference on Social Stability "The Challenge of Technology Development"*, Vienna.

Kulba V.V., Malugin V.D., Shubin A.N. (2002). Globalization – Information Control Processes. In: *Proceedings of the 4th International Conference on Complex Systems, pp. 21-25,* Samara, Russia.

Shubin A.N., Kulba V.V., Tsyganov V.V. (2003). Globalization As a Kind of Law of Historical Development of the World Community: Merits and Demerits. In Preprints Volume 10th IFAC Conference on "Technology and International Stability". SWIIS'03, Waterford, Republic of Ireland, July 3-4, 2003, pp. 80-84.

ELSEVIER
IFAC
PUBLICATIONS
www.elsevier.com/locate/ifac

MODELLING STOCK MARKET VIA FUZZY RULE BASED SYSTEM

Hakan Aksoy°,
Kemal Leblebicioglu

*Senior Portfolio Manager, Mutual Funds Division, Koc Portfolio°,
Department of Electrical & Electronics Engineering, METU*

Abstract: A rule based fuzzy logic model is implemented to forecast the monthly return of the Istanbul Securities Exchange 100 (ISE100) Index by combining technical, financial and macroeconomic analysis. Starting with the technical analysis, an index level observation (Aksoy and Saglam, 2001) by using classifier systems is used as a long term input. Another technical analysis rule using moving average is also modified as a short period input. In the financial analysis that borrows from the methodology of the Altman's bankruptcy prediction analysis, a model (Aksoy, 2003) for predicting next period's stock return is developed with the logistic regression by using 81 financial ratios. The weighted average of the logit score of the stocks is used to as an input. Finally, macroeconomic data is gathered in three main groups: real economy, FX market and TL market. The data is ruled and modeled within the period from 1996 to 2002 and optimized with steepest descent learning algorithm. Furthermore, the model is also tested for optimal investment decision in 2003 and the model's suggestion performs better than the return of the repo and the ISE100 Index. *Copyright © 2004 IFAC*

Keywords: Stock Market, Bankruptcy Prediction, Logistic Regression, Learning and Fuzzy Logic.

1. INTRODUCTION

In the past decade, fuzzy systems have been used with conventional techniques in many scientific applications and engineering systems, especially in system theory. Fuzzy sets, introduced by Zadeh (1965) as a mathematical way to represent vagueness in linguistics, are different from the classical set theory. In a classical nonfuzzy set, an element of the universe either belongs to or does not belong to the set. A fuzzy set is a generalization of an ordinary set in that it allows the degree of membership for each element in a unit interval. In order to quantify the vague rules, as characteristic of fuzzy systems, an expert, data set and a model are needed. Fuzzy logic can be perceived as the bridge between the knowledge of humans and their expectations by using the information set with mathematical calculations. Thus, fuzzy rules represent the people's vague knowledge in the system. Transformation methods are applied to train this knowledge. The theory of fuzzy is widely used in control theory. There exist two major types of fuzzy controllers: Mamdani (1974) and Takagi-Sugeno (1985). Mamdani (1974) uses fuzzy sets and Takagi-Sugeno (1985) uses linear functions. Fiordaliso (1998) finds appropriate rules in

the framework of chaotic time series forecasting. He proposes a Takagi-Sugeno fuzzy controller to evaluate the rules in the inference process by using gradient descent algorithm. The results of his study point out the advantage of using a nonlinear "combining forecast" method. Lee and Antonsson (2000) also works on the Takagi-Sugeno Fuzzy System with evolutionary computation approach and finds a locally Pareto optimal set of solutions. Nomura, Hayashi and Wakami (1992) works on the optimization process for fine-tuning of the input data. One of the acceptable ways of fine-tuning in fuzzy logic is the optimization by steepest descent algorithm to minimize the error. Not only the membership functions but also the rule generation in fuzzy logic system should work properly. Furthermore, some time series applications are also available in the fuzzy logic theory literature. Economakos (1979) estimates the demand for electrical power by using linguistic terms of the needed electricity. For the finance application, Shnaider and Kandel (1989) constructs a model to forecast corporate income tax revenue by using moving averages of the corporate tax revenue and per capita GNP for Florida. Besides, Song and Chissom (1993) analyzes the theoretical part of the fuzzy time series modelling. They fuzzify the data, develop the time series model, and calculate the output. Then, Song and Chissom (1994) applies a time invariant forecasting model by using different defuzzifying methods. Meanwhile, Chen (1996) works on the same model by changing the arithmetic operations in the rule generation part. As a detailed macroeconomic modelling example in fuzzy systems, Kooths (1999) develops a macroeconomic model to realize an alternative to conventional expectation hypothesis. The experience and rule based expectations can be used in forecasting behaviour that is characterized by explicit rule orientation (theory foundation), vague formulation (bounded rationality) and learning process (acquisition of experience). By using rules, he builds an experimental macroeconomic model involving money supply, interest rate, exchange rate, currency reserves, net capital imports, price increase, unemployment rate and technical progress. Human reasoning can also be modelled as if the thought process is described by the application of fuzzy logic (LeBaron, Arthur and Palmer 1999). Traders are capable of handling a large number of rules for mapping market states into expectations. Linn and Tay (2001) shows this by allowing agents to have the ability to compress information into a few fuzzy notions, which they can in turn process and analyze with fuzzy logic. Their work also analyzes the fuzzy inductive reasoning and nonlinear dependence in security returns by using the outputs of artificial stock market environment. In order to understand the price movements of the stocks with effect and reaction mechanism, a laboratory environment should be designed as in the models of the computational economics. The laboratory asset markets permit the controlled manipulation of the rules and procedures. One of the famous experimental stock markets, the Santa Fe stock market investigates market efficiency and price convergence with rational expectation asset pricing model. The model is described in detail in Arthur, Holland, LeBaron, Palmer & Tayler (1997) and LeBaron, Arthur and Palmer (1999). This model attempts to integrate the training mechanism into a well-defined economic structure, along with inductive learning using a classifier-based system. One feature of this model is that it allows agents to explore a fairly wide range of possible forecasting rules (Chen and Yeh 2001). The results on this market show that the artificial stock market is able to replicate certain empirical problems observed in real markets. Consequently, stock prices have specific properties that make the prediction more complex because of the strong resemblance to random walk processes (Lo and MacKinlay, 1988). A successful prediction algorithm does not have to provide predictions for all points in the time series. The important measure of the success is the generated profit where the algorithm produces predictions and the explanatory power of the model above the standards. This study supports the idea of the

possibility to identify these movements and predict future changes. Our argument against the efficient market hypothesis is constructed because of the time delay between the point when new information enters and the point when the information has been evaluated and a new equilibrium price has been established. From this standpoint, the controversy revolves around how the word "immediately" should be interpreted in the efficient market hypothesis definition. Being an academician and a trader, I believe that the answer to this problem is a matter of perception and time differences of traders and the academicians. In other words, traders may be in general faster than academics in the market. To characterize the ISE100 Index, one may implement fuzzy models, explaining to some extent predictable behaviour patterns. The fuzzy models for the ISE100 Index in this study have been used for the first time in the literature. By using the rules in fuzzy logic method, the model can be constructed by an expert (myself) and optimized with steepest descent learning algorithm. The adjustment process for prices of the stocks and the index should be taken long enough. Thus, daily and weekly periods can be quite short for the adjustment of the fast movements in prices. Besides, macroeconomic data are generally released on a monthly basis in general. The independent variable can be taken as the monthly return of the ISE100 Index for the modelling.

2. MODELLING

One of the main advantages of using fuzzy logic is that it more closely models the kind of reasoning that a person engages in when dealing with issues or elements that are not precisely defined and that involve aspects of degree or judgment. However it may not be possible to think in terms of sharp boundaries for modelling. Thus, the disadvantages of fuzzy logic systems are the same as those found in traditional knowledge based systems: Someone has to write rules, which means that expert knowledge has to be available and formalized. These systems can not learn on their own nor can they adapt to changing market conditions, except by manually rewriting the rules, adjusting the membership functions or other specific rule-finding methodologies. According to professional investors, the stock markets are generally affected by four main criteria (Virtual Trading, 1995): Stock Market Factors, Macro Economic Factors, Risk Factors and Inter-Market Factors. In the model designed for the ISE100, the inter-market factors (effects of other markets) are included in macroeconomy part. Thus, the models in the following pages generally consist of three main parts; technical analysis (for the risk factors), fundamental analysis (for the stock market factors) and macroeconomy. The design of the model is constructed according to the relations of the variables between each other. Firstly, in the technical analysis section the index level observation (Aksoy and Saglam 2000), which is used to analyze the ISE100 Index in the long-run, and short term technical analysis, which is used to analyze the momentum of the ISE100 Index, are used as the variables to represent the first and dynamic section of the model. Secondly, the fundamental analysis section contains the results of the logit scores derived from data of the stock markets balance sheets and income statements by using 81 financial ratios after running logistic regression in SPSS. The final score is the weighted average of the companies' separated logit score, which represent the firms' financial statements in the medium and long run. Finally, the macroeconomy section in the model has three main parts: Real Economy, FX Market and TL Market. Real economy shows the fundamental changes in Turkey. These are GNP, industrial production index and capacity utilization ratio. Capacity utilization ratio is an indicator for the industrial production and industrial production is an indicator for GNP. The FX Market consists of current account deficit, which is an important item of the balance of payments, FX reserve and FX rate. Current account deficit and FX reserve variables measure the amount of foreign exchange, while FX rate is used for the valuation of these assets. The aim of this part is to quantify the resistance of currency crises in the economy. TL Market consists of primary budget deficit, repo and DIBS after inflation adjustments. This last part as an alternative investment section in TL terms other than

the stock market, shows how good the TL market expectation is. Consequently, the 12 inputs used in the models are summarized in the table below.

Table 1 Input Variables

V1	Long Term Tech. Analysis	Long Term Risk (Aksoy&Saglam,2001)
V2	Short Term Tech. Analysis	Moving Average (for 10 working days)
V3	Fundamental Analysis	Discriminant Analysis for Turkish Companies
V4	GNP	% Change in Gross National Product (y-o-y)
V5	Industrial Production	% Change in Industrial Production (y-o-y)
V6	Capacity Utilization	% Change in Capacity Utilization (y-o-y)
V7	Balance of Payments	Current Account Deficit
V8	FX Reserves	Weekly change in FX Reserve of the Central Bank
V9	FX Rate	US Dollar/TL Parity
V10	Budget Deficit	Budget Deficit/Wholesome Prices (y-o-y)
V11	Repo Rates/Inflation	Monthly Average of Repo Rate over ex-ante Inflation
V12	DIBS/Inflation	Gov. Bills&Bonds Index over ex-ante Inflation

As shown in Figure 1, by using the first two of the variable, another variable will be produced, named as the Technical Analysis Variable. The third variable will be used to estimate the output directly. From the fourth variable to the sixth variable, the Real Economy Variable will be produced, as the first middle variable. From the seventh variable to the ninth variable, the FX Market Middle Variable will be produced. From the tenth variable to twelfth variable, the TL Market Variable will be produced as the last middle one. By using these three middle variables, the Macroeconomy Variable will be produced. Finally, the output will be calculated with the Technical Analysis, Fundamental Analysis and Macroeconomy Variables.

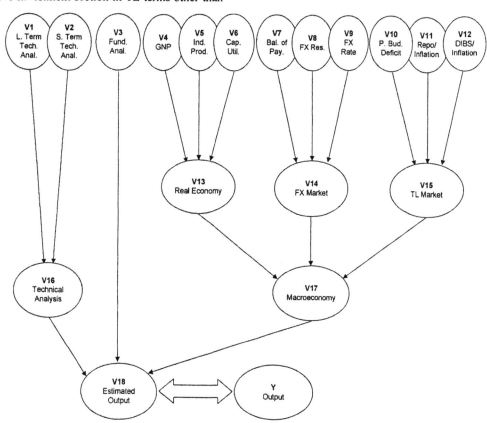

Fig. 1. The Fuzzy Logic Model

For each variable the maximum, minimum, average and the standard deviations are analyzed. By using these values with expert opinion, the borders of the fuzzy membership functions of the variables are determined. All of the variables have five membership functions. If there are two variables with five membership functions for each, the number of rules will be 25. The number of rules will be 125 for three variables with five membership functions. After having the variables with membership functions, the rules should be produced for the last step of the modelling. In Figure 1, there are 12 input variables and the other 6 middle variables are produced from these 12 variables. One of the 6 middle variables is produced from two different fuzzified variables (25 rules for each variable). The other 5 variables are produced from three different fuzzified variables (125 rules for each variable). Finally, there should be 650 rules for the overall model. For the structure of the data set, when the data is released, the variable is adapted according to the recent data in the next day. The monthly return of the stock market index is in daily frequency. However some other input variables are not in daily frequency. The data set is constructed in an overlapping way because the variables in the data set are announced daily, weekly, monthly and quarterly. Like the economists or strategists' points of view in portfolio management companies, when a variable in the data set is changed, the estimation of the model for the next day will be adapted to these changes. Besides, the weightings of some variables for the output are more effective. Thus, the rules are generated according to their priority with respect to the expert (myself)! The result of the model by using the rules with fuzzy logic algorithm in Matlab is 5.02%, which is the sum of the squares of the difference of the real returns from the estimated returns divided by the real returns of the stock market for the period between 1996 and 2002.

3. LEARNING

After having 5.02% R square value, the question that should be asked is: Is there any improvement in the accuracy? The data set covers the period starting from the beginning of 1996 and to the ending of 2002. The next work is to find the optimum stock market model, which is developed by changing the border points of the membership functions by using steepest descent algorithm. By minimizing the square of the error, which is the difference of real stock market returns from estimated stock market returns, the inner five border points of the membership functions are changed. Firstly, this process is done for the last three variables (technical analysis, fundamental analysis and macroeconomy) and the output. Secondly, the membership functions of the 12 input variables are changed according to the same algorithm. Thirdly, the same process is done to the last three variables and the output again. Finally, this process repeats until the convergence happens. The

new model is a little different from the first model with membership functions of the 12 input variables, the last 3 variables and the output. The learning of the model is with %11,8 R square value as seen in the following figure, which is the sum of the squares of the difference of the real returns from the estimated returns divided by the real returns of the stock market. Since the fluctuation in stock market is significant, peak points are not successfully estimated. However, the correlation analysis will be different as in Figure 3 and the model seems to be more successful with 34.4% correlation.

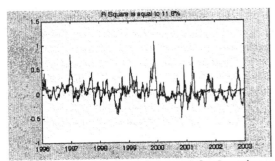

Fig. 2. Actual and Estimated Output After Learning

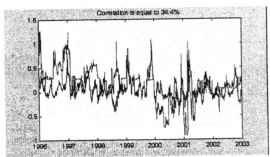

Fig. 3. Correlation of Actual and Estimated Output

4. TESTING THE MODEL

It will be better to compare these models using the data that are not used in the optimization process. By using the data in the period between 2003 and 2004, the accuracy of the model is better than the model development period with 17.6% R square as seen in the following figure. Besides, the correlation of the actual and estimated output is 52.4%, which is considerably high in terms of stock market results.

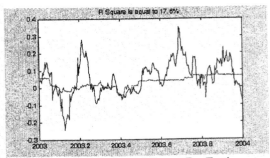

Fig. 4. Actual and Estimated Output For Testing Period

The in-sample R square result of the fuzzy logic model is 11.8% and the out of sample R square result of the model is 17.6%. The results are comparable with the findings in the literature. Besides, the Istanbul Stock Exchange is one of the stock markets with the highest degree of volatility in the world. Because of this volatility, the modelling accuracy can be distorted. Furthermore, the data used and tested for the modelling the ISE100 Index covers three main crises and other political and financial crisis during the period. After the devaluation in Russia, the Turkish economy was also hit due to its strong financial relations with Russia. The banking crisis in November 2000 and the currency crisis of the Turkish Economy in April 2001 can be considered as the other two major crises during the period. As a result, not only the stock market movements but also the macroeconomic fundamentals are very volatile in Turkey. Being one of the emerging markets in the world, the forecast of the stock market may be possible since the market is not efficient. While comparing the accuracy of the models, it should be kept in mind that it is harder to design models for emerging markets than for developed markets due to the significant volatility difference.

Finally, for the 1-year out of sample testing period, there are around 21 working days in a month and 21 different time series outputs can be calculated in monthly frequency of this 1-year period. By using bootstrap re-sampling method, these 21 different outputs can be used to analyze the results of the model statistically. To test the performance of the model, it will be particularly useful in helping to determine entry and exit times for periodic trading or position holding by using the results of the output (which the monthly returns of the ISE100 Index). If the forecasted information suggests that the next period will be an up in the market, investors enter a long position. The reverse would involve to stay in TL position or to sell the stocks. The profitability of

those trades that are executed should be substantial. Besides, there should be four different types of benchmarks to compare the results of the model; TL benchmark, USD benchmark, Repo benchmark and Stock Market Index benchmark. Thus, the corresponding benchmarks are based on the returns of the USD, based on the returns of the repo and also on the returns of the ISE100 Index. If the benchmarks are achieved successfully, it can be said that the model works good enough. To justify how well the model is working statistically, the distribution of the data should be known. By using bootstrap re-sampling method, statistical tables can be constructed and used for testing the model's suggestion. As a result, after the end of the one-year period (from the beginning of 2003 to the beginning of 2004) the investor will have reached his new portfolio amount. Thus, the three hypotheses to be tested will be:

- Hypothesis 1: Does the difference between investing in the model and fully investing in the ISE100 Index yield a positive return?
- Hypothesis 2: Does the difference between investing in the model and fully investing in USD yield a positive return?
- Hypothesis 3: Does the difference between investing in the model and fully investing in Repo yield a positive return?
- Hypothesis 4: Does investing in the model yield a positive return?

These hypotheses are tested with the results of the data in the period between 2003 and 2004. Since the model is built in the period between 1996 and 2002 (inclusive), the model and the data for the testing period are independent from each other. Furthermore, the data is overlapping. Since there are 21 working days in a month, there exists a set of 21 series to test for the four hypotheses above.

Table 2 Statistical Results of the Hypothesis

Hyp	μ_i	σ_{μ_i}	Z_{μ_i}	$C_{i,0.98}$	$C_{i,0.95}$	$C_{i,0.90}$	$C_{i,0.80}$	p	H_0
1	0.1322	0.094	1.41	[-1.51,1.32]	[-0.91,0.92]	[-0.62,0.67]	[-0.35,0.42]	0.98	NO
2	1.4132	0.156	9.04	[-1.68,1.12]	[-1.09,0.82]	[-0.75,0.61]	[-0.41,0.40]	1.00	NO
3	0.5227	0.127	4.10	[-1.74,1.09]	[-1.17,0.81]	[-0.77,0.60]	[-0.40,0.40]	0.99	NO
4	1.0046	0.149	6.76	[-1.71,1.09]	[-1.15,0.81]	[-0.78,0.60]	[-0.41,0.40]	0.99	NO

Finally, the answers to the four hypotheses are positive and in favour of the fuzzy logic model. The difference of the means and their standard deviations are given in Table 2. All of the mean differences are positive enough with respect to their standard deviations. Thus, it is statistically proved that the suggestion of the model has positive returns in USD terms and higher returns than the ISE100 Index and the return of the repo by using bootstrap re-sampling method. The reasons for better results are: a) the ISE

is not a developed and an efficient market; thus, predicting future prices is easier than is the case with other developed markets; b) rule based fuzzy logic modelling with a large enough data set improves the explanation of the future movements. However, we also state that the testing period of one year is not enough to feel confident on this out of sample results. The model is built by using the past data and there is no guarantee for the future performances of the model. Inevitably, in all prediction models there is a risk of underestimation and overestimation. In order to minimize forecast error, all available information as to the predicted variables should be exploited with best mathematical tools and/or econometric models. Thus, fuzzy logic model is a useful tool in this aspect. We use this technique in order to forecast the monthly return of the ISE100 Index. Definitely, our aim is not to give a clue or a hint for the future behaviour of the stock market index. The aim in this modelling technique is solely to try this new methodology and to compare with other mathematical methods. Other than this academic motivation, we have no intention to obtain practical and strategic results for investors in the Istanbul Stock Exchange.

Since all the stock markets in the world are very volatile and the Istanbul Stock Exchange has one of the highest fluctuations, modelling these extreme increases or decreases is hardly possible. In spite of this difficulty, the fuzzy logic models can be developed and new rules can be added to the previous models. As for the future work, one may consider that the existing rules for the fuzzy models may not be good enough to explain the real outputs. There may be some other rules to explain the behaviours more explicitly and the models can be developed and can be redesigned better with new rules. Similarly, there may be some new variables to better explain the characteristics of the behaviours. In modelling of this study, there are no lagged variables of the inputs. There may be one or more period lagged variables to extend the models' modelling perspectives. Besides, the data set can be enlarged by the introduction of some new variables such as population growth or GNP per capita. Additionally, incorporating "combining forecast" method with fuzzy logic may yield profitable results in the framework of chaotic time series forecasting. As in Fiordaliso (1998), the different forecasting tools can be combined to perform better results with fuzzy logic.

REFERENCES

Aksoy, H., İ. Sağlam (2001). Sınıflayıcı (Classifier) Sistem ile İMKB'de Yeni Bir Anomali Gözlemi. *Working Paper, Department of Economics, Boğaziçi University*, 15.

Aksoy, H. (2003). A Logistic Regression Analysis in the Istanbul Stock Exchange. *Unpublished Working Paper.*

Arthur, W.B., J.H. Holland, B. LeBaron, R. Palmer and P. Tayler (1997). Asset Pricing under Endogenous Expectations in an Artificial Stock Market. *The Economy as an Evolving Complex System II*, Edited By W.B. Arthur, S.N. Durlauf and D.A. Lane, Addison-Wesley, pp: 15-40.

Chen, S.M. (1996). Forecating Enrollments Based on Fuzzy Time Series. *Fuzzy Sets and Systems*, **81**, pp: 311-319.

Chen, S., C. Yeh (2001). Evolving Traders and the Business School with Genetic Programming: A New Architecture of the Agent-based Artificial Stock Market. *Journal of Economic Dynamics and Control*, **25**, pp: 363-393.

Economakos, E. (1979). Application of Fuzzy Concepts to Power Demand Forecasting. *IEEE Transactions on Systems, Man and Cybernetics*, **9**, pp: 651-657.

Fiordaliso, A. (1998). A Nonlinear Forecasts Combination Method Based on Takagi-Sugeno Fuzzy Systems. *International Journal of Forecasting*, **14**, pp: 367-379.

Kooths, S. (1999). Modelling Rule and Experience Based Expectations Using Neuro Fuzzy Systems. *Paper in the 5th International Conference of the Society for Computational Economics*, Boston. USA.

LeBaron, B., W.B. Arthur, and R. Palmer (1999). Time Series Properties of an Artificial Stock Market. *Journal of Economic Dynamics and Control*, **23**, pp: 1487-1516.

Lederman, J., R.A. Klein (1995). Virtual Trading: How Any Trader with a PC can Use the Power of Neural Nets and Expert Systems to Boost Trading Profits. Probus.

Lee, C.-Y., E.K. Antonsson (2000). A Pareto Optimal Approach to Takagi-Sugeno Fuzzy System Synthesis. *Unpublished Working Paper*, Caltech, USA.

Linn, S.C., N.S.P. Tay (2001). Fuzzy Inductive Reasoning, Expectation Formation, and the Behaviour of Security Prices. *Journal of Economic Dynamic and Control*, **25**, pp: 321-361.

Lo, A.W., A.C. MacKinlay (1988). Stock Market Prices do not Follow Random Walks: Evidence from A Simple Specification Test. *Review of Financial Studies*, **1**, pp: 41-66.

Mamdani, E.H. (1974). Application of Fuzzy Algorithms for Simple Dynamic Plant. *Proc. IEE*, **121**, pp: 1585-1588.

Nomura, H., I. Hayashi, and N. Wakami (1992). A Learning Method of Fuzzy Inference by Descent Method. *Proceedings of the IEEE International Conference on Fuzzy Systems*, pp: 203-210.

Shnaider, E., A. Kandel (1989). The Use of Fuzzy Set Theory for Forecasting Corporate Tax Revenues. *Fuzzy Sets and Systems*, **31**, pp: 187-204.

Song, Q., B.S. Chissom (1993). Forecasting Enrollments with Fuzzy Time Series – part I. *Fuzzy Sets and Systems*, **54**, pp: 1-9.

Song, Q., B.S. Chissom (1994). Forecasting Enrollments with Fuzzy Time Series – part II. *Fuzzy Sets and Systems*, **62**, pp: 1-8.

Takagi, T., M. Sugeno (1985). Fuzzy Identification System and Its Applications to Modelling and Control. *IEEE Transactions on Systems, Man, and Cybernetics*, **15**, pp: 116-132.

Zadeh, L.A. (1965). Fuzzy Sets. *Information and Control*, **8**, pp: 338-353.

www.elsevier.com/locate/ifac

ANALYSIS OF SOCIO-ECONOMICS FACTORS IN POLISH REGIONAL STRUCTURE

Jerzy Holubiec, Grażyna Petriczek

*Systems Research Institute, Polish Academy of Sciences,
01-447 Warsaw, Newelska 6 ,Poland
tel. 0-22 36- 44- 14, fax. 022-37-27-72, e-mail: petricz@ibspan.waw.pl*

Abstract: In the paper the concept of partition of Poland territory consisting of vojvodships into group with the similar socio-economic characteristics is presented. This concept is based on the method of determining and selecting the homogeneous groups from the data set describing the analysed phenomenon. The data set describing vojvodships in Poland has the form of matrix consisting of rows number corresponding to vojvodships number and columns number depended on the number of considered socio-economic factors. The analysis of partition of Poland territory with respect to various socio-economic characteristics was performed for years 1998, 2000, 2001. In 1999 year new partition of Poland territory was performed. Until 1999 year Poland territory consisted of 49 vojvodships. After administration reform Poland territory contains 16 vojvodships. *Copyright © 2004 IFAC*

Keywords: Regional modelling, hypotheses, data set homogeneity, statistics, statistically stable boundaries between two sets

1. METHOD OF SET DIVISION INTO HOMO-GENEOUS SUBSETS

The essential element of the method is a criterion for testing data set homogeneity hypothesis. The criterion function has a form of statistics U, which has the χ^2 distribution. The method consists in iterative partition of non homogeneous set into two parts. If number of this division increases, than the process of successive partitions can lead to the existence of statistically unstable boundaries between adjacent homogeneous subsets.

The subsets aggregation is based on adequately formulated hypothesis testing concerning stability of the boundary between the two sets.

Thus, the data set division algorithm contains two basic fundamental steps :

- the set partition procedures into homogeneous, disconnected subsets, that enable the primary set division
- the procedures that examine the stability of boundaries between these subsets.

This two - steps algorithm is as follows:

1.1 Set division into homogeneous, separated subsets

The basic idea underlying the partition is principle of the equivalence of random variable that have the same distribution.

Let $S = \{ s_1 , s_2 , ..., s_n \}$ denotes a data set.

Assume that to each element $s \in S$ corresponds a random variable ξ_s with the distribution function $F_s(x)$.

Let E^s denotes the set of random variables ξ_s and R denotes the set of random variable values x.

Def.1. (Kildyshev and Abolentzev, 1978) Random variables ξ_{s_1}, ξ_{s_2} are equivalent if for each two elements $s_1, s_2 \in S$ the following relation is fulfilled :

$$F_{s_1}(x) - F_{s_2}(x) = 0 \quad \text{for any } x \in R \qquad (1)$$

The above condition states that a random variable set E^s can be disconnected into equivalence classes, and subsequently, the set S has the form :

$$S = S_1 \cup S_2 \cup ... \cup S_k \quad k \geq 1 \qquad (2)$$

If $k = 1$ then the set S is homogeneous.
If $k > 1$ then the set S is non homogeneous and the relation (2) describes this non homogeneity.
On the basis of the conception of random variable equivalence one can formulate the homogeneity definition.
Def.2. (Kildyshev and Abolentzev, 1978) The set of random variables $E^{S_1} \subset E^S$ is homogeneous if the following condition is fulfilled:

$$F_{s'}(x) - F_{s''}(x) = 0 \quad \text{for any } s', s'' \in S_1 \qquad (3)$$
$$\text{and } x \in R$$

By contradiction to the homogeneity condition (3) one obtains the non homogeneity definition.
A set of equivalent random variables is homogeneous. In order to formulate homogeneity hypothesis testing criteria one assumes that the random variables are independent and normally distributed with the probability density function:

$$f(x) = \frac{1}{\sqrt{(2\pi)^k}} |\Sigma_s|^{-\frac{1}{2}} \exp\left(-\frac{1}{2}(x - m_s)^T \Sigma_s^{-1}(x - m_s)\right) \qquad (4)$$

where:
m_s - the vector of expected values of the random variable ξ_s
Σ_s - the covariance matrix with $[k \times k]$ dimension
$|\Sigma_s|$ - the determinant of covariance matrix.

Moreover, if one assumes that the considered variables have the same covariance matrices then the homogeneity condition takes the following form: (the H_0 hypothesis)

$$H_0: \qquad m_{s'} = m_{s''} \text{ for all } s', s'' \in S$$
$$\text{subject to:} \qquad \Sigma_{s'} = \Sigma_{s''} \qquad (5)$$

Defining on the set of all partitions of S into two subsets S_1 and S_2 the function:

$$\delta(S_1, S_2) = \frac{1}{n_1} \sum_{s \in S_1} m_s - \frac{1}{n_2} \sum_{s \in S_2} m_s \qquad (6)$$

where: n_1, n_2 - number of elements of S_1 and S_2 respectively,

one obtains a homogeneity index of k - dimensional variable set.
On the basis of index (6) the homogeneity hypothesis can be formulated as follows:

$$H_0: \qquad \delta(S_1, S_2) = 0 \qquad (7)$$
$$\text{for any pair } (S_1, S_2) \text{ belonging to all}$$
$$\text{partitions of S into two subsets.}$$

Let n - be a number of observations and k - be a number of considered characteristics describing the analyzed phenomenon.
The set of all observations of k - dimensional random variable is a matrix

$$\begin{bmatrix} x_1 \\ x_2 \\ \cdot \\ \cdot \\ \cdot \\ x_n \end{bmatrix} = \begin{bmatrix} x_{11}, & x_{12}, & \cdots, & x_{1k} \\ x_{21}, & x_{22}, & \cdots, & x_{2k} \\ \cdot & \cdot & \cdots & \cdot \\ \cdot & \cdot & \cdots & \cdot \\ \cdot & \cdot & \cdots & \cdot \\ x_{n1}, & x_{n2}, & \cdots & x_{nk} \end{bmatrix} \qquad (8)$$

The matrix (8) is a realization of k - dimensional normally distributed random variable ξ_s, with mean value m_s and the same diagonal covariance matrices.
The H_0 hypothesis (5) testing is performed as to compare two samples.
That result is the partition of matrix (8) into two various, disconnected parts containing n_1 and n_2 rows, respectively.
A criterion function for the H_0 hypothesis (7) testing is constructed using the maximum likelihood method.
The statistical estimation of k - dimensional partition $\delta(S_1, S_2)$ is given by the random variable $\tilde{\xi}$ of the form:

$$\tilde{\xi} = \frac{1}{n_1} \sum_{s \in S_1} \xi_s - \frac{1}{n_2} \sum_{s \in S_2} \xi_s \qquad (9)$$

Each component appearing in the expression (9) is a random variable with the following distribution parameters: the expected values (m'_j, m_j) and the variances $(\sigma_j^2 / n_1, \sigma_j^2 / n_2)$.
Assuming that the H_0 hypothesis given by condition (5) is true, one can formulate the likelihood function of the form: (Fisz, 1967; Mardia, et al., 1979)

$$L(x, m) = \frac{1}{\sqrt{(2\pi)^k}} \left(\prod_{j=1}^{k} c_j^2\right)^{\frac{1}{2}} \exp\left[-\frac{1}{2}\sum_{j=1}^{k} \frac{x_j^2}{c_j^2}\right] \qquad (10)$$

where: c_j^2 - the variance of the random variable $\tilde{\xi}$ described by the following relation:

$$c_j^2 = \frac{\sigma_j^2 (n_1 + n_2)}{n_1 n_2} \qquad (11a)$$

$\bar{\sigma}_j^2$ - sample variance of the random variable ξ_{s_j}

$\bar{x}_j$ - observation values of j-th component of the random variable $\tilde{\xi}$ determined as follows:

$$\bar{x}_j = \frac{1}{n_1} \sum_{s \in S_1} x_{sj} - \frac{1}{n_2} \sum_{s \in S_2} x_{sj} \qquad (11b)$$

The sample variance can be calculated from the relation:

$$\bar{\sigma}_j^2 = \frac{1}{n_1 + n_2 - 1} \left[\sum_{s \in S_1} x_{sj}^2 + \sum_{s \in S_2} x_{sj}^2 - \frac{1}{n_1 + n_2} \left(\sum_{s \in S_1} x_{sj} + \sum_{s \in S_2} x_{sj} \right)^2 \right] \quad (11c)$$

It can notice that the form of the likelihood function depends on the exponent $\sum_{j=1}^{k} \frac{\bar{x}_j^2}{c_j^2}$

Substituting the expressions (11a) - (11c) to the exponent one obtains the following criterion function $U(S_1, S_2)$:

$$U(S_1, S_2) = \frac{\dfrac{1}{(n_1 + n_2) n_1 n_2} \sum_{j=1}^{k} \left(n_2 \sum_{s \in S_1} x_{sj} - n_1 \sum_{s \in S_2} x_{sj} \right)^2}{\sum_{j=1}^{k} \left(\sum_{s \in S} x_{sj}^2 - \dfrac{1}{n_1 + n_2} \left(\sum_{s \in S} x_{sj} \right)^2 \right)}$$

$$S = S_1 \cup S_2 \qquad (12)$$

The statistics U has the χ^2 - distribution with k degrees of freedom.

From the maximum likelihood principle and the properties of function L(,) it follows that the H_0 hypothesis can be accepted if the following condition holds :

$$U(S_1, S_2) \le \chi_{\alpha,k}^2 \qquad (13)$$

for any pair (S_1, S_2) belonging to all partitions of set S

where: α - significance level
k - degree of freedom

If the above condition is not satisfied then the H_0 hypothesis should be rejected; the set is nonhomogenous and can be partitioned into two disconnected subsets.

The determination of statistics U for all pairs (S_1, S_2) and examination of inequality (13) for a large number of observations are complicated and causes computational difficulties.

To simplify the H_0 hypothesis testing it was assumed, that the observations { X_1 , X_2 , ... , X_n } are ordered increasingly with regard to most significant characteristics.

In this case it is sufficient to verify the hypothesis H_0 only for the pairs (S_1, S_2); where the subset S_1 consists of l - elements and the subset S_2 consists of the remaining n-l elements.

In this case, the statistics U has the form:

$$U(l, n-l) = \frac{\dfrac{n-l}{n(n-l)l} \sum_{j=1}^{k} \left((n-l) \sum_{i=1}^{l} x_{ij} - l \sum_{i=l+1}^{n} x_{ij} \right)^2}{\sum_{j=1}^{k} \left(\sum_{i=1}^{n} x_{ij}^2 - \dfrac{1}{n} \left(\sum_{i=1}^{n} x_{ij} \right)^2 \right)}$$

$$\text{for } l = 1, 2, \dots\dots, n\text{-}1 \qquad (14)$$

From the above consideration, it follows that the homogeneity hypothesis H_0 is accepted if the following inequality:

$$U(l, n-l) \le \chi_{\alpha,k}^2 \quad \text{for } l = 1, \dots., n\text{-}l \quad (15)$$

is fulfilled.

If the above inequality is not satisfied for any row l-th then the H_0 hypothesis should be rejected ; the set is nonhomogeneous. In this case the alternative hypothesis is assumed:

$$\begin{aligned} H_1 : \quad &\delta(S_1, S_2) \ne 0 \\ &U(l, n-l) > \chi_{\alpha,k}^2 \end{aligned} \qquad (16)$$

The partition of nonhomogeneous data set into homogeneous, disconnected subsets is based on the acceptation of the alternative nonhomogeneous hypothesis (H_1 hypothesis).

From the maximum likelihood method follows that the H_1 hypothesis can be accepted if the L function achieves its maximum value. Thus, the exponent occurring in this function should be minimized.

By appropriate transformations the minimization problem leads to the maximization of statistics U(l, n-l) and it can be written as follows: (Hołubiec and Petriczek, 1997)

$$\max_{l} U(l, n-l) \qquad (17)$$

From the condition (17) it follows that the likelihood function achieves the maximum for such a partition of nonhomogeneous data set into two parts, that gives rise to maximum of statistics U(l, n-l).

The condition (17) establishes the theoretical base of the method for set partitioning into two homogeneous, separated subsets.

1.2 The subsets aggregation - hypothesis on unstable intergroup boundaries

The presented method of set division consists in the iterative partitions of nonhomogeneous set into two parts. If number of these partitions (iteration number) increase that successive divisions can cause the existence of statistically unstable boundaries between adjacent homogeneous subsets.

The statistic stability of boundaries between subsets can be examined by comparing multi- dimensional means. If multi - dimensional means of two compared groups are statistically equivalent than it can be assumed, that the statistically unstable boundary exists, so the boundary can be removed and the sets can be aggregated into one homogeneous set.

The examination of the intergroup boundaries stability leads to verification of suitably formulated hypothesis.

Let m_i - denote multi - dimensional mean of i-th subset.$(i = 1,...., M)$.

The hypothesis H_0 on statistic unstable intergroup boundary has the form:

$$H_0: \qquad m_i - m_{i+1} = \{0, 0, \ldots\ldots, 0\} \quad (18)$$

If the H_0 hypothesis holds, than the boundary between two sets is statistically unstable and the sets can be connected.

The rejection of the H_0 hypothesis of the form (18) enables acceptation of the alternative hypothesis (on boundaries stability) and it means that there exist essential stable intergroup boundaries.

The examination of boundaries can be performed successively for all considered subsets.

A criterion function for the H_0 hypothesis testing is constructed using the maximum likelihood method. As the result, one obtains the U statistics of the form:

$$U(S_i, S_{i+1}) = \cfrac{\dfrac{n_i + n_{i+1} - 1}{(n_i + n_{i+1})n_i n_{i+1}} \sum_{j=1}^{k}\left(n_{i+1}\sum_{s \in S_i} x_{sj} - n_i \sum_{s \in S_{i+1}} x_{sj}\right)^2}{\sum_{j=1}^{k}\left(\sum_{s \in S} x_{sj}^2 - \dfrac{1}{n_i + n_{i+1}}\left(\sum_{s \in S} x_{sj}\right)^2\right)}$$

$$S = S_i \cup S_{i+1}, \qquad i = 1, ..., M \qquad (19)$$

where:
M - the number of subsets
k - the number of considered characteristics
n_i, n_{i+1} - number of elements of S_i, S_{i+1}, respectively

It can be proved that under some assumptions the H_0 hypothesis testing leads to inequality of the form:

$$U(S_i, S_{i+1}) \leq \chi^2_{\alpha,k} \qquad i = 1, ..., M \qquad (20)$$

If the condition (20) holds, than one can assume that the boundary between subsets S_i and S_{i+1} is unstable.

In this situation these subsets can be interconnected and the hypothesis of the stability between subsets $(S_i \cup S_{i+1})$ and S_{i+2} is tested.

If the condition (20) is not satisfied then the boundary between sets S_i and S_{i+1} is stable (essential) and these sets can not be connected.

2. MODELLING REGIONAL STRUCTURE

The presented algorithm consists of two steps:
a) The initial partition of input set into separate (disconnected) homogeneous subsets. For this purpose one tests the homogeneity hypothesis of the form (15); if H_0 is rejected, then the nonhomogeneity hypothesis (16) is examined. The H_1 hypothesis testing leads to the U statistics maximization problem (17).
b) The examination of stability of the boundaries between subsets obtained by the initial partition. It results in testing hypothesis on the boundary stability. Depending on the case whether the H_0 hypothesis of the form (20) is true or not, the two adjacent subsets can be connected or remain disjoined.

The presented algorithm was used for selecting the homogeneous groups of vojvodships in Poland described by a set of characteristics.

It means, that the vojvodships belonging to the same group have similar feature; for example the socio-economic characteristics or development dynamic characteristics. The selection of the characteristics describing vojvodships depends on applications.

The algorithm was used for analyzing the Polish regional structure for 3 years - 1998, 2000, 2001.

In 1999 administration reform was performed and the new regional structure was introduced.

Until 1999 the Polish regional structure consisted of 49 vojvodships and after new partition there are 16 vojvodships.

The following cases were considered:
a) the vojvodships described by 3 characteristics (population, employment, investments)
b) the vojvodships described by 5 characteristics (population, employment, investments, industry production, number of flats)
c) the vojvodships described by 9 characteristics (population, employment, investments, industry production, number of flats, agricultural grounds, construction production, unemployment, accommodation)

The analyzed statistical data (Statistical Year-Book, 1998, 2000, 2001) concern 3 years: 1998 – when Poland territory consisted of 49 vojvodships and 2000, 2001 – when Poland territory contained 16 vojvodships.

In all cases the number of people is assumed to be the most significant characteristic. The χ^2 distribution value for 3 degrees of freedom equals 7.815; for 5 degrees of freedom it is equal to 11.07 and for 9 degrees of freedom equals 16.919. The significante level α= 0.05.

The input data set is ordered increasingly with regard to the significant feature assumed.
The input data set is represented as matrix with: the number of rows corresponding to the number of vojvodships and the number of columns corresponding to the number of characteristics considered.
For each year the partition of input data set into homogeneous, separable groups of vojvodships were obtained.

➢ for 1998 for two cases (3 characteristics and 5 characteristics) after 10 iterations one obtains the complete partition of input data set into 11-th homogeneous groups of vojvodships. The subsets obtained from 1-st step procedure have stable boundaries.

For 9 characteristics after 11 iterations one obtains the initial partition of input data set into 12-th homogeneous, disconnected subsets (1-st step of algorithm).
From the analysis of stability of the boundaries between subsets (step 2 -nd algorithm) follows that the subsets number 2, 3 can be connected; the boundaries between subsets are unstable. Finally one obtains 11-th groups of vojvodships.

From the analysis of obtained subsets results that number of characteristics has not caused any significant change in number of subsets; only the attachment of particular vojvodships to groups has varied.
Tables given below present partition of Poland territory for 1998 year into homogeneous, separable groups of vojvodships described by 3, 5 and 9 characteristics respectively.

Table 1. Partition of 49 vojvodships into groups
1998- 3 characteristics

1- subset	2- subset	3- subset
Chelmskie	Leszczynkie	Koninskie
BialskoPodlaskie	Sieradzkie	Suwalskie
Lomzynskie	Ostroleckie	Zamojskie
	Przemyskie	Elblaskie
	Skierniewickie	Pilskie
	Slupskie	Krosnienskie
	Wloclawskie	
	Ciechanowskie	

7 - subset	8 - subset	9 - subset
Nowosadeckie	Bielskie	Lubelskie
Rzeszowskie	Szczecinskie	Lodzkie
Radomskie	Opolskie	Kieleckie
Olsztynskie		Wroclawskie
Czestochowskie		Bydgoskie

10 - subset	11 -subset
Krakowskie	Warszawskie
Poznanskie	Katowickie
Gdanskie	

From the analysis of obtained results for 3 and 5 characteristics one can perform following conclusion:
• for 16 vojvodships belonging to subsets 1, 8 9, 10, 11 increasing number of characteristics has not caused any changes in the attachment of particular vojvodships to groups.
• for 2 – subset in the case of 5 characteristics follows partition of this group into two subsets, which are given in below:

Table 2. 1998 - 5 characteristics. The partition subset
number 2 into two groups

2 - subset	3 - subset
Leszczynki	Skierniewickie
Sieradzkie	Slupskie
Ostroleckie	Wloclawskie
Przemyskie	Ciechanowskie

• for the 5 characteristics Gorzowskie vojvodship belong to 3-group
• for 5 characteristics the changes occur in subsets 5, 6, 7 and it is presented in Table 3.

Table 3. 1998- 5 characteristics. The subset in which
occur the changes

5 - subset	6 - subset	7 - subset
Plockie	Tarnobrzeskie	Bialostockie
Jeleniogorskie	Piotrkowskie	Kaliskie
Legnickie	Siedleckie	Walbrzyskie
Koszalinskie	Torunskie	Nowosadeckie
	Zielonogorskie	Rzeszowskie
	Tarnowskie	Radomskie
		Olsztynskie
		Czestochowskie

• for 9 characteristics the changes occur in subsets number 1, 2, 3, 4, 5, 6, 7 and it is presented in Table 4.

Table 4. 1998- 9 characteristics. The subset in which
occur the changes

1 - subset	2 - subset	3 -subset
Chelmskie	Lomzynskie	Koninskie
BialskoPodlaskie	Leszczynskie	Suwalskie
	Sieradzkie	Zamojskie
	Ostroleckie	Elblaskie
	Przemyskie	Pilskie
	Skierniewickie	Krosnienskie
	Slupskie	Gorzowskie
	Wloclawskie	
	Ciechanowskie	

4 - subset	5 - subset	6 - subset
Plockie	Tarnobrzeskie	Walbrzyskie
Jeleniogorskie	Piotrkowskie	Nowosadeckie
Legnickie	Siedleckie	
Koszalinskie	Torunskie	
	Zielonogorskie	
	Tarnowskie	
	Bialostockie	
	Kaliskie	

7 - subset
Nowosadeckie
Rzeszowskie
Radomskie
Olsztynskie
Czestochowskie

➤ for the years 2000 and 2001 (after administra
tion reform) homogeneous groups with 3, 5 and 9
characteristics considered were taken into account.
The of characteristics did not influence on the number of homogeneous groups.
For the years 2000 and 2001 4 homogeneous groups
with stable boundaries between subsets were obtained. Table 5 present partition of Poland territory
for years 2000 and 2001 into homogeneous, separable groups of vojvodships described by appropriate
3, 5 and 9 characteristics.

Table 5. Partition of 16 vojvodships into groups – for
the years 2000, 2001 for 3 , 5 and 9 characteristics

1 - subset	2 - subset
Lubuskie	Zachodnio-Pomorskie
Opolskie	Kujawsko-Pomorskie
Podlaskie	Podkarpackie
Świetokrzyskie	Pomorskie
Warminsko-Mazurskie	Lubelskie

3 - subset	4 - subset
Lodzkie	Slaskie
Dolnoslaskie	Mazowieckie
Malopolskie	
Wielkopolskie	

It can be seen that increasing number of characteristics has not caused any significant change in number
of subsets ; only the attachment of particular vojvodships to groups has varied.
The determined vojvodships groups form regional
structures with similar characteristics.

3. REFERENCES

Fisz, M. (1967). Rachunek prawdopodobieństwa
i statystyka matematyczna. PWN, Warszawa.
Kildyshev, G.S., and J.A. Abolentzev (1978). Mno-
gomernye gruppirovki. Izd. Statistika, Mosk-
va
Mardia, K.V., J.T. Kent and J.M Bibby (1979). Mul-
tivariate Analysis. Academic Press,, London
Hołubiec, J. and G. Petriczek (1997). Homogeneity
Algorithm For Modeling Regional Structure.
In: *Proceedings of the International Confer-
ence on Methods and Models in Automation
and Robotics MMAR* (S. Domek, Z. Emirsa-
jłow, R. Kaszyński, Ed.). **Vol.1**, pp. 397-402.
Technical University of Szczecin, Szczecin
Statistical Year-book 1998, 2000, 2001. Główny
Urząd Statystyczny, Warsaw

SELF-ADJUSTING OF ATTRACTION CHANNEL FOR SHORT-TERM SUBSTRUCTURES BY ABRUPT INCREASE OF STOCK MARKET POTENTIAL

Kleparskiy V.G.

V.A.Trapeznikov Institute of Control Sciences of Russian Academy of Sciences
117997 Moscow, Profsouznaja 65
Fax: (095)420-20-16, E-mail: kleparvg@ipu.rssi.ru

Abstract: The study of short-time substructure influence on adaptation of attraction channel of such dynamical system as stock market was carried out. The fractal dimension and intermittency exponent were estimated using probability description of the system trajectory changes. Period doubling bifurcation was observed for short-term substructures as a result of sharp forcing caused by financial rating change. *Copyright ©2004 IFAC*

Keywords: attraction channel, self-adjusting, dynamic system, stock market, short-term substructures, bifurcation.

1. INTRODUCTION

Recently, considerable study has been given to complex hierarchical systems, low-level cascades of which can introduce corrections into top-level control actions based on self-made estimations of new reality. Evolution of such systems is governed not only by nonlinear responses to top-level control but also by inter-level time/space distributed feedback as well. As a result, one can expect arising both short-scale and long-scale time/space corrections during process of system's self-organization. Therefore, one should also expect evolution variations associated both with randomness of trajectory within a "channel" of top-level decisions and with changes of the "channel" itself due to system self-adjusting to a new reality. A temporal behavior of such cascade-like system will correspond not only to the motion along a smooth manifold (motion to a fixed point, or to a limit cycle, for example) but to "infinitely fragmented" fractal set. So, the problem of short-scale chaos influence on adaptation of the cascade hierarchical system is now being put in the forefront.

The possibility of the solution this problem is determined by basic feasibility of the modernized Kolmogorov theory of cascade dynamic structures (see, for example, Nelkin (1996), Cross and Hohenberg (1993)) that reflects major regularities of structure formation in systems driven away from equilibrium. Based on this theory, the self-adjusting of attraction channel can be studied during the time-evolution of such hierarchical cascade system as, for example, a stock market. The choice of the stock-market system as a study-object is conditioned by accessibility of data on attraction channel evolution. Such data can be derived by using probability description of the system trajectory changes. To reduce the degree of complexity of the investigated system the study was carried out close to instability point. The instability was originated by abrupt increase of effective potential due to Moody's rating change.

2. BASIC CONCEPTS AND PROBLEM POSING

The theoretical starting point of evolution analysis is usually based on a nonlinear differential equation

$$\frac{du}{dt} = F(u, \Phi(u)) = F(u(t)), \quad (1)$$

Here $u = u(t)$ is the system's state, and $\Phi(u)$ is the feedback law. The variable $u = u(t)$ belongs to a linear space H called the phase space, and operator F is a mapping of the phase space H into itself.

In the general situation, the functional F depends on some physical parameter λ $F(u) = F_\lambda(u).$

This parameter λ is some "measure" of system driving away from equilibrium, i.e. some "measure" of control forcing. For fluid media, as example, such forcing parameter λ is Reinolds number.

It was pointed out (see, for example, Temam (1988)), that with increase of forcing (i.e. at sufficiently large λ ($\lambda > \lambda_H$)) a Andronoff-Hopf bifurcation can occur. So, the system's behavior becomes time periodic with a limit cycle attractor.

At this $u(t) - \varphi(t) \to 0$ as $t \to \infty$,

where φ is a time-periodic solution of (1), of period $T > 0$.

With further increase of deviation from equilibrium, i.e. for larger λ, $\lambda_H < \lambda < \lambda_T$, invariant tori can appear, i.e., for $t \to \infty$,

$$\varphi(t) = g(\omega_1 t, ... \omega_n t),$$

where g is periodic function with period 2π in each variable and the frequencies $\omega_i = 1/T_i$ are rationally independent numbers. So, at this stage, the system motion is governed by modes with discrete frequencies ω_i, that is not characteristic of full chaotic behavior.

Finally, for $\lambda > \lambda_T$, the system reaches the last stage, where $u(t)$ looks completely space/time random for all time. This transition to chaos from invariant tori phase may occur, according to Ruelle-Takens approach to turbulence arising (see, for example, Ruelle (2001)), after two-three Hopf bifurcations due to doubling-of-period phenomena. So, the motion of the system for $\lambda \sim \lambda_T$ must become chaotic with double-period bifurcations from one time-stretch of torus-type trajectory to another. The complicated convolutions of trajectory mean the transition of the system dynamics to a chaotic (strange) attractor. The complexity of a system evolution can be "measured" through the fractal dimension of system attractor. The higher the fractal dimension of the attractor is, the more complicated becomes the system motion.

For complex hierarchical system one cannot hope to provide a detailed theoretical picture of the chaotic dynamics. So even in a small system chaotic motion can only be analyzed statistically. Therefore we must adapt statistical methods to estimate the fractal dimension of strange attractor for a given system.

Such adaptation can be based on the recent modifications of Kolmogorov's theory (see, for example, Nelkin (1996), Frish (1997)). It retains the notion of scaling in inertial range of turbulent flows, as a result of energy transfer through the cascade structures down to smaller scales by nonlinear term in the Navier-Stokes equation.

The existence of turbulent-like cascade structures of varying time scale (short-, middle- and long-term) is essential characteristic of stock market evolution (see, for example, Lux and Marchesi (1999)). This turbulent-like behavior has been first emphasized by Ghashghaie, et all (1996) due to similarity between probability density functions (PDFs) of price variations in stock market and velocity variations in turbulent flows. Such similarity makes it possible to suggest that the underlying dynamics of stock-market system is given by the Navier-Stokes-type equations. The stock-market system can thus be considered as multifractal cascade system (Kleparskii (2001)). Since such cascade system never becomes stationary, the stock market evolution can be interpreted, from nonlinear physics viewpoint, as a chain of Andronoff-Hopf bifurcations. The fractal dimension d_f of such attractor channel must increase with time in the case of chaos increase. Therefore measuring of the fractal dimension d_f (following the analysis of intermittency exponent μ that value is determined by the dissipation intensiveness) can be usefully applied to the study of the self-adjusting of attraction channel on various stages of system evolution. Special interest presents the study of the fractal dimension and intermittency for stages of stock-market evolution with abrupt financial rating changes. One can assume that on this stage of evolution - close to instability point – the investigated system reduces the degree of complexity. So one can also assume that the picture of doubling-of- period phenomena can be easily revealed at abrupt changes of system effective potential.

3. METHODS OF SOLUTION

The fractal dimension d_f of the attractor for the self-affine trajectory of stock-market index can be defined according to Feder (1988) as

$$d_f = 2 - \alpha, \qquad (2)$$

where α is the self-affine scaling exponent. The value of scaling exponent α can be estimated using power-low scaling of high-order moments M_n of PDF as it was first emphasized by Kolmogorov (see, for example, Nelkin (1994), Frish (1997)). The mentioned above analogy between stock-market and turbulence gives the possibility to use the relation

$$M_n = \left\langle (\Delta R)^n \right\rangle \sim (\Delta t)^{\xi_n} \qquad (3)$$

Here ΔR is the stock-market index difference, Δt is the observation interval for the index-price time series, and ξ_n is the short-time structure scaling exponent.

Obtained experimental dependencies for the moments $M_n \equiv \langle(\Delta R)^n\rangle$ vs the observation intervals Δt (in log-log scale) can be used to estimate the structure scaling exponents ξ_n for various n-order of PDF moment. Knowledge of the structure scaling exponent ξ_n makes it possible to determine the self-affine scaling exponent α with use of the relation

$$\xi_n = \alpha n + \gamma_n , \qquad (4)$$

where γ_n is the multifractal exponent. The value of γ_n can be defined, according to theory of turbulence (see, for example, Nelkin (1994), Frish (1997)), as:

$$\gamma_n = \mu n \, (3-n)/18, \qquad (5)$$

where μ is the intermittency exponent, that value is caused by local dissipation intensity in cascade dynamic structures as it was pointed out by Nelkin (1994), Frish (1997)).

4. RESULTS AND DISCUSSION

To reveal the actual regularities of the attraction channel self-adjusting the temporal price-variations ΔR of the Sberbank-Russia stocks were investigated from October 7 to October 10 2003. The selected to study time-interval has abrupt price-increase (nearly 14% as it can be seen in Fig.1) of the Sberbank-index caused by Moody's rating change that took place on October 8.

To obtain the d_f - and μ - values, the day-averaged PDF-moments M_n of the Sberbank-Russia index were determined for n = 2, 4, and 6. The observation intervals Δt were ranging from 15 to 120 min. The everyday analysis of the time series of the Sberbank index allows to obtain the day-averaged structure scaling exponent ξ_n as a slope of short-term segments of $\lg M_n(\lg \Delta t)$ curves for given values order n. The duration of monofractal segments of $\lg M_n(\lg \Delta t)$ curves is interpreted as life-time $\Delta\theta$ of hidden short-term substructures (Kleparskii (2001)).

In Fig. 2 and Fig. 3 the averaged values of the attractor dimension d_f and the intermittency exponent μ vs the effective existence duration $\Delta\theta$ of short-term

dynamical substructures are presented. The obtained results show that the d_f- and μ-values have visible periodical components with periodicity L $\approx$ 30 min at October 7 (day before Moody's rating change). The short-term substructures with smaller d_f - values have smaller μ-values. This can be explained by smaller dissipation intensity in substructures with less chaotic behavior. Already from 11:45 a.m. of October 8 - time of rating change - to 10:45 a.m. October 9 the d_f-values run up to minimum as compared with previous days. The μ -values also run up for shortest-term substructures that can be interpreted as decrease of dissipation intensity.

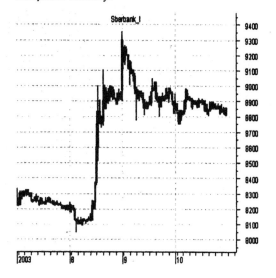

Fig. 1 Sberbank-Russia stocks changes from October 7 to October 10, 2003

It was also observed a minimal scattering for existing dynamical substructures. Such behavior can be caused by shrinking of attraction volume (as it usually suggested for dissipative systems) and by decrease to a minimum of number of positive Lyapunov exponents. The latter can be caused by an unidirectional drift of the market participants during rapid stocks price-increase.

The data presented in Fig. 3 make it possible to estimate the small amplitude periodicity of d_f–values with wave-length L $\approx$ 60 min for October 8. The obtained results can be interpreted as doubling of the period of d_f–changes as compared with previous day, October 7.

To explain the obtained results in framework of nonlinear dynamics one should remember that the forcing parameter λ in systems described by Navier-Stokes-type equations is given by excitation to dissipation ratio. So, the sharp increase of excitation that take place in case of abrupt stocks-price increase (due to Moody's change) leads to the increase of forcing parameter λ. The λ increase results in the Hopf bifurcation to next period of Feigenbaum cascade (due to doubling-of-period phenomena). After the Hopf bifurcation has occurred the initial (on October 7) conditions of nonlinear transfer of effective potential through cascade system is decomposed. One should remember that nonlinear transfer of energy in a direction which drives the system towards equilibrium, i.e. towards large wave-number (see, for example, Nelkin (1994), Frish (1997)) is the main reason of arising and existence of cascade structures in turbulent-like systems. As the dissipation increases with a square of wave number, the period doubling of life-time $\Delta\theta$ of short-term substructures results in decreasing of dissipation in stock-market system. As a result, the nondissipated (in shortest-term substructures) potential can be accumulated in stock-market system. This assumption corresponds to the rule of the energy distribution among its modes for longitudinal energy spectrum $E_{11}(k) \sim k^{-5/3}$ within turbulent flows (see, for example, Nelkin (1994), Frish (1997)).

On the next time-stretch of the stock-market trajectory (October 9 and 10), for the take-profit stage, one can observe the decay of d_f-periodicity with wave-length $L \approx 60$ min and arising of short-term substructures with $L \approx 30$ min. Such behavior can be explained as a results of decay of rating-increase excitation and, consequently, of decrease of forcing parameter λ. The latter must leads to Hopf bifurcation with twofold decrease of the previous period.

So, the Hopf bifurcation with double-increasing/decreasing of the period can be suggested as basic mechanism of self-adjusting of attraction channel for short-term substructures at abrupt increase of stock-market potential. It can be also observed the appearance of substructures with negative μ-values,

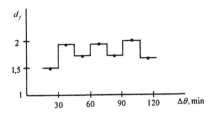

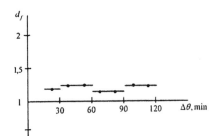

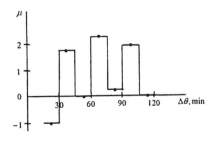

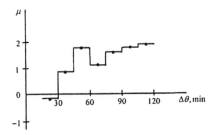

Fig.2. The changes of the attractor dimension d_f and intermittency exponent μ versus $\Delta\theta$ for short-term substructures. The studied stage includes October 7 (open 8260, closed 8190) and October 8 till 11:45 a.m. (8050.58). Total number of intervals N = 38 at $\Delta t = 15$ min.

Fig.3. The changes of the attractor dimension d_f and intermittency exponent μ versus $\Delta\theta$ for abrupt increase of Sberbank index from 11:45 a.m. of October 8 (8050.58) to 10:45 a.m. of October 9 (9350). Observation interval $\Delta t = 15$ min. N = 30.

i.e. with negative values of the transferred potential. This phenomenon can be attributed to removal of the effective potential from the stock-market system.

5. CONCLUSION

The study of scaling properties for cascade structure self-organized system of the stock market was carried out. The fractal dimension and intermittency exponent were estimated using probability description of the system trajectory changes. The existence of short-term substructures with various values of life-time $\Delta\theta$ was revealed. The period doubling bifurcation was detected for the life-time $\Delta\theta$ of short-term substructures in the presence of sharp forcing caused by financial rating change. In succeeding days, for the take-profit stage, twofold decrease of the previous period, i.e. the bifurcation of the period halving was observed for the life-time $\Delta\theta$. So, a chain of Hopf bifurcation with double-increasing/decreasing of the period can be suggested as a basic mechanism of self-adjusting of attraction channel for short-term substructures to maintain the ordered development of the system.

REFERENCES

Cross M.C., Hohenberg P.C (1993). Pattern formation outside of equilibrium. *Rev. Mod. Phys.*, **v.65, No.65, p.II**, p.851-1090.

Feder J. (1988). *Fractals* Plenum Press, N.-Y.

Frish U. (1997). *Turbulence. The legasy of A.Kolmogorov.* Cambridge University Press.

Ghashghaie S., Breymann W., Peinke J. *et al.*(1996). Turbulent cascades in foreign exchange markets. *Nature*, **V.381**, p.767-770.

Kleparskii V.G.(2001) Multifractality and self-Adjustment of the attraction channel of stock market. *Automat. and Remote Control*, **V.62, No.4**, p.607-616.

Lux T., Marchesi M.(1999) Scaling and criticality in a stochastic multi-agent model of a finansial market. *Nature*, **V.397**, p.398-400.

Nelkin M. (1994) Universality and scaling in fully developed turbulence. *Adv. Phys.*, **V.43, No.2**, p.143-181.

Ruelle D. (2001) *Hasard et chaos*. R&C Dynamics, Moscow-Igevsk, 2001 (in Russian).

Temam R. (1988) *Infinite-dimensional dynamical systems in mechanics and physics.* Springer,N.-Y.

TECHNOLOGY DEVELOPMENT AND ETHICAL DECISION MAKING: IDENTITY FACTORS AND SOCIAL CONSTRUCTION

M.A. Hersh

Electronics and Electrical Engineering,
University of Glasgow, Glasgow G12 8LT, Scotland.
Tel: +44 141 330 4906. Fax: +44 141 330 6004. Email: m.hersh@elec.gla.ac.uk

Abstract: This paper investigates the use of feedback block diagrams to explore the relationships between identity, values and behaviour, as well as how these relationships, together with power dynamics, impact on technological and social development. The models are introduced by a discussion of values and virtue ethics and the relationship between technology and society. In addition a model of behavioural, and attitudinal change of individuals, organisations and societies is presented in terms of single, double, triple and quadruple loop action learning. Copyright © 2004 IFAC

Keywords: Technology, identity, values, minority status, modelling, feedback

1. INTRODUCTION

Technology development is one of the most important factors in shaping modern society, both in the richer industrialised countries, which fairly quickly experience new technologies, and the poorer so called developing countries, where access to new technologies is more restricted. Many of the scientists and engineers who are involved in the research, development and implementation of these new technologies often still consider themselves to be purely problem solvers and pay less attention to the nature of the problems they are solving, who has set them and whose interests the results will serve.

However there is also growing awareness of the importance of ethical decision making in science and engineering and interest in the development of tools to support it. This includes the codes of ethics or professional conduct, developed by many science and engineering societies (Martin et al., 1996; Hersh, 2000) and a variety of ethical theories, principles and methodologies. In many cases such theories and methodologies can be used to structure problems and highlight issues, but value judgements will be required to support ethical decision making.

Approaches such as the Johari window can be used to support engineers and other professionals in determining their own values (Stapleton et al., 2003).

However the technique will need to be modified in order to take account of the values of engineers who experience social exclusion, either for identity reasons such as being female or black, or due to approaches which are not part of the engineering mainstream. In this paper engineering modelling techniques and, in particular, feedback block diagrams will be used to explore the relationships between identity, values and behaviour, as well as how these relationships together with power dynamics impact on technological and social development. In deriving the models tradeoffs have been made between complexity i.e. deriving all relevant factors and comprehensibility i.e. having a model which is simple enough to be meaningful. This is the first in a series of papers in which engineering modelling techniques are used to explore the relationships between science, society, technology and power factors and some of the models will be refined and further developed in subsequent work.

The paper is set out as follows: Section 2 discusses values and virtue ethics and obtains a model for the relationship between behaviour, identity and values. Section 3 considers the role of single, double, triple and quadruple action learning in supporting change. Section 4 presents models of the relationships between science, technology, society and power dynamics and the role of marginalized or minority individuals and views in achieving genuine innovation and change.

2. MODELLING BEHAVIOUR, IDENTITY AND VALUES

2.1 Values

In some cases the requirements for ethical action in a particular situation are very clear. However in others applying different ethical theories, philosophies or approaches will clarify the issues, but value judgements will still be required to support decision making. There are a number of different sources of values, including religion, politics, humanist or other non-religious philosophies, education, family and friends, culture and the society you are living in. There also seem to be considerable differences between the ethical values of different societies and a wide range of different codes of values. However this does not mean that all possible codes of values are acceptable and ethical.

Unfortunately specifications to be met by ethical codes of values or tests to determine which sets of values are ethical have not yet been devised. Common elements have been noted in the value systems of very different societies and groups. For instance it has been suggested (Kluckholm, 1955) that every culture has a concept of murder and distinguishes murder from other types of killing which are not considered murder, and that every culture has some regulations about permitted and forbidden sexual behaviour. However the significance may be in the details, where there are often very great differences, rather than in the superficial commonality. Even within one society there are often significant differences of values, as evidenced by hectic debates about abortion, euthanasia and capital punishment.

2.2 Virtue Ethics

There are a number of different theories of ethics. Virtue ethics (Oakley, 1998) will be discussed here due to the feedback relationship between conduct and the development of 'virtuous' character. It supports actions which build good character. It differs from some other types of ethics, such as deolontological (concerned with duties and obligations) and consequentalisist ethics (concerned with consequences) in that the focus is the effects of the

action on the person carrying it out (and the relationship between action and character) rather than on the results of the action or particular obligations and rules. Thus virtue ethics assumes that the main ethical question concerns desirable character. It is based on the premise that a person with moral virtues is more likely to behave ethically than someone who purely follows rules. Behaviour often has an impact on character. Therefore virtue ethics is concerned both with the expected behaviour of a person with particular virtues and the type of behaviour which will promote the development of these virtues. There then remains the issue of the list of relevant virtues, which will depend on a number of factors, including culture, age and gender.

The strength of virtue ethics is its recognition that conduct has an effect on the person. This gives a feedback system, as illustrated in figure 1, in which ethical conduct has an effect on character and the development of virtues and these virtues lead to further ethical behaviour. Virtue ethics is also consistent with spiritually motivated approaches to ethics, since it could be considered to encourage personal and spiritual development through ethical behaviour. However even 'virtuous' people sometimes make mistakes or do things they regret and there is no universally accepted understanding of a 'virtuous' person, since, as discussed above, concepts of virtue depend on factors such as gender and culture.

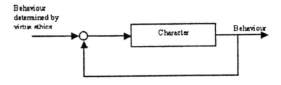

Figure 1, Ethical Behaviour and Virtuous Character

3.1 Model of Behaviour, Identity and Values

The simple model in figure 1 can be used to develop the model of behaviour, identity and values in figure 2. It will be assumed that identity status, including factors such as gender, race and ethnicity, class, sexuality, age and nationality, contributes to determining your values and that these values will determine your perspective on virtue ethics. The values of the dominant social group generally have an important role in influencing your values. There are a number of different ways of representing this influence and it is represented here as 'noise' to highlight the fact that it may be counter to the values, particularly of outsiders, and have the effect of distorting them. In addition there may be a struggle to maintain these values against outside pressure. The resulting virtue ethics then generates ethically desirable (virtuous) behaviour which is fed into real world dilemmas and situations and leads to actual behaviour.

This actual behaviour is also influenced by previous positive and negative experiences which generate fears, concerns and/or self confidence. Comparing actual and ethically desired behaviour can be used to modify actual behaviour by feeding the difference into the real world dilemmas and situations. Actual behaviour has an effect on character, either directly or through comparison with the desired virtuous behaviour. This modified character then generates a new set of values which can be fed back and

compared with the values generated by your identity, leading to a modification in your view of virtue ethics. This new set of values can then be fed back into the identity module and may lead to a change in how you perceive your identity or how it is perceived by the dominant social group or the minority group(s) you identify with. This is consistent with the fact that identity is not necessarily fixed but may be constructed through discourse and could be the temporary outcome of the powers, regulations and experiences encountered by an individual (Karreman et al., 2001).

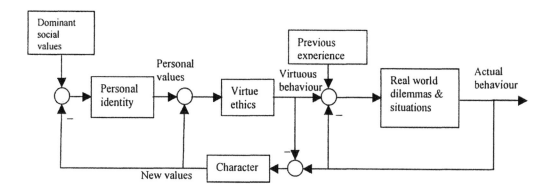

Figure 2, Model of Behaviour, Identity and Values

3. MULTI-LOOP ACTION LEARNING

There are a number of different theories, philosophies and methods which can be used to support ethical decision making (Babcock, 1991; Beauchamp, 2001; Beauchamp et al., 1978; Kuhse et al, 1998; Madu, 1996; Martin et al, 1996). However, once decisions have been made on what is ethical in the particular context, it will be necessary to implement them. In some cases action will require the involvement of other people and/or organisations and there may be institutional and other barriers. There is thus a need for methods for overcoming barriers to ethical action and persuading individuals and organisations of the value of such action. Some of the available methods have been categorised (Nielson, 1996) as single, double and triple loop action learning. Quadruple loop action learning can also be added, as shown in figure 3.

The following distinctions can be made (Nielson, 1996):

- Single loop action learning is about changing behaviour, rather than learning about ethics and changing values.

- Double loop action learning involves changes in values (generally of individuals) as well as behaviour.

- Triple loop action learning involves changes in the underlying tradition or ethos of the organisation, as well as changes in values and behaviour.

- Quadruple loop action learning additionally involves changes in the ethos or tradition of the surrounding society. Alternatively it involves changes in the underlying tradition with reference to the nature of the organisation, in addition to its practices.

In terms of a simple example:

- Single loop action learning could lead to measures to increase the proportion of disabled people recruited to senior positions, for instance due to fear of legal action on the grounds of disability discrimination, without any increase in awareness of the ethical responsibility to recruit more disabled people or a change in values.

- Double loop action could lead to a change in ethical values by some individuals in the organisation with a recognition of the ethical responsibility not to discriminate against disabled people, in addition to practical measures. This ethical commitment is likely to make the practical measures more effective than they would be otherwise.

- Triple loop action could lead to a change in the ethos of the organisation with a recognition of the value to the organisation and its ethical responsibility to employ more disabled people at a senior level. This could be accompanied by measures to overcome structural barriers and make the organisation attractive as a place of employment for disabled people.

- Quadruple loop action could lead to a change in the ethos of the wider society with a commitment to the value of diversity in society and ensuring equality and lack of discrimination. Disabled people would be considered one of the many diverse groups which enrich society. Measures would be taken to remove structural barriers and make all environments attractive and accessible to the whole population, including disabled people.

Alternatively, the difference between triple and quadruple loop learning can be illustrated as follows:

- Triple loop learning involves a particular organisation examining and discontinuing its behaviour of defrauding the Ministry of Defence by charging for materials that have not been used on the Ministry of Defence project.

- Quadruple loop learning involves the organisation examining the whole context of military contracts and deciding to have nothing to do with them.

Real technological and social change requires triple and quadruple loop learning. It has been suggested that real change or innovation happens at the margins. This is where women and minorities are situated. Therefore triple and quadruple loop learning is required to change the ethos of organisations (triple loop learning) and of society as a whole (quadruple loop learning) to allow us all to benefit from the knowledge and expertise that is currently sited at the margins. Nielson (1996) suggests some methods that can be used to achieve triple loop learning, but they will not always work. The problem of quadruple loop learning or changing deep seated attitudes in society as a whole is generally even more difficult to resolve. In practice change occurs slowly and not necessarily linearly. It should also be noted that once such change occurs the margins will have shifted and new sources of creativity and learning will be required to achieve significant change and innovation. This gives an iterative process which should converge to a state in which significant change is no longer desirable.

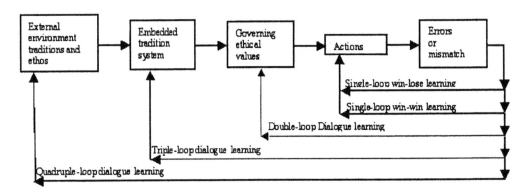

Figure 3 Single, Double, Triple and Quadruple Loop Learning

4. TECHNOLOGY AND SOCIETY

There has been considerable discussion of the relationship between society, technology and science, but power relations have rarely been mentioned explicitly in mainstream advanced technology literature. One perspective considers technology to be neutral in itself and its consequences to be determined solely by the nature of particular applications. An almost diametrically opposed perspective, technological determinism (Ellul, 1954; Winner, 1977), considers technology to be all-powerful.

In the strongest versions of this perspective technology totally determines the future directions of society in ways that are not possible to resist. Although useful, both these perspectives are too simplistic. In particular they ignore the power relations and dynamics that effect choices about what technology is developed, how it is used and in whose interests it is deployed. These are highly complex processes that are difficult to address according to the positivism underpinning current engineering research (Jervis, 1997). Technology design and development are influenced by existing power structures and contribute to developing and further institutionalising particular structures (Baudrillard, 1999; Borgman, 1984).

Figure 4 illustrates some of the relationships between science, technology, society and power dynamics. Rather than either technical or social determinism, it is assumed that there are feedback relationships between technology and society and that developments in both society and technology influence existing power relationships, as well as science. In order not to overcomplicate the model a number of factors have not been shown. For instance, the unconscious, or deliberate attempts to impose the economic, political and ideological structures in which this technology developed have not been included in the model. This can be considered a form of colonisation through technology, which is subtler, but no less insidious than previous attempts (Banerjee, 2001).

There has been some discussion in the previous section of the fact that change occurs at the margins. This means that real technological innovation which occurs in new directions, rather than purely continues with more of the same, requires the involvement of women and minorities.

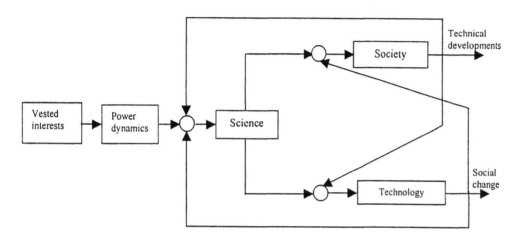

Figure 4, Feedback Relationships Between Power Dynamics, Science, Technology and Society

However there are gatekeeping processes which are used to maintain orthodoxy through restriction of access to resources and publication in respected journals to individuals who are considered to conform and who present ideas or projects within the canon. As a consequence, indigenous knowledge, for instance, of edible plants, is disappearing or even being suppressed, since it is not recognised as valid or authoritative (Ilkkaracan and Appleton, 1995). The mechanisms by which this occurs are different in different contexts and include the lack of transparency and gender and race discrimination in the peer reviewing process for academic journals and the deliberations of research councils and other funding bodies. Gatekeeping processes can then be considered to act as a filter on innovation from minority or marginalised researchers and paradigms. Figure 5 illustrates the way in which creative development and innovation at the margins is to a certain extent in competition with more mainstream developments which tend to result in more limited change within existing paradigms.

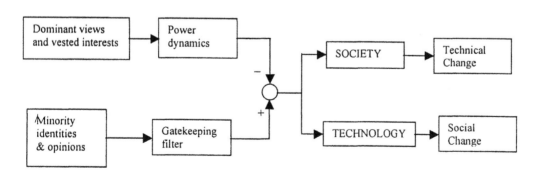

Figure 5, Innovation: The Role of Minority Identity and Views

Facilitating the involvement of minorities and minority opinions and paradigms will require organisational and societal change i.e. triple and quadruple loop learning, in terms of the multi-loop learning model presented in the previous section. Combining the multi-loop learning model with the models of society, technology, change and innovation in figures 4 and 5 could have interesting conclusions, but this will form the subject of further work.

5. CONCLUSIONS

This is the first in a series of papers using engineering modelling techniques to explore the relationships between science, technology, society and power dynamics. The focus in this paper has been the use of feedback and block diagrams to explore the relationships between identity, values and behaviour, as well as how these relationships and power dynamics impact on technology and society. These models have been introduced by a discussion of values and virtue ethics and the relationship between technology and society and the impact of power dynamics on this relationship.

Drawing on the work of Nielson (1996), a model has been presented of multi-loop individual, organisational and societal learning and change. It has also been argued that real innovation requires the involvement of minority and marginalised individuals and paradigms and that this will require change at the third and fourth loop level. These arguments will be taken further and the relationship between the different models further developed in subsequent work.

REFERENCES

Banerjee, R. (2001). Biodiversity, biotechnology & intellectual property rights: unpacking the violence of 'sustainable development', *19th Standing Conference of Organisational Symbolism*, Dublin.

Babcock D. L. (1991). *Managing Engineering and Technology*, Prentice-Hall.

Baudrillard, J. (1999). *The Consumer Society: Myths and Structure*, Sage: Thousand Oaks.

Beauchamp, T.L. (2001). *Philosophical Ethics*, McGraw Hill Higher Education.

Beauchamp, T.L. and W. LeRoy (eds.) (1978). *Contemporary Issues in Bioethics*, Chapters 1 and 2, Dickenson Publishing Co. Inc.

Borgman, A. (1984). *Technology and the Character of Contemporary Life: A Philosophical Enquiry*, Univ. Chicago Press.

Ellul, J. (1954). *La Technique ou l'Enjeu du Siècle*, Librairie Armand Colin, Paris.

Hersh, M.A. (2000). Environmental ethics for engineers, *Engineering, Science and Education Journal.*, **9(1)**, 13-19.

Ikkaracan, I. and H. Appleton (1995). *Women's Roles in Technical Innovation*, Intermediate Technology Publication.

Jervis, R. (1997). *System Effects: Complexity in Political & Social Life*, Princeton University. Press, NJ.

Karreman, D. and M. Alveson (2001). Making newsmakers: , *Organisational Studies*, **22(1)**, 1-24.

Kluckholm, C (1955). Ethical relativity: sic et non, *Journal of Philosophy*, **55**, 663-667.

Kuhse, H. and P. Singer (eds.) (1998). *A Companion to Bioethics*, chapters 1-12, 40 and 41, Blackwell Publishers.

Madu, C. N. (1996). *Managing Green Technologies for Global Competitiveness*, Quorum Books.

Martin, M.W. and R. Schinzinger (1996). *Ethics in engineering* (3rd ed.), McGraw-Hill

Nielson, R.P. (1996). *The Politics of Ethics*, Oxford University Press.

Oakley, J. (1998). A virtue ethics approach, In: Kuhse, H. and P. Singer (eds.), *A Companion to Bioethics*, 86-97, Blackwell Publishers.

Stapleton, L. and M.A. Hersh (2003). Exploring the deep structure of ethics in engineering technology: *SWIIS '03*, Waterford, Ireland.

Winner L. (1977). *Autonomous Technology, Technics-out-of-control as a Theme in Political Thought*, MIT.

Acknowledgements: I would like to thank Prof Mike Johnson of the University of Strathclyde for drawing the graphics.

THE GLOBAL VALIDITY OF ETHICS: APPLYING ETHICS TO ENGINEERING AND TECHNOLOGY DEVELOPMENT

Balázs Bitay * and Dietrich Brandt**

* Industrial Engineering and Management
University of Technology and Economics
Budapest, Hungary
balazs.bitay@innovationcentre.hu

**Dep of Computer Science in Mechanical Engineering (ZLW/IMA)
University of Technology (RWTH), Aachen, Germany
brandt@zlw-ima.rwth-aachen.de

Abstract: Recently several attempts have been made to define and implement new codes of ethics in dealing with technology today and tomorrow. Thus the *global validity of Ethics* under the impact of new Technologies has become visible as an issue widely discussed within engineering, philosophy and society. It requires increasingly the awareness of our own responsibility towards all humans as well as towards the natural environment and towards future generations. The issue has been taken up by the European Commission in recent statements and projects. As an example, it has been suggested to create a network supported by a systematic information tool, on ethical issues in science and technology. It would offer access to information in various languages on legislation, codes of conduct, best practices, and the ethical debates taking place in different European countries. These and other examples of such strategic approaches towards ethics of technology will be described and discussed in this presentation. *Copyright © 2004 IFAC*

Keywords: Business, engineering, ethics, ethics networks, Hippocratic Oath

1 INTRODUCTION: THE QUESTIONS OF ETHICS

"The Principles of a Global Ethic:
Our world is experiencing a fundamental crisis: A crisis in global economy, global ecology, and global politics. The lack of a grand vision, the tangle of unresolved problems, political paralysis, mediocre political leadership with little insight or foresight, and in general too little sense for the commonwealth are seen everywhere: Too many old answers to new challenges" (Declaration, 1993).

What is right or wrong to do? How do we decide about right or wrong? What is the *Ethics of our Actions*? These are the questions that come up when thinking about Ethics. In this presentation, we are going to focus on *Technology Ethics* and its connection to business, in particular.

Not every situation from which we gain short term win, is sustainable with regard to interpersonal or intercultural settings. The questions of how to decide and how to act, are the questions of ethics. But not only philosophers have to take up the challenge answering them. It is the duty of every person him- or herself to consider and discuss processes of everyday decision finding, be it in engineering, entrepreneurship and business, science and research, university teaching or any other activity. (Rose, 2003)

The far aim is to develop a common frame of reference, a code of *universal ethics*. The ethical way of acting means: unlimited liability towards our actions in the broadest sense. It involves - without being complete - environmental consciousness, human-centredness in technology development and other fields of action; responsibility in our actions toward next generations, social responsibility; furthermore sustaining, growing and passing on civilisation and knowledge, accepting mistakes but striving towards excellence; freedom of actions, and accepting everybody as being equal.

2 THE HISTORY OF ETHICAL CODEX

2.1 The Hippocratic Oath - classical version

It is amazing how early humankind started to think about ethics. The first well-known written example which relates to ethics, dates back to Hippocrates. He created *the Ethical Code of Medicine* which has had a tremendous effect on medicine and science. By creating this set of principles, he turned medicine from a superstitious, magic activity - as it was believed to be in those times - into science. And by doing so he started a process which has led towards the wide acceptance of sciences.

"I swear by Apollo Physician and Asclepius and Hygieia and Panaceia and all the gods and goddesses, making them my witnesses, that I will fulfil according to my ability and judgement this oath and this covenant: To hold him who has taught me this art as equal to my parents and to live my life in partnership with him, and if he is in need of money to give him a share of mine, and to regard his offsprings as equal to my brothers in male lineage and **to teach them this art - if they desire to learn it - without fee and covenant; to give a share of precepts and oral instruction and all the other learning to my sons and to the sons of him**...

I will neither give a deadly drug to anybody who asked for it, nor will I make a suggestion to this effect. Similarly I will not give to a woman an abortive remedy. Whatever houses I may visit, I will come **for the benefit of the sick**, remaining free of all intentional injustice, of all mischief and in particular of sexual relations with both female and male persons, **be they free or slaves. What I may see or hear in the course of the treatment** or even outside of the treatment in regard to the life of men, which on **no account one must spread abroad**, I will keep to myself, holding such things shameful to be spoken about.

If I fulfil this oath and do not violate it, may it be granted to me to enjoy life and art, being honoured with fame among all men for all time to come; if I transgress it and swear falsely, may the opposite of all this be my lot."

In the following we will examine to what extent the Hippocratic Oath lives up to the ideas we have about ethics nowadays, as mentioned above:

"to teach them this art - if they desire to learn it - without fee and covenant; to give a share of precepts and oral instruction and all the other learning to my sons and to the sons of him" This part relates to passing on civilisation and knowledge.

"I will neither give a deadly drug to anybody who asked for it, nor will I make a suggestion to this effect. Similarly I will not give to a woman an abortive remedy." It means: being responsible toward next generations, and human-centredness.

"for the benefit of the sick... be they free or slaves." Freedom of actions, and accepting everybody as being equal.

"What I may see or hear in the course of the treatment ... on no account one must spread abroad" Human rights, not giving out personal information which is very much in the centre of discussions nowadays.

"If I fulfil this oath and do not violate it, may it be granted to me to enjoy life and art, being honoured with fame among all men for all time to come" Finally, this relates to maturity: in the sense of leadership, not to misuse any power we may have.

As we can see, Hippocrates was a real pioneer in the field of ethics, and many of his original thoughts are still followed nowadays. This document has laid the foundation of, and is still the basis of the current ethical code of medicine. Now, it is interesting to examine how did (or did not) ethical standards change in the course of the last 2000 years.

2.2 The Hippocratic Oath - modern version

The new and more modern version quoted here in abridged form, was written in 1964. The author was Louis Lasagna, Academic Dean of the School of Medicine at Tufts University, and it is used in many medical schools today.

"I swear to fulfil, to the best of my ability and judgement, this covenant: I will respect the hard-won scientific gains of those physicians in whose steps **I walk, and gladly share such knowledge as is mine with those who are to follow...** I will remember that there is art to medicine as well as science, and **that** warmth, sympathy, and understanding may outweigh the surgeon's knife or the chemist's drug.

I will not be ashamed to say "I know not," nor will I fail to call in my colleagues when the skills of another are needed for a patient's recovery. I will respect the privacy of my patients, for their problems are not disclosed to me that the world may know. Most especially must I tread with care in matters of life and death. If it is given me to save a life, all thanks. But it may also be within my power to take a life; this awesome responsibility must be faced with great humbleness and awareness of my own frailty. Above all, I must not play at God.

I will remember that I do not treat a fever chart, a cancerous growth, but a sick human being, whose illness may affect the person's family and economic stability. My responsibility includes these related problems, if I am to care adequately for the sick. I will remember that I remain a member of society, with special obligations to all my fellow human beings, those sound of mind and body as well as the infirm."

When comparing the classical and modern versions, we note the following aspects:

"I walk, and gladly share such knowledge as is mine with those who are to follow." It is interesting to see the effect of the increased need for financial welfare, the idea of sharing knowledge has stayed on, but doing it "without fee and covenant" is not in the text anymore! In the same time, the awareness of financial issues is also shown in the following quote:

"I do not treat a fever chart, a cancerous growth, but a sick human being, whose illness may affect the person's family and economic stability"...."Warmth, sympathy, and understanding may outweigh the surgeon's knife or the chemist's drug": This part proves the increasing need for human-centredness.

"I will not be ashamed to say "I know not," nor will I fail to call in my colleagues when the skills of another are needed for a patient's recovery." This aspect may come from the broadened scope of medicine and the trend of world economy toward specialisation: one cannot get hold of all the information and knowledge needed, and we all have to be aware of sharing them. Leaders have big responsibility to confess that they cannot know everything and do *not have to know* everything. Leaders as humans are allowed to make mistakes. It is not ethically acceptable, though, not to confess mistakes and not to learn from them, or to hide information, endangering the welfare of all when doing so (we may think of atomic disasters, as an example). The last part of this oath also relates to this

aspect: "I will remember that I remain a member of society, with special obligations to all my fellow human beings".

3 ENGINEERING ETHICS

3.1 The Institution of Civil Engineers, U.K., 19th Century

Being engineers and scientists, we briefly introduce the engineering ethics which was designed many centuries later. An example of what we may regard as a *Code of Ethics of Engineers*, is the poem which was found by the Quaker Edmund C Hambly in 1995, the youngest President the Institution of Civil Engineers had ever had. He struggled too hard, suffered a heart attack and died, half-way through his presidency. Only weeks before he did so, half-way across India, visiting a construction site in the middle of nowhere, as part of his presidential visit to meet engineers in India, he found this text on the wall of a site engineer's hut, (cited from Jim Platts: Meaningful Manufacturing, 2003):

"I take the vision which comes from dreams
and apply the magic of science and mathematics
adding the heritage of my profession
and my knowledge of Nature's materials
to create a design.
I organise the efforts and skills of my fellow workers
employing the capital of the thrifty
and the products of many industries,
and together we work toward our goal
undaunted by hazards and obstacles.
And when we have completed our task
all can see
that the dreams and plans have materialised
for the comfort and welfare of all.
I am an Engineer.
I serve mankind
By making dreams come true."

It is an ethical code of a total different profession than medicine, and it comes from a country with a totally different background. But it basically shares the same concerns as the Hippocratic Oath:

"Fellow workers" and "together we work toward our goal" refers to the freedom of actions, being equal. "I serve mankind by making dreams come true": this idea involves growing civilisation, and human-centredness. "For the comfort and welfare of all" goes very much in line with being socially responsible, including responsibility towards the next generations.

We are more and more realising nowadays that "I serve mankind by making dreams come true" has to be carefully considered: people have various kinds of needs which today's business and engineering world can easily satisfy, but satisfying such needs does not always contribute to *sustainability*. Thus it is no longer enough for engineers to satisfy these needs, but they have to carefully think about the future impact of their inventions.

3.2 The Association of Engineers, Germany, 20th Century

Another example is the *Fundamentals of Engineering Ethics* suggested by the *Association of Engineers VDI*, Germany, in 2002. They offer to all engineers, as creators of technology, orientation and support as they face conflicting professional responsibilities. Some paragraphs from this document are following here:

"Engineers carry both individual and shared responsibilities. They are responsible for their professional actions: to the community, to political and societal institutions, to employers, customers, and technology users.
Engineers are aware of the societal, economic and ecological context: usability and safety, the welfare of the citizens, and the lives of the future generations.

In cases of conflicting values, engineers give priority: to the values of humanity over the dynamics of engineering, to issues of human rights over technology implementation and exploitation, to public welfare over private interests, and to safety and security over functionality and profitability of their technical solutions.

Engineers, however, are careful not to adopt such criteria or indicators in any dogmatic manner. They seek public dialogue in order to find acceptable balance and consensus concerning these conflicting values." (VDI, 2002)

Again we find in these quotations the same elements of striving for communication and co-operation among equal partners as in the previous documents. In particular the last paragraph deals explicitly with the issue of finding balance and consensus in the public dialogue – a serious indicator for changes of views within the engineering and science community. Here it is important to point to the individual and shared responsibilities, and the usability and safety issue, which are interrelated in terms of technology. As an example, in case of complex technology failure (like a space shuttle accident) it is hard to find out why the failure occurred: was it the material, the way it was put together, or human failure, or a combination of more factors. This is meant by the concept of *shared responsibilities*.

3.3 The Memorandum Information and Communication

The application of these concepts to *engineering action* is described in the following section of this paper as it was discussed during the World Engineers Convention, Hannover, Germany, 2000. This discussion led to the Memorandum *Information and Communication*. Here some points are quoted from this memorandum:

Global versus Regional Development: Law and Governance
The main trend today is towards globalisation: all communications and transactions take place within world-wide dimensions. To make the world a really global world, however, we need to reconcile national laws at the international level as well as mechanisms to enforce them, without succumbing to any dominant perspective. Hence we need to contribute to more political control of technology-triggered developments through making more information available to all citizens.

Entrepreneurship on Different Scales: Economics and Business
Networking on the *global* "macro" scale is leading to both strong economic co-operation and mutual dependencies of large enterprises and countries. The liability and responsibility of global enterprises is no longer towards any specific country or people. There seems to be no control or governance possible through any single country. In parallel, *regions* are challenging traditional national politics by developing their own political momentum. We need to create awareness for, and adapt policy to the joint design and implementation of technological, political and organisational renewal.

Data Availability versus Data Security: Transportation and Processing of Data
Today, all information on the technological networks is available to everybody. But the misuse of the web and the breaking of data security are well known. Insufficient data reliability, trustworthiness and dependability are increasingly becoming a global problem. Data availability and data security are contradicting challenges on both the technical and organisational level. A new security culture needs to be developed concerning all developers and users.

The Ethics of Multimedia Information and Internet-based Action
The calling-up and exchanging of information and pictures have proved their importance and necessity in personal life as well as in many fields of research, business, politics etc.. There is, however, the freedom of storing and sending all those pictures which symbolise the harmful or abusive side of human life (e.g. pornography, racism, violence and violent

games etc.) It is unethical to transmit consciously and purposefully abusive, wrong and misleading information. Hence the individual responsibility of the engineer needs to be reinforced by a professional code (like the Hippocratic Oath). Furthermore we need to discuss the ethics of information and pictures in view of the *cultural pluralism* of countries, their different traditions and value systems while avoiding to establish any *one value system* across the world. (Memorandum, 2000)

The quotations demonstrate how engineers today are aware of the needs and challenges put forward by Ethics, and the claims for universal validity of values implemented in engineering. In each of the quotes, the attempts become visible to organize engineering activities as co-operative endeavours based on *communication and negotiation among equals*. The activities are meant to also include *society* on equal terms. We all know that such endeavours are only successful as far as the principles of this communication are taken seriously, and we all know how again and again these endeavours fail. Nevertheless the attempts are signals of attitude changes which in the long run may become common practice.

4 BUSINESS ETHICS TODAY

4.1 Russia today

Here follows an example from Russia, presented at the conference on *Business Ethics* in Saint-Petersburg, May 18-19, 2000, by Yelena Dotsenko, representative of the Consortium of Legal Information Codex:

a) Regarding our staff: adherence to the corporate culture of business conduct; consideration for each individual employee's potential; fair compensation for work output; special benefits; creation of opportunities for professional growth etc.
b) Regarding our customers: the prohibition of directly or indirectly misleading our customers with distorted information about product quality and the manipulation of independent research data.
c) Regarding our competitors: the prohibition of receiving undue advantages over other legal systems market participants. Competition should be based on increased work quality and professionalism within legal regulations.
d) Regarding our partners: the prohibition of violating the legal systems market stability and the legal rights and interests of other market participants. There should be no damage to the interests of clients or other participants in this

market. Trust, credibility and liability should be guaranteed. (Dotsenko, 2000)

It is a widely shared view in the world that Russia has not been performing well in the fields of ethics; the current trends seem to be positive, though. But it is interesting to see that the document is rather based on, and formulated in a way to state *what not to do*. This may mean that Russia is realizing the need to catch up first to the minimum standards, which proves a realistic, long-term approach.

4.2 Western business ethics

It is all very well for consultants to urge entrepreneurs to run business ethically. They don't have to live with the consequences. Do they even know the consequences?

In the course of our ethics considerations, we have run across some very provocative questions concerning ethics and its bottom line. Here are a few:

For Owners/Managers:
1. Can we afford to do business ethically? Won't we lose business to external competitors?
2. How do we ensure that our employees will follow the company's ethical policies?
3. How do we react to unethical behaviour on the part of employees?

For Employees:
1. Will our superiors back us if we choose to behave ethically?
2. What do we gain from ethical behaviour?
3. Will our co-workers (internal competitors) gain competitively if we decide to behave ethically?

These are difficult questions, and they must be answered to the satisfaction of both management and employees before any organization can do business ethically. We do not presume to know how to answer these questions. Yet we can assist in determining answers.

4.3 What about actions?

All theory gains impact only when it is put into practice. Frank Vogl, Senior Ethics Resource Center 'ERC' Fellow, came up with some suggestions how to realize ethical actions in the field of business: what are the core strategies needed for achieving it? According to him, corporations need to consider five sets of actions:

1. Learning a lot more about global business ethics.
2. Developing effective in-house corporate global business ethics strategies.
3. Creating internal corporate coalitions involving the corporate Ethics Office, Human Resources,

Government Relations, Corporate Communications and Legal Department, to implement new global business ethics programs.
4. Securing dialogue with civil society groups in all the countries in which you operate.
5. Monitoring the impact of your new global business ethics strategies, and monitoring the rising expectations around the world of public expectations of corporate behaviour. (Vogl, 1998)

We cannot afford *not to do business ethically* if the organizational goal is long-term success. It is also counter-productive to assume that competition is unethical.

Marvin Bower (2003) from McKinsey suggests that there are three primary advantages over competitors whose ethical standards are lower.

1. A business of high principle generates greater drive and effectiveness because people know that they can do the right thing decisively, they can rely on ethical principles. So decisions can be met faster.
2. A business of high principle is more likely to attract high caliber people, thereby gaining a basic competitive and profit edge.
3. A business of high principle develops better and more profitable relations with customers, competitors, and the general public because it can be counted on to do the right thing at all times and in the same time they build favourable image. The general public is also more likely to be open-minded towards its actions.

5 CONCLUSIONS: SUGGESTIONS BY THE EUROPEAN COMMISSION

Recently the European Commission has taken up the issue of Ethics to be discussed, and developed further across all European countries. Here follow two short quotations from their *Science and Society Action Plan* which symbolize the direction of thoughts in Europe although there is still a long way to go before these perspectives will be part of daily practice in engineering and business.

"Put responsible science at the heart of policy making: Most policies have a scientific and technological dimension and decisions must be supported by transparent, responsible opinions based on ethical research. It is therefore necessary to strengthen the ethical basis of scientific and technological activities, to detect and assess the risks inherent in progress, and to manage them responsibly on the basis of past experience...

As recommended by the European Parliament, researchers, business circles, standard-setters and social players need to be encouraged to enter into a public dialogue across Member States and the Candidate Countries on the new leading-edge technologies as soon as they begin to emerge. This will enable responsible choices to be made, supported by the appropriate policies and implemented at the right time. The European Group on Ethics has helped guide the Community policies on culturally sensitive ethical questions in science.

(The European Group on Ethics in Science and New Technologies is an independent, pluralist and multidisciplinary body which has been set up by the European Commission to give advice on ethical aspects of science and new technologies in connection with the preparation and implementation of Community legislation or policies)

Networks of ethical committees will be fostered at both national and local levels. The aim will be closer co-operation and a more effective exchange of experience and best practice." (European Commission, 2002, p 8 and p 21, 22)

REFERENCES

Bower, M. (2003): *Business Ethics.* McKinsey, 2003

Declaration (1993): Declaration Toward A Global Ethic. The Parliamant of the World's Religions, Chicago 1993.

Dotsenko, Y. (2000): Making the Choice to be Ethical. Conference on Business Ethics, St. Petersburg, May 18-19, 2000: www.ethicsrussia.org.

European Commission (2002): Science and Society Action Plan. Office SDME 06/62, Bruxelles http://www.cordis.lu/science-society.

Hambly, C. (1995) - in: Platts, M.J. (2002): Meaningful Manufacturing. Yorkshire Press, p 22

Lasagna, L. (1994): pbs.org/wgbh/nova/doctors/oath_modern.html.

Memorandum (2000): *Information and Communication - The Memorandum. Congress Information and Communication.* VDI World Engineers' Convention, June 19-21, 2000, Hannover, Germany.

Rose, C. (2003): The Concept of Universal Ethics in Global Networking. In: *Human-Centred System Design - First: People, Second: Organisation, Third: Technology. 20 Case Reports (*Brandt, D. (Ed.)), ARMT 42, Aachen 2003, 173-180.

VDI - The Association of Engineers (2002): *The Fundamentals of Engineering Ethics.* Dusseldorf, Germany, 2002.

Vogl, F. (1998): Introductory Global Business Ethics Remarks. Meeting of the Ethics Resource Center's Fellows Program, Nov 5/6, 1998: http://www.ethicsrussia.org/voglremarks.html.

www.elsevier.com/locate/ifac

A COMPLETE SOLUTION FOR MOTION PICTURE RESTORATION: SCANNING, PROCESSING, RECORDING

G. L. Kovacs*, I. Kas, S. Manno

*Computer and Automation Research Institute, Hungarian Academy of Sciences,
1111-Budapest, Kende u. 13-17. Hungary, gkovacs@sztaki.hu
and Technical University of Budapest and University of Pécs

L. Czuni, T. Sziranyi, A. Hanis

*Department of Image Processing and Neurocomputing, University of Veszprém, Hungary
8200 -Veszprém, Egyetem u. 10. Hungary, czuni@almos.vein.hu*

Abstract: A digital restoration system (**DI**gital **MO**tion Picture **R**estoration System for **F**ilm Archives (DIMORF)) is introduced, including all main stages of restoration, as scanning, processing and recording. The system produces high-resolution restored film as final result to meet the request of film archive specialists. The film-scanner and the -recorder work with a maximum resolution of 6K. The scanner uses line cameras, while the laser recorder has a rotating optics to project RGB laser-light onto the film positioned on the inner surface of the recorder drum. The PC-based digital processing unit is responsible for restoration. Built-in film correction includes: stabilization, de-flicker, blotch detection, scratch removal, de-noising, color manipulation, de-fade effects. Interactive user support and intelligent, adaptive algorithms help semi-automatic operation. *Copyright © 2004 IFAC*

Keywords: film restoration, scanner, recorder, software, network, artificial intelligence

1. INTRODUCTION

In the Hungarian National Film Archives (HNFA), similarly to other film archives of the world, there are thousands of films to be saved from final perdition. DIMORF is a product of efforts to create a relatively low cost system capable to handle all degradations, especially those typical in the HNFA. The main objectives were to create a system that meets current and future needs of film archivists:

- To handle all major types of film errors
- To create an open interface for future algorithms
- To reach 6K resolution as an upper limit for high-resolution films
- To make the application user friendly and to support the work of operators with easily understandable digital report files
- To enable future data-mining and intelligent process control where XML data structures record all user interactions

Special requirements of the HNFA (and probably of other archives), regarding the restoration of a large number of archive films, are the following:

- Usually there is only one remainder copy of an artwork in rather bad or critical physical condition. This means that only one scanning procedure is possible.
- Due to the long restoration time (at high resolution) continuous operator intervention is not possible. The components of the systems must operate parallel in semi-automatic mode, with occasional manual interventions for fine-tuning.
- The sequence of the restoration process elements can be greatly varying but it is roughly according to the following schedule:

  a) Visual investigation to decide the degree of degradation and the necessary pixel resolution since most archive films do not require 6K.

Library work - to know what to expect;

Physical cleaning and some repairing;

Determining and setting the scanning parameters (lighting, physical scratch removal by liquid-gate, application of infrared scanning);

Scanning.

b) Estimating and planning the necessary restoration tasks.

Defining main cuts, marking main degradation places and features (approximate position, noise level, scratches, etc.);

Starting automatic restoration.

c) Periodic manual interventions: supervision of film reconstruction, fine tuning of parameters, ROI selection, defining further batch filtering.

d) Last step of the restoration cycle: final qualification leads to new iterations or to the decision to finish the digital restoration work and to transmit data to the recorder.

e) Preparation for film writing: the method of frame enlargement (blowup to 6K with special methods giving natural grainy sensation) is to be determined, final grading according to the raw film material.

- All steps of the restoration process are recorded in XML files to achieve traceability and reversibility. With the help of these data the restoration can be reproduced at different scales and further analysis is possible.

Many of the above requirements are related to the problem of high costs of manual film restoration, the most relevant problem of the restoration industry (Kovacs, 2002). That is why the proposed algorithms and software solutions support an open, expandable, and well-structured system driving towards built-in intelligence and automatisms.

2. SYSTEM DESCRIPTION

2.1 Looking Around

As there is a real request to restore and save old movies serious efforts were put into research and development to get appropriate scanners, film writer units and restoration software solutions. Most of these solutions in Japan, in Western Europe and in the USA are very specific (as the CINEON system of KODAK used in the United States) and no detailed information is available on them, and they are not sold as systems, only sometimes as system units (scanner, recorder, plenty of software), but these are extremely expensive. Only the FRAME system (Joanneum, 2002) offers services – on high prices. There is a joint European R&D project (PrestoSpace, 2003), which recently started, and offers partially similar solutions to those we are fighting for. When the DIMORF project was

started almost 3 years ago we made calculations, according to which – partially due to our cheaper labor prices – our total project costs are less than the restoration costs of 5-10 movie films using the FRAME or any similar service or system.

2.2 Previous Work in Analogue Film Copying

A film copy machine was developed by our team to save films by means of copying them onto a new safety film material, making only simple restoration (Kovács, 2002-3). It applies high diameter vacuum-wheel film driving, instead of perforation needles, while the precise, CCD sensor controlled, repositioning of the transfer optics compensates film shrinkage. A new type of digital sound restoration system was developed at the same time (Bisztray, 2002).

2.3 System Architecture

The system consists of 5 main components described below.

Very high-resolution film scanner. The film scanner utilizes 3 units of 6K line CCDs (manufactured by Kodak) with top sensitivities at 450, 550, and 650nm. Light is supplied with a halide incandescent lamp through fiber optics. Pixel data are transferred in three independent data channels to achieve reliable very high bandwidth transfer at 6K resolutions. Besides the line sensors, a low-resolution (PAL) camera is built in for fast online monitoring of the scanning process. Scratch removal is supported with an infra sensor sensible to light from 700nm to 1300nm. High resolution scanning of a 35mm frame is below 10sec and at a 3 times reduced resolution (the same device in a different operating mode) it is below 3sec. Additional CCD sensors are responsible for the detection of the position of perforations.

Digital film restoration workstation(s). To achieve high compatibility and low cost, film restoration algorithms are implemented on the Intel-based PC architecture running MS-Windows operating system. Software applications support multithreading technology thus multi-CPU platforms are proposed. The digital restoration project and its operations are hierarchically organized into Projects, Tasks, Jobs and Filters. This opens the gate for further extensions to multi-platform processing with the help of Gigabit Ethernet network.

Sound processing unit. Sound is processed independently from images via proprietary digital optical technology (Bisztray, 2002). Conventional sound processing is also possible for further sound reconstruction.

Film recorder. In the film writer a high-speed rotating mirror projects modulated laser beam onto unexposed film positioned on the inner surface of the recorder drum. The film is fixed in the inner arc (of the drum) by vacuum to guarantee mechanical stability and accuracy. Laser light is produced with Point Source RGB lasers at 405, 532, and 640nm.

Information storage and data network. Storing high-resolution film data requires huge amount of disk space. To keep costs low RAID controlled EIDE disks are used in a dual CPU PC host. All devices are connected to this Terabyte capacity host via Gigabit Ethernet network.

2.4 Recent status of the work

The scanner and the digital restoration unit are already in laboratory service while the film recorder is just being manufactured at the time of submitting this paper. Moreover the restoration software is in everyday use as we take part in the restoration of the first Hungarian color movie the Ludas Matyi (1949). This work is supported by the HNFA and our working partner is the Hungarian Film-laboratory Ltd. with several years of experiences in film restoration. This joint project demonstrates all problems of an old, worn out nitro-cellulose film with about 3% color-content, with several physical failures, etc. on one hand and the strengths of some of our restoration software solutions on the other hand.

3. DIGITAL RESTORATION

DIMORF has the ability to correct several types of film errors, as:
- Film vibration - Flickering - Blotch removal
- Noise - Scratches - Color fading

Recent results of restoration algorithms have been presented in Kovács L. (2002) and Kato (2002). Great impact has been made to achieve automatic operations. For this purpose film analysis is supported by cut detection and by film indexing based on color information and motion activity. Data representation in XML is also aiming to create well-defined and controllable processes.

In the next sections automatic stabilization, indexing, cartoon rendering is discussed in some details. The restoration environment is also introduced at some level.

3.1 Automatic stabilization

Eliminating the vibration of a degraded image sequence causes difficulties in an automatic film restoration system. The vibration is usually caused by improper film transportation during the recording,

copying or the digitization process. We have developed an automatic method for image stabilization consisting of two main steps: estimating vibration then correction by drifting the whole frame. Earlier stabilization algorithms are unsuccessful in cases of multiple motions and human interaction is necessary to achieve satisfactory results. Our algorithm is automatic, robust on noisy films and avoids false results for most difficult sequences.

The observed vibration of films can be very complex since camera ego-motion can also cause vibration having serious effects on the 2D projection of 3D sceneries. There are cases where the whole image cannot be tested for estimating the stabilization parameters, since the scene may contain multiple, complex object motion. Consequently, recent systems need manual selection of a base point or an object for adequate stabilization. This manual task is time consuming so the aim of the proposed method is automatic vibration stabilization without any manual interaction. This is achieved by means of automatic ROF (Region of Fixation) selection in the image sequence. The motion information of these selected ROFs will be applied for the stabilization of the images.

We combined the phase correlation motion estimation method in Hanis (2003) with a top-bottom image splitting algorithm based on motion activity. This motion estimation method is relatively insensitive to fluctuations in image intensity (flicker and blotches are very typical for archive films). For finding ROFs in a scene the images are divided into sub-regions in a quad-tree manner. If the motion observed on whole image can be well characterized with global motion then the proposed method uses only the first level of the quad-tree (i.e. the whole image). In case of complex scenes images are divided into sub-regions due to local objects motions. Analyzing the motion trajectories of regions (the leaves of the quad-tree) we can find a ROF or groups of ROFs, which describe adequately the dominant motion of the whole image. These ROFs are valid for a predefined length of app. 1 - 0.5 seconds (depending on the constancy of the film). Next step is the filtering of the motion of ROFs to estimate the ideal noise-free motion and then comes the repositioning to obtain a stable sequence

3.2 Film indexing

Computer intelligence – sophisticated decision making support - assists in film indexing. To support the semi-automatic configuration of filters the systems builds up a knowledge database, which contains the parameters of manual user interactions (such as fine tuning). These data are correlated to the different shots of a movie and thus the system can automatically propose adaptive filter browsers for supporting the operator's decision making.

One of the most popular approaches to presenting video content is the structured modeling approach: the video sequences are first divided into shots and then each shot is summarized by one or more frames called representative frames (r-frame), Hanis (2003). Several techniques exist to segment video, we use a reliable combined method to detect hard cuts and editing effects (fade, dissolve, wipe). Hard cuts, dissolves and wipes are detected by analyzing intensity edges between two consecutive frames and fades are identified by investigating the shrinkage of the intensity histogram of each frame. After shot extraction r-frames are determined for each shot then color and motion information are calculated. In the current stage two types of content based indexing can be used for finding similar shots within or between films. In both cases representative frames to be indexed are selected according to the similarity of frames' histogram to the histogram of the average frame of the shot. Our present technique assigns one r-frame to each shot but it can be extended to determine more pro shot with the help of a threshold. This threshold is lower than the value used for detecting cuts. Frames with higher histogram difference value than the threshold but lower than cut detection threshold are chosen as r-frame.

3.3 The restoration environment

The overview of the digital restoration software environment is in Fig. 1 (a). Digital filters are responsible for the removal of film errors. The implementation of a filtering algorithm must follow the specific rules of a filter interface definition. This enables DIMORF to have future filters as plug-ins simply added to the system. These filters are organized into *jobs*, which are grouped into tasks. This enables the operator to define well-separated (and possibly) parallel restoration processes. All information that can be described textually (and does not exceed a certain amount of size) is stored in XML files. XML has the advantage to describe information in well-defined structures (and it is also notable that with simple transformations <<such as XSLT>> information can be transformed into readable format).

Fig. 1 (b) illustrates the use of XML data structures in multi-level film restoration. For many types of degradation (f. e. color fading, flickering) film analysis is accomplished on a low-resolution version of the input sequence. Results of the analysis are recorded in XML files and interpreted by the restoration algorithms for applying on the high-resolution images.

4. SOME INTERESTING SOLUTIONS IN THE FILM-RECORDER

The output unit of the DIMORF film restoration system has some unique solutions, which are worth to be discussed in details. This output unit (or film-writer,

or recorder, or printer) is able to produce 35-millimeter movie pictures using RGB lasers as light sources.

Film recording is done with a maximum of 6.000-points/line resolution which corresponds to a point size of 4 microns. The speed of film writing is 10 sec/frame and 24 frames are written in the same time, however this number can be programmed to be smaller. The longest film in the machine can be 600 meters. The equipment is able to record color and black and white movies, using positive or negative technology as well.

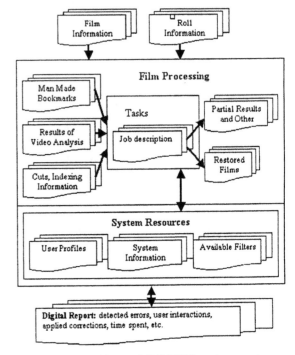

Figure 1. (a) Main objects of DIMORF environment

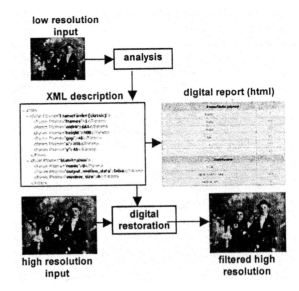

Figure 1. (b) Multi-level restoration

The wavelengths of the lasers are R (640 nm), G (532 nm) and B (405 nm). The color depth is 12 bits linear, or 10 bits logarithmic. Recently the following formats are accepted: 16-bit RGB TIFF, logarithmic CINEON, or our homemade linear LGMF.

The recorder is based on a rotating optics construction where the film to be exposed is moving on the inner arc of a drum structure. The high speed rotating optics is in the geometric center of the arc to project the focused, modulated laser beam onto the film to be exposed. The focus of the projected laser beam is always on the surface with emulsion of the film along the inner surface of the drum to guarantee a steady patch size. Synchronization is provided by a high-resolution position signal source on the axis of the rotating lens. The rotating optic is making stepwise movements parallel to its axis to make line-by-line recording of the film frames according to the resolution.

LaserGraph Motion

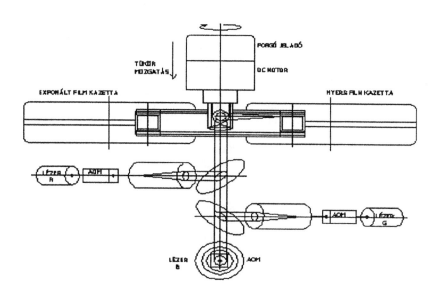

Figure 2. A sketch of the film-recorder

The film is held on the inner arc of the drum by means of vacuum, to guarantee mechanical stability and accuracy. After recording of a set of frames the film-moving unit moves the film as the positioning unit gives the control information, and after the new vacuuming the recording starts again. Positioning is done by the signals of 24 pieces of photodiodes getting red laser light. The diodes are positioned in-line with the film perforation.

A well-separated part of the optics contains the laser light sources, the fast (10 MHz) acusto-optic modulators (AOM) the projecting and light-forming lens-system and the dichroic mirror system.

The RGB lasers are responsible for color mixing and they have their own 12-bit modulators. The lens between the lasers and AOM produces the appropriate size light beams in the modulator crystal for the prescribed switching speed. The afocal system after the AOM increases the beam size to such measure that the beam could be led into the beam-combining system and into the outgoing beam increaser as a quasi-parallel beam. This beam-combining system consists of two dichroic mirrors.

The second lens of the first afocal system just after the AOM is a rubber optic. This makes it possible that the three different color beams produce equivalent size of patches after mapping. Fine-tuning of the patches is possible, too. The second, outgoing afocal system controls the beam-sizes at the same time and to the same direction, as every second part of it is a zoom lens.

The adjustable gray filter assists to fix the basic values of the power of the appropriate beams. The outgoing plate projects some percents of the beams to photo diodes to be able to see the values of the outgoing light power and to control the unwanted deviations using a PC.

When the lasers were chosen we had to take into account the following:
-   Color sensitivity of the tools used at the incoming unit

- Color sensitivity data and intensity values of the outgoing films (KODAK EASTMAN EXR 50D Film 5248 and 5245 negative, and KODAK VISION Color Intermediate Film 5244 and 5242 intermediate positive).

Naturally it was not possible to reach a perfect match. This was approached by software compensation and by means of controlling the power reserves.

Both hardware units (the film-scanner and film-writer) are upgradeable to work with an 8K resolution, too.

The information corresponding to the 3 colors are accepted from the background memory by means of separate computers equipped with interfaces and a PCI bus. A separate computer controls the equipment. This computer has a user-friendly interface to be used by the operator when communicates with the system. The user interface works under Window's xx to assist in testing, making the necessary settings and adjustments, and to control the work. The operator can select the information concerning the film to be recorded and can check the necessary data, as title, director, date, serial number of the roll, etc. He/she makes the bookkeeping of working files, etc. The operator can send messages to the system and to the archives, too. All these are provided in a comfortable, user friendly way.

5. CONCLUSIONS AND FUTURE WORK

We have introduced a complete digital film restoration system consisting of both hardware and software components. The DIMORF project aims at restoration at high-resolution with semi-automatic functions on PC platform to obtain a relatively low cost solution for film archives. The film scanner and recorder work at resolution of maximum 6K, which is adequate for the highest quality requirements. The digital restoration processes advocate multi-level restoration and digital report files by XML description. Multi-level restoration, where image analysis and synthesis are separated and can operate at different scales, decreases computational costs tremendously when working on high-resolution image sequences.

The color transmission of the scanner and the modeling of some possible aging effects of films are examined. At the current stage we are investigating the color transfer model of the film processing chain including the film recorder under production. The developed system is capable of structured data collection for future data-mining applications to increase system automation and intelligence.

6. ACKNOWLEDGEMENT

This paper is based on the research supported by the project NKFP - 2/049/2001 of the Ministry of Education, Hungary.

7. REFERENCES

Bisztray, F., Erdélyi, G., Feketü, J. Manno, S. and Méder, I. (2002) .Method and device for the correction of errors of the injured or scuffed sound recording on sound-films. Patent P0201132, Hungarian Patent Office, Budapest, 2002.

Hanis, A., Szirányi, T. (2003). Measuring the motion similarity in video indexing, Proc. Of the 4th EURASIP, Zagreb, 2003, pp. 507-512.

JOANNEUM (2002) – FRAME System http://www.joanneum.ac.at/iis/projects/frame

Kokaram, A, et al. (2002). Robust and Automatic Digital Restoration Systems: Coping with Reality, Proc. of IBC, 2002. pp. 405-411.

Kato, Z., Ji, X., Szirányi, T., Tóth, Z., Czúni, L. (2002). Content-Based Image Retrieval Using Stochastic Paintbrush Transformation. Proc. of ICIP'2002, IEEE, Rochester, 2002. pp. 944-947.

Kovács, G.L., Kas, I. (2002). Some Problems of a Digital Motion Picture Restoration System, In: Proc. of the VIPromCom 2002, IEEE/EURASIP International Symposium on Video/Image Processing and Multimedia Communications (ISBN953-7044-01-7, IEEE Catalog No. 02EX553, ed. M. Grgic), 16-19 June 2002, Zadar, Croatia, pp. 127-132.

Kovács, Gy. et al. (2002-3) DIMORF, Proc. of 2nd and 3rd International Conference on Film Restoration, Budapest, 2002 and 2003.

Kovács, L. and Szirányi, T. (2002). Creating animations combining stochastic paintbrush transformation and motion detection. Proc. of 16th ICPR, Quebec, Canada, 2002, vol. 2, pp. 1090 to 2002.

Kuglin, C. D. and Hines, D. C. (1975) The phase correlation image alignment method. Proc. of International Conference on Cybernetics and Society, IEEE, 1975. pp. 163 - 165.

Licsár, A., Czúni, L. and Szirányi, T. (2003). Adaptive Stabilization of Vibration on Archive Films. Proc. of CAIP, 2003. pp. 230-237.

PrestoSpace (2003), EU FW VI. Project – http://prestospace.org

www.elsevier.com/locate/ifac

ASSEMBLY MECHATRONIC SYSTEM WITH FLEXIBLE ORGANISATION

Roman Zahariev, Nina Valchkova

Central Laboratory of Mechatronics and Instrumentation
Bulgarian Academy of Sciences
"Acad. Bonchev" str. Bloc. 1, 1113, Sofia, Bulgaria

Abstract: The Assembly Mechatronic System is described from the point of view of the system organisation in relation to an executing technological assembly process. The created system offers a modular construction; therefore it is possible to attach and to optimise the number and type of necessary executive modules according to the set-up task. The assembly robot is used as a basic organising element, equipped with a turret for instruments (multifunctional end-effectors), a sensor system with force sensors, which is implemented in the wrist of the robot and a visual (television) system with CCD camera. The process of constructing the informational environment of the system in which it works is also described. The authors reveal the realisation of a part of the standard assembly operations and show in practice the result of the activity of the sensor systems and the specification of the model of the working environment used. At the end, conclusions and evaluations for the work of the described Assembly Mechatronic System and considerations for its future development are suggested. *Copyright © 2004 IFAC*

Keywords: mechatronics, robotics, assembly systems, sensors.

1. INTRODUCTION

The main reason for the creation of such an Assembly Mechatronic System is the synthesis and the testing of a global approach toward the organisation of its work based on the main principles, which are derived and appropriate for the specific conditions under which it
has to work. The high level of indeterminateness of many parameters of the environment, has to be taken into consideration, this implies the possibility of making an intelligent decision for achieving a determined target.

As it is known, it is necessary every precise assembly robot to fulfil its task in an environment consisting of peripheral devices and arrangements, which ensure the performance of the technological process. This question is underestimated in many cases as the whole attention is directed toward the assembly

robot, its characteristics and elements. Of course these questions are important, but they do not cover all of the necessary information for the creation of an assembly system with sufficiently large abilities. But, if we speak about an intelligent system with elements of artificial intelligence, which is ready to fulfil the necessary acts in an unclassified stochastic environment in order to achieve the set target function, which is the discussed case, then this question is very important.

Also an interesting fact is that such system is an example of a mechatronic system, in which are combined and work synergistically, as one whole, the main principles of mechanics, electronics, informatics and mathematics, in order to built up a medium of knowledge, which to be used as an appropriate tool and to be ready to fulfil the previously determined high and ambitious targets.

The Assembly Mechatronic System is created on the basis of the following principles:
- Flexibility, the system can easily be used for different assembly applications.
- Openness, giving the possibility for compatibility with other systems or devices, using accepted standards, in order to allow the system to be easily upgraded.
- Modularity - the system must be designed from relatively autonomous modules which can be combined in different ways.
-"Black box" approach – it can be applied to different levels of the system, so that the various systems and subsystems have minimum knowledge.
- The algorithms that are used in the system gather and process experience from the environment using fuzzy elements. These algorithms have to be used so that the system can receive higher degree of knowledge. (Albus *et al.*, 1983).

2. STRUCTURE OF THE SYSTEM

The Assembly Mechatronic System consists of the following subsystems:
Main modules (shown in Fig.1):
■ Assembly Robot REM 10 type SCARA with its own modular control system.
■ Six-positional multifunctional end-effectors,

which allows a given instrument to be chosen easily.
■ Force processing system with force-torque six component sensor, which is built in the wrist of the robot.
■ Visual processing system with CCD camera and VIDICON type camera.
■ Master control system of the whole configuration.

Additional modules:
■ Robot - manipulator, along with its control system.
■ Technological periphery, necessary for the fulfilment of the particular assembly process.

A basic organising module in the Assembly Mechatronic System is the assembly robot. It performs all precise and delicate operations during the process of assembly. It is equipped with a six-position turret, which allows with a rapid turn to take its position and to use one of the six assembly instruments or grippers (according to the technological process). Along with this in the wrist of the robot is implemented six-component force-torque sensor, realising the force feed-back from the working area. (Zahariev *et al.* 1995).

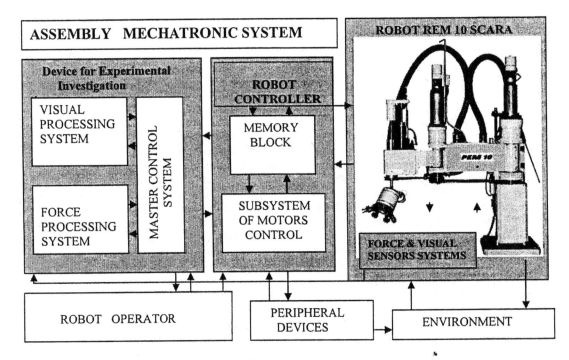

Fig.1. Structure of the Assembly Mechatronic System.

A CCD camera is also built in near the end effectors, which allows to be observed parts of the assembly process during the realisation of a local strategy. The system has one more video sensor. This is a video camera VIDICON, situated above the working area, and giving the opportunity to obtain a general view of the whole system and the movement of the objects in it.

The force and visual feed-back use different computers for the purposes of processing and obtaining their information. The computers have an ON-LINE communication with the controller of the robot and with the upper level of the main computer. The main computer co-ordinates the activities of all components of the Mechatronic system as it decomposes the targets and tasks at a technological level. It has data about the modules connected to the system and makes decisions about the organisation of the work. Using the force and visual feed-back, it has the possibility to observe the execution of the main programme, which it has created. The main computer can also interfere in case of divergence from the programme. Along with this activity, the model of the world is upgraded (the working model). This process is directly related to the information coming from the force and visual feed-back. Along with this visual information is also used, created with the help of a CAD system. The data for the geometric and kinematic characteristics of the modules used in the system are obtained through images created OFF-LINE in a computer, which uses a problem-oriented CAD system. On the basis of a hierarchically ordered model of the working environment, an intelligent decision is taken for the optimal control of the resources of the mechatronic system, in order to fulfil the target function.

3. ORGANISATION OF THE WORKING ENVIRONMENT

Let the concrete objects, which are representing the assembly technology, be considered as a technological space T, in which every technological task has to be realised through a sequence of orders to the executing mechanisms of the system, determined by the entered programme. Such a programme is considered as a function F_t, defined in T and depending on time (Korn *et al*, 1968). The technological space includes the following projection hyper-plains:

■ Hierarchical decomposition of the targets (tasks), denoted by T_g

■ Hierarchy when collecting the information from the sensor system - T_d

■ Hierarchy in the environmental model - T_m.

In every one of these projection surfaces the technological programme F_t has a projection representation - namely the sequence of orders in the corresponding surface. They will be denoted correspondingly with F_{tg}, F_{td}, and F_{tm}. As these three projection surfaces are defined for one and the same interval of time, they can be represented as parallel passing processes.

- Decomposition of the tasks - hyper-surface T_G

The hierarchical structure of this hyper-surface is the following:

■ First hierarchic level (the lowest one) is built of two sublevels:

■ Level of coordination transformations with servo control

■ Level of coordination transformations, in which either the direct or inverse kinematics or dynamic problems are solved.

■ Second is the level of elementary motions (primitives). The motions are decomposed into elementary ones or into trajectories of the co-ordinate system of the work place or of the co-ordinate system of the working tool. Such primitives can be, for example, "move to point X" or "close gripper" or "achievement of set-up orientation", etc.

■ The third level is the level of the elementary tasks. Each task is decomposed into elementary motions - primitives, which are given as problems to the second level. For example the task "Move the object A from position X into position Y" is decomposed into a sequence of elementary motions, according to the present state of the working environment.

■ The fourth level is the level of control of the automation of the workstation. At this level control of the robot, the tools, the supplying systems and the different sensors is realised.

■ The fifth level is the level of control of the automated technological cell. There the responsibility is taken for control of the production of several technological groups according to the different details. That level communicates directly with the systems for control of the separate robots, and according to the concrete technological group, different programmes are generated.

- Collection of data from the sensor systems - hyper-surface T_d.

For each level from the hyper-surface T_G there is a correspondence with a service level in T_d, which ensures the necessary information for making decisions and control. The whole stream of information from all sensor systems is segmented and for each level in T_g, the necessary information is released for regulation of the motions, for recognition of images, scenes and planning of paths with no accidents. (Zahariev 1995).

- The module, which provides sufficient information for position and velocity, corresponds to the first level in T_g This data is then used for the servo regulators and for the co-ordinate transformations.

- The second level of Ts gives tactile proximity information.

- The third level processes the information for whole objects.

- The forth level is the information interdependencies among the objects from a geometric and kinematic point of view.

- On the planning level the required formation is related to the location of the transporting and feeding devices, tool magazines, the lot sizes and others.

4. PRACTICAL EXPERIMENTS

As an illustration of the activities of the Assembly mechatronic system, some practical laboratory experiments are shown - for example the most commonly occurring assembly operation, the so called peg/hall insertion. The process of making a decision can be illustrated during that experiment as shown in Fig.2& Fig.3. The decision is made also with the help of a heuristic procedure. In the technological space T, which is in fact n-dimensional phase space, with the increase of the number of the variable states "n" (phase variables) and with the increase of the a priory information, coming from the sensor systems, T_d can be seen as a sharp increase of the probability for an optimal solution (Zahariev et.al. 1995, 1996). Fig. 2 shows the dependencies of the insertion force as a function of time of the process of peg/hall insertion, when the values of the translation error "x" and angular error "α" are different. For example, these values are $X_1 = 0.4$ mm, $\alpha_3 = 20$' and $X_1 = 0.6$ mm, $\alpha_3 = 30$'. The graphics are obtained during the realisation of the described operation with clearances of 0.02 mm.

The results when the robot is connected to the system with sensor feed-back for real time decision taking are shown in Fig.3. Evidently when comparing the data from Fig.2 and Fig.3 it can be concluded that under one and the same conditions, the value of the insertion force decreases by as much as 3 to 4 times.

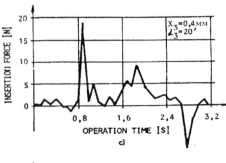

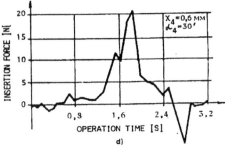

Fig.2 Process of peg/hall insertion when the robot is not connected to the system with sensor feed-back for decision taking

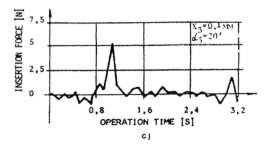

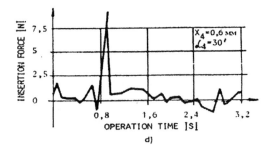

Fig.3 Process of peg/hall insertion when the robot is connected to the system with sensor feed-back for decision taking.

5.CONCLUSION

By the use of a mechatronic approach for the organisation of the working environment the maximum homogeneous and synchronisation of the Assembly Mechatronic System can be achieved. Thus an opportunity is given to answer more completely the set-up requirements toward the work of the system. The same is valid for the organisation of the internal informational environment, as an image of the working environment. The degree of adequate synthesis of the model of the working environment is increased, and also the possibilities for making an optimal decision in every moment of the technological process are increased.

The experiments with the created system show that during the process of making an optimal solution there exists a heuristically determined threshold of the number of the phase variables (the state variables), which are included in the model of the working environment. This process is synthesised with enough degree of adequacy from a practical point of view, this the reason why it is not necessary to examine the case of Markovian processes, where the number of variables, n, tends to infinity. Thus it is not necessary to use a very complicated and sophisticated mathematical approach for the optimisation of the decision making, which would cause difficulties to the system.

REFERENCES

Albus J.S., C.R.McLean, A.J.Barbera., M.L.Fitzgerald. (1983). Hierarchical Control for Robots in an Automated Factori. *Pr. 13-th ISIR*, Chicago. USA.pp. 252 - 259.

Korn G. A., T. M. Korn. (1968) *Mathematical handbook*. McGraw - Hill Book Company, NY.

Zahariev R., B.Stoyanov. (1995). Intelligent Automated Cell for Assembly of Lamps. *First ECPD International Conference on Advanced Robotics and Intelligent Automation*. Athens. Greece. Sept. 6-8. pp.173-178.

Zahariev R. (1995). Robotized Assembly System for Intelligent Automation. First *ECPD International Conference on Advanced Robotics and Intelligent Automation*. Athens. Greece, Sept. 6-8. pp.336-341.

Zahariev R., N.Valtchkova. (1995) Mechatronical System for applied research and diagnostics of Industrial robots on the base of sensor fusion. *Pr. of 2-nd International Workshop on Mechatronics Design and Modelling*. Ankara. Turkey. November 13-17. pp.135-139.

Zahariev R., B. Stoyanov. (1996). Experimental Robotised Assembly System in Condition of Undetermined Environment. September 26-28. Vienna, Austria, pp.560-566.

ELSEVIER

IFAC

PUBLICATIONS
www.elsevier.com/locate/ifac

A MULTI AGENT SUPPORT SYSTEM FOR MODELING, CONTROL DESIGN AND SIMULATION

Ivana Budinská, Baltazár Frankovič, Tomáš Kasanický, T.-Tung Dang

utrrbudi, utrrfran utrrtung@savba.sk, kasanicky@neuron.tuke.sk
Intitute of Informatics SAS,
Dúbravská cesta 9,
841 08 Bratislava

Abstract: The paper describes a support system for modeling, control design and simulation
of production systems. The system is based on Multi Agent technology. The system is
intended for experts and technical staff from manufacturing enterprises, and for students of
technical schools as well. It might be used for solving critical situations in quazi real time,
and/or for learning and training technical staff and students of technical universities.
Copyright © 2004 IFAC

Keywords: Multi Agent Systems, production systems, structural database, case based
reasoning

1. INTRODUCTION

The basic idea of building a support system for modeling, control and simulation of discrete and continuous production system raised from the need of having a simple tool for technical staff, and managers of production systems to help them in some critical situations when they need very quickly to react on changed environment, to set up new controlling values, etc. The system enables to build a case library for various types of systems, either discrete or continuous with the help of tolls that may be integrated in the system directly (built in tools), or connected via http. In a critical situation the system searches in database for some similar case and return the user a similar solutions. On the basis of user's requirements the found solution is adapted on the new situation and after approving by user, it is entered to the database as a new case. Such a behavior of the system enables to enlarge system capabilities. The other system functionalities enable learning and training of students and technical staff in modeling, design of control and simulation of production system.
Section 2 describes some related publications and works. Section 3 introduces architecture of the system. System interfaces are described in Section 4, and a method for building a case library is presented in Section 5. Section 6 is addressed to ontology utilization in building the

system. An example of similarity computation is given in Section 7. Section 8 contains conclusion and some ideas for future work.

2. RELATED WORKS

Many publications are devoted to utilization of multi agent systems in manufacturing. In (Balasubramanian et al., 1995) a multi-agent architecture for the integration of design, manufacturing, and shop floor control activities is presented. In the project Pabadis (Penya et al., 2002), Java™-based and Jini™-technology with mobile and residential software agents are combined. This will enable industry to minimize time and expenditure related with setting up new machines or changing existing ones. Many publications are devoted also to problems in industrial control teaching (Flochova, Hruz, 2003).

3. ARCHITECTURE OF THE SYSTEM

The system consists of three basic modules; each of them works in two regimes. There are modeling module control module and simulation module (See Fig.1: Architecture). Each of modules works in a regime of searching in database of ready to use cases on the base of

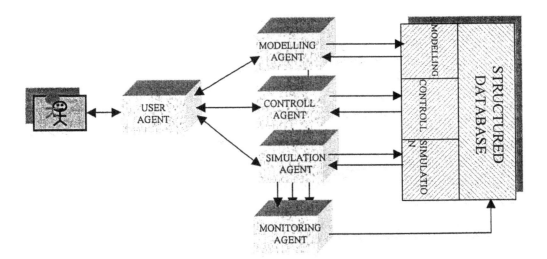

Fig. 1: Basic architecture of the system.

Cased Based Reasoning (CBR), and in a regime of connection to appropriate tools, which enable modeling, design of control and simulation of production systems. The basic algorithm of CBR is in Fig. 3.
The core of the system is created by Generic Block, which consists of the following agents (Dang et al, 2003):
User Agent (UA) – interposes communication between a user and the system. UA transfers user's requirements to the system's agents and on the other hand return solution from the system's agent to the user in a human readable form. There is one UA per user in the system.
Modeling Agent (MA) – after receiving information about the process, search for appropriate modeling tools and algorithms on the basis of previous cases. It returns suggested algorithms and ask for more precise information according to a chosen algorithm. The final decision on which algorithm and/or tool has to be chosen is up to the user. Finally, Modeling agent returns a model of described production process.
Control Agent (CA)- receives a finally chosen model with all necessary attributes defined. On the basis of the model, CA searches for an appropriate control algorithms in the database. Through PAA it negotiates with the user and finally chooses appropriate control algorithms. The user has to choose an algorithm from the suggested ones and specify all needed values for that. CA returns control algorithm for the process according user's specifications.
Simulation Agent (SA) – is responsible for simulation of control for the chosen model and control algorithm with the aim to help the designer to assess the proposed solution.
Monitoring Agent (MoA) - follows the system behavior after applying the recommended method for designing the control. If all the requirements are satisfied, then MoA updates the database by newly achieved results. That means the system stores all solutions for the next reuse and application. Otherwise, the MA and CA have to repeat their calculations.

Basically agents work according the following scheme:

{what user?
  what system?
        Search for appropriate methods
        Identify available tools
        Select the best tool (CBR method)
Create model}

After specification of the user and the system, the agent search in the database for appropriate methods and tools. The important element of the support system is also Structured Database, in which historical situations; modeling, controlling and simulation tools and methods are stored. Structured database contains description of methods and tools via their attributes, and enables searching according users requirements. The database has been developing with a strong relationship with user's questionnaire.

4.   THE SYSTEM'S INTERFACES

User's Interface: Provides intelligent interface that enables logging into the system, user profile creation, and user requirements specification. User requirements are specified through so called "Questionnaire". There are several levels of questionnaires.   User fills in **an introductory questionnaire**. It contains questions about what user wants to do in the system and general characteristics about the concrete production system. The introductory questionnaire is the basis for a user profile creation. Depending on the introductory questionnaire filling in, the second level of questionnaire is generated for the user. **The second level questionnaire** specifies more in details the production system and the methods for modeling, control design and simulation. According filled in the second level questionnaire; a first evaluation of user's requirements is done. The User Agent contacts appropriate Agent in the Generic Block (Modeling Agent, Control Agent, or Simulation Agent), and send information from the questionnaire to the Agent. Contacted Agent will act according a message from User Agent. For instance, if User Agent contacts Control Agent, it also says what kind of control system a user wants to design. The user would choose it from predefined options in the questionnaire. Control Agent would search for appropriate regulator (discrete or continuous), stability criteria, etc. Then Control Agent generates **a third level questionnaire,** and asks the user for more precise definitions of parameters, sampling period, etc. The third level questionnaire is intended for advanced users, however the system provides a guide and help for answering all questions in the third level questionnaire. The last level of questionnaire presents very specific information about the concrete control algorithm. The last level of questionnaire is tightly connected to a control design tool and requires a very precision definition of all parameters, which is required by the concrete tool to suggest a control algorithm, or regulator.

Interface between Generic Block and external application:
The system supposes utilization of several tools for modeling, control design and for simulation. In order to connect these tools with the system, an interface must be defined. The external tools, or applications can be located generally anywhere on the net, and they are called by the system through XML RPC.

Interface between the system and the structured database:
This is a very important interface, which enables query the database; insert new cases in the database and update solutions in the database. There is no choice for user to enter the database directly, but through Monitoring Agent and through appropriate agent for modeling, control design and simulation.

5. MODELING, CONTROLLING AND SIMULATION METHODS AND TOOLS ONTOLOGY

Ontology is a very useful framework for organizing knowledge. In order to reason about manufacturing processes, modeling, control and simulation methods and tools, ontology of manufacturing processes and modeling, control and simulation tools is being built. Some ontology editors provide technological support to most of the ontology lifecycle activities and have extensible architecture, letting one to add new modules to provide more functionality to the environment. Examples of these environments are Protégé (Noy et al, 2000), developed by the Stanford Medical Informatics at Stanford University, and OntoEdit (Sure, 2002), and developed by AIFB in

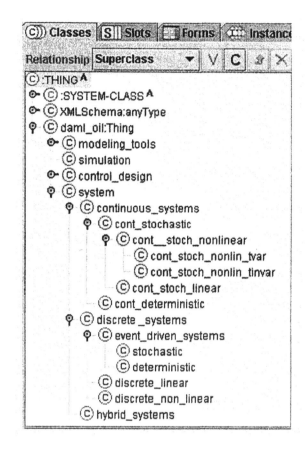

Fig. 2: An example of ontology in Protégé.

Karlsruhe University. In addition, these tools allow one to express the ontologies in Semantic-Web languages such as RDF(S) and OWL.

The application of the OWL format for ontology for the agent system is relatively new. One advantage of owl ontology is the availability of tools that can reason about it. Tools provide generic support that is not specific to the particular subject domain. Building useful and reliable

reasoning system is not a simple effort. Constructing ontology is easier. Constructing ontology in owl enables to benefit from third party tools based on the formal properties of the OWL language.
Ontology for the described system consists of four main classes:
Class System
Class Model
Class Control
Class Simulation
Class System contains three subclasses:
continuous_systems, discrete_systems, hybrid_systems
(See Fig. 2). Three subclasses have inherited attributes from the class System, but each of them contains some new specific attributes. Also other classes contain some subclasses. For instance class Models contains modeling tolls and methods for continuous systems and for discrete systems as well.

6. STRUCTURED DATABASE CREATION

One of the most suitable ways to organize case data in the database, is to create a case library. Each case is characterized by a set of attributes. Generally, the number of attributes and their meaning may vary from case to case. It is important that each case in the database is associated to a solution within a set of solutions.

Suppose a set of cases of one class – concrete manufacturing system with a discrete or continuous production process and all attributes that specifies that system, represents a case. Each case is described by a set of its attributes.

$$C = \{c_1, c_2, ...\}$$

$$c_i = \{a_{i1}, a_{i2}, ... a_{in}\}; \qquad a_{ij} \in A$$

A is a set of attributes that can be recognized in any case. If an attribute is not related to the concrete case, a value is set to zero.

Let's define a set of solutions:

$$S = \{s_1, s_2, ...\}$$

$$s_i = \{el_{i1}, el_{i2}, ...\}$$

A Case Base is defined as a space of case descriptions and case solutions:

$$CB = C \times S$$

User defines a case through attributes and the system starts to search for similar cases in the CB.

Let a current case is defined as follows:

$$cc = \{ca_1, ..., ca_n\}$$

The agent looks for similar cases to a current one by using a method for similarity assessment.

The general problem of similarity assessment is according to (Bergmann, 2002):

$$sim((cc_1, ..., cc_n)(c_1, ..., c_n)) = \Phi(sim_1(cc_1 c_1), ..., sim(cc_n c_n))$$

A traditional similarity measures for individual attributes is measured by a function Φ that is monotone increasing in every argument.

For attribute-value representation a couple of methods can be used to measure a similarity degree. A method of weighed attributes is used to assess similarity between a current case (that user has defined) and a case in the CB in this project. The weights allow expressing the importance of all attributes.

Two different methods to evaluate attributes are used: user specific weights, which are given as a part of user's requirement through user's questionnaire; and case specific weights (weights are specified as a part of structured database).

7. A NUMERICAL EXAMPLE OF SIMILARITY CALCULATION

In this Section a numerical example of similarity computation between two cases is given. Let's have a case c_i, which possesses a number of attributes a_k (where k=1, ..., n).

Given two cases c_i and c_j we can define a similarity value S_{ij} between them as a weighted sum of the similarity values between the single attributes of the two cases. So if s_k^{ij} is the similarity value between the attribute a_k of the case c_i and the attribute a_k of the case c_j, and if w_k is the weight that the a_k attribute has in the calculation of the case similarity, then S_{ij} is given by $S_{ij} = \Sigma_k w_k s_k^{ij}$

Computation depends on the values that attributes can get. Generally attributes can be numerical and non-numerical. In case of non-numerical values, each attribute can get predefined values. For instance in the class *System*, attributes are: *continuous, discrete, hybrid*. The second level of attributes is *linear, non-linear*.

If only cases with exactly the same values of these attributes can be similar, the weight assigned to these properties is therefore the highest possible: $w_{ct} = 100$

Similarity function can have two values:

$$s_k^{ij} = \begin{cases} 1, if\ a_k^i = a_k^j \\ 0, otherwise \end{cases}$$

In case, where it is not important that the values must be exactly the same, similarity function can get more values. It must by specified by an expert, which attributes can be assess as similar and with which scores. In case of numerical attribute values, the similarity degree has to be assessing by the difference between two values. In order to simplify similarity computation, three-value system is used to evaluate the similarity degree: *good, bad, and very bad*. Each of these values is bounded according to experience. Only case with a *good* similarity degree between all attributes can be chosen as a similar system.

An example:

Let's have three case according Table1. For simplicity it is assumed, that similarity function for all these attributes can get only two values (0,1). In fact, some of these attributes need more complicated similarity function.

Table 1: An example: Three cases described by attribute values and weights of attributes.

w	Case1	Case2	Case3
100	Discrete	Discrete	Discrete
100	Event_driven	Event_driven	Event Driven
100	Deterministic	Deterministic	Stochastic
98	Manufac_trans_system	Transport_system	Transport sysstem
98	Fixed_transport_line	Fixed_transport_line	Fixed_transport_line
75	One_way_transport	Multi_way_transp	Multi_way_transp
70	Fixed_tasks	Non_fixed_tasks	Non_fixed_tasks
60	Noncrash_opt_control	Noncrash_nonopt_cont	Noncrash_nonopt_cont
60	Regarding_jobs	Regarding_jobs	Regarding_jobs

Similarities are computed as it was described:

$$s_{12} = 100 \cdot 1 + 100 \cdot 1 + 100 \cdot 1 + 98 \cdot 0 + 98 \cdot 1 + 75 \cdot 0 +$$

$$70 \cdot 0 + 60 \cdot 1 + 60 \cdot 1 = 518$$

$$s_{13} = 100 \cdot 1 + 100 \cdot 1 + 100 \cdot 0 + 98 \cdot 0 + 98 \cdot 1 + 75 \cdot 0 +$$

$$70 \cdot 0 + 60 \cdot 1 + 60 \cdot 1 = 418$$

Cases 1 and 2 are more similar as 1 and 3.

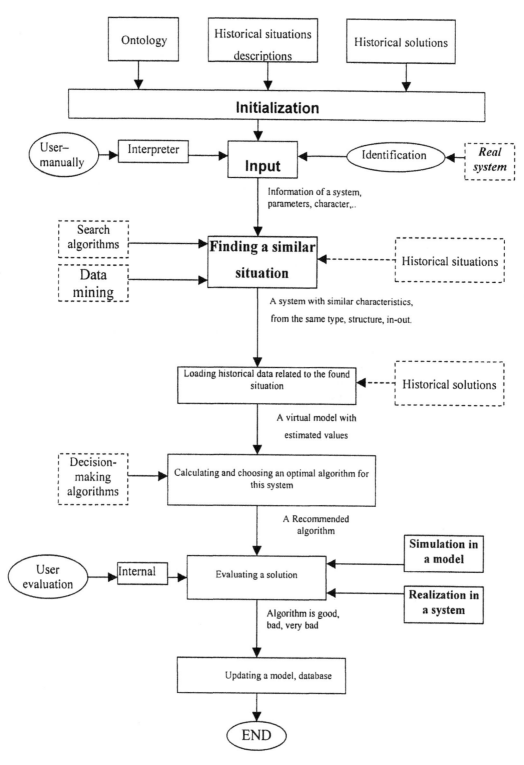

Fig. 3: Basic algorithm of reasoning that is used in the system.

8. CONCLUSION

The described support system uses Case Based Reasoning (CBR) to find a solution that matches the best to the user's requirements. The structured database is built independently for three block of the systems; Modeling, Control Design, and Simulation, however all these databases are related to each other. The future work is concentrated on testing CBR methods in the system, using more methods to evaluate similarity degree for case retrieval and to design methods for a case adaptation.

9. ACKNOWLEDGEMENT

This work was supported by Science and technology Assistance Agency under the contract No. APVT-51-011602.

10. REFERENCES

Balasubramanian S.and D.H. Norrie (1995) A Multi-Agent Intelligent Design System Integrating Manufacturing And Shop-Floor Control, http://ksi.cpsc.ucalgary.ca/DME/AnAgent.html

Bergmann R (2002) *Experience Management*, Lecture Notes in Artificial Intelligence Series, vol. 2432.Springer-Verlag

Dang T.Tung, Frankovič B., Budinská I. (2003) Case-based reasoning applied for CAS-Decision system, In *Proc. of the* 2nd IFAC Conference on Control System Design. Bratislava, Slovakia, September, 7-10, 2003. Š. Kozák, M. Huba (Eds.). Bratislava, Slovakia, Compact Disc, Section C: DEDS Control System Design.

Flochová, J., Mudrončík, D.: (2003) Industrial control software teaching at the Department of Automatic Control Systems FEI STU. *ECC' 03 European Control Conference*. Cambridge, UK: September 1-4

Noy N.F., Fergerson R. W., Musen M. A.. The Knowledge Model of Protégé-2000: (2000) Combining Interoperability and Flexibility. *2nd International Conference on Knowledge Engineering and Knowledge Management*. Lecture Notes in Artificial Intelligence, vol. 1937, Springer.

Penya Y. K., Sauter T.: (2002) Network Load Imposed by Software Agents in Distributed Plant Automation, *Proceedings of INES 2002*, Croatia

Sure Y., Erdmann M., Angele J., Staab S., Studer R., Wenke D.. (2002) OntoEdit: Collaborative Ontology Engineering for the Semantic Web. *First International Semantic Web Conference (ISCW'02)*. Lecture Notes in Computer Science, vol. 2342, Springer.

ELSEVIER

IFAC
PUBLICATIONS
www.elsevier.com/locate/ifac

AUTOMATIC COLOR DECISION USING FUZZY CONTROL

Takashi Mitsuishi*, Koji Saigusa*, Noriya Kayaki** and Masayuki Kochizawa*

**Department of Spatial Design and Information Systems,
Miyagi University
**Department of Economics and Business Administration,
Kagoshima Prefectural College*

Abstract: This paper proposes the fuzzy system for color construction. This system is divided into two parts. The 1st system is supposed to consider the color preference and decide the color from surroundings. The usual fuzzy production rule is applied to this system. On the other hand the 2nd system is supposed to consider the influence of color on the human and adjust the color decided by the 1st system. Therefore the predictive fuzzy control is applied to the 2nd system. The usefulness of this coloration system is shown by numerical example. *Copyright © 2004 IFAC*

Keywords: Feed forward control, Fuzzy inference, Non linear control systems, Expert systems, Computer-aided design.

1. INTRODUCTION

In 1965 Zadeh (Zadeh, 1965) introduced the notion of fuzziness. And then Mamdani (Mamdani, 1974) has applied it to the field of control theory using what is called Mamdani method. This method is one of the ways to numerically represent the control given by human language and sensitivity, and it has been applied in various practical control plants.

Generally, the colors in the living environment have been made decision by some factors, safety and design, and user's preference. Thus, the need of the coloring system that evaluates the complicated mentality and physiology by the scale of feeling because the living environment is the field expected a peace of mind for people. There are many researches treating human's ambiguous sensitivity using fuzzy logic on the color scheme, for example a coloring system about the landscape using fuzzy logic by Terano et al (Terano, *et al.*, 1991; Nakanishi, *et al.*, 1992). But there is not a coloring decision system that thought the liking of the patient in the sickbed environment.

The authors are studying about automatically color construct system using a fuzzy logic, which analyzes psychologically the reason of color decision (Mitsuishi *et al.*, 2003). The methodology of a coloring decision and the outline of a coloring system are shown. In this paper, the color is determined by fuzzy system that has two processes. In the first process, the hue, luminosity and chroma

of color are temporarily decided by state evaluation fuzzy control in consideration of the situation and state for the object of color scheme. This process that decides the color by sensibility of human does not take effect of color scheme into consideration. Therefore, next process using predictive fuzzy control considers what effect does the color scheme brings about, and decides three attributes of the color again. This predictive fuzzy control plays the role of performance function. Since the authors are researching into 2nd system now, the detail of 2nd system could not be shown in this paper. Simple application of 1st system is described in the end of this paper.

2. FUZZY THEORETICAL APPROACH TO COLOR SCHEME

Color is composed of three attributes of hue (tinge), luminosity (brightness of color) and chroma (clearness of color). The numerous expression of color is possible by each continuous order and is symbolized with the table color system of Mancel etc (Chijiiwa, 2001). However, it cannot make it special to one of color for example "red" that said intangibly because the relation of the attributes of color is unclear. It can become possible to formulate certain color by considering each of hue, luminosity and chroma as fuzzy set, because color is one of the expressions including language expression similar, ambiguity (Sagawa, 1999; Sagawa, 2000).

3. COLOR DECISION BY CONDITION

The 1st system (first process) is given as following:
1st System:

Temporarily Decision of Hue
IF x_1 is $A1_{i1}$ and x_2 is $A1_{i2}$ and ... and x_k is $A1_{ik}$
  THEN u_1 is HU_i ($i = 1, 2, ... , l$)

Temporarily Decision of Luminosity
IF x_1 is $A2_{i1}$ and x_2 is $A2_{i2}$ and ... and x_k is $A2_{ik}$
  THEN u_2 is LU_i ($i = 1, 2, ... , m$)

Temporarily Decision of Chroma
IF x_1 is $A3_{i1}$ and x_2 is $A3_{i2}$ and ... and x_k is $A3_{ik}$
THEN u_3 is CH_i ($i = 1, 2, ... ,n$)

Here, x_j ($j = 1, 2, ... , k$) are premise variables and are the value of factor for decision of color scheme. $A1_{ij}, A2_{ij}, A3_{ij}$ ($i = 1, 2, ... ,l$ or m or $n, j = 1, 2, ... , k$) are the fuzzy sets for premise variables. Moreover consequent variables u_1, u_2, u_3 are the values of the hue, luminosity and chroma respectively. And let HU_i, LU_i, CH_i ($i = 1, 2, ... , l$ or m or $n, j = 1, 2, ... , k$) be fuzzy sets that shows the fuzziness of three attributes of the color. l, m and n are the number of fuzzy production rules, and k is the number of premise variables.

The outputs of previous system (hue u_1*, luminosity u_2* and chroma u_3*) are calculated by approximate reasoning. In this paper, Product-Sum-Gravity method (Mizumoto, 1990) is applied to approximate reasoning because of simplicity of calculation.

Approximate reasoning of hue:

$$F(u_1) = \sum_{i=1}^{l} \left\{ HU_i(u_1) \cdot \prod_{j=1}^{k} A1_{ij}(x_j) \right\}$$

Here, $A1_{ij}(x_j)$ and $HU_i(u_1)$ ($i = 1, 2, ... ,l, j = 1, 2, ... , k$) are grade of fuzzy set $A1_{ij}$ and HU_i. To calculate the defuzzification, the center of gravity method is applied.

$$u_1* = \frac{\int u_1 F(u_1) du_1}{\int F(u_1) du_1}$$

Output of Luminosity:

$$u_2* = \frac{\int u_2 \sum_{i=1}^{m} \left(LU_i(u_2) \prod_{j=1}^{k} A2_{ij}(x_j) \right) du_2}{\int \sum_{i=1}^{m} \left(LU_i(u_2) \prod_{j=1}^{k} A2_{ij}(x_j) \right) du_2}$$

Output of Chroma:

$$u_3* = \frac{\int u_3 \sum_{i=1}^{n} \left(CH_i(u_3) \prod_{j=1}^{k} A3_{ij}(x_j) \right) du_3}{\int \sum_{i=1}^{n} \left(CH_i(u_3) \prod_{j=1}^{k} A3_{ij}(x_j) \right) du_3}.$$

4. SECDOND COLOR DECISION SYSTEM

The 2nd color decision system is introduced in this section. This inference method is what estimates the results which are calculated by 1st color construction systems. A predictive fuzzy control is applied to this method toward a cost function and the adjustment.
2nd System:

Final Decision of Hue
IF (u_1' is HU_j' $\rightarrow z_1 = f(u_1')$ is $B1_j$ and u_1* is $C1_j$)
  THEN u_1' is HU_j' ($j = 1, 2, ... , l'$)

Final Decision of Luminosity
IF (u_2' is LU_j' $\rightarrow z_2 = g(u_2')$ is $B2_j$ and u_2* is $C2_j$)
  THEN u_2' is LU_j' ($j = 1, 2, ... , m'$)

Final Decision of Chroma
IF (u_3' is CH_j' $\rightarrow z_3 = h(u_3')$ is $B3_j$ and u_3* is $C3_j$)
  THEN u_3' is CH_j' ($j = 1, 2, ... , n'$)

Similarity to 1st system, u_1', u_2' and u_3' are the values of the hue, luminosity and chroma respectively. HU_j', LU_j' and CH_j' ($j = 1, 2, ... , l'$ or m' or n') are fuzzy sets of them and may differ from 1st system's ones. z_1, z_2 and z_3 are the values of fuzzy objective evaluation. They are constructed by the functions f, g and h. Here these functions are objective evaluate functions. These objective evaluate functions estimate an influence of the outputs of these approximate reasoning based on the fuzzy rules. On the other hand Bk_j and Ck_j ($j = 1, 2, ... , l'$ or m' or $n', k = 1, 2, 3$) are the fuzzy sets for z_1, z_2 and z_3 , and Ck_j ($j = 1, 2, ... , l'$ or m' or $n', k = 1, 2, 3$) are the fuzzy sets which estimates the outputs of 1st system.

When the values of hue u_1*, luminosity u_2* and chroma u_3* from 1st systems to 2nd systems, the outputs of these systems are calculated by approximate reasoning similarity to 1st system. The calculation of the outputs of three attributes is shown as following:

$$y_k = \frac{\int u_k \sum_{j=1}^{p} \left(Bk_j(z_k) \cdot Ck_j(u_k*) \cdot X_j'(u_k') \right) du_k'}{\int \sum_{j=1}^{p} \left(Bk_j(z_k) \cdot Ck_j(u_k*) \cdot X_j'(u_k') \right) du_k'}$$

where $k=1, 2, 3$, $X=HU$ ($k=1$), LU ($k=2$), CH ($k=3$) and $p=l'$ ($k=1$), m' ($k=2$), n' ($k=3$). These hue y_1, luminosity y_2 and chroma y_3 are the final outputs of this system that the authors proposed. Then the color based on fuzziness is decided by composition of three attributes.

5. APPLICATION TO REAL EXAMPLE

In this section, using an idea and framework mentions in the previous section, the example of application will be described. Since it is difficult to construct the functions *f*, *g* and *h* in 2nd system, which indicates the value affected by the color scheme, then unfortunately the application model using 2nd system is not introduced in this paper. This system is just going to be examined now. Then real application model using only 1st system is described below.

5.1 Determination of premise variables and setup of its membership functions (fuzzy sets)

The premise variables are selected from the result of a questionnaire. The questionnaire was carried out about the factor that determines the color scheme for the students at the Miyagi University. As the result, the age and the temperature selected by many students are chosen from some factors, contrast between color schemes, brightness, season, feeling, sex, etc. Since it is easy to evaluate numerically, the authors selected them.

Four fuzzy sets (VY : very young, Y : young, O : old, VO : very old) are decided for the age because of simplicity. The membership functions of them are shown in Fig.1.

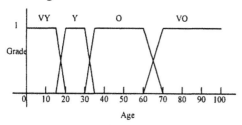

Fig. 1 Membership functions of the age

As the fuzzy sets about the temperature, two sets (C: cool, W : warm) are constructed in consideration of the climate in Japan. They are decided according to generally knowledge of Japanese. It is possible to construct the membership function automatically (Oda, 1992). The membership functions of C and W are shown in Fig.2.

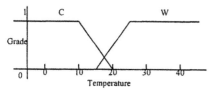

Fig. 2 Membership functions of the temperature

5.2 Set-up of consequent fuzzy sets

The following fuzzy sets of three attributes of hue, luminosity and chroma are decided according to the opinion of color scheme specialist Saigusa. He is one of authors of this paper and a wood-block print artist.

The fuzzy sets about the hue are assumed to be five sets based on his experiences and knowledge. The other ones about luminosity and chroma are assumed to be three fuzzy sets, low, middle and high. The ranges of membership functions of them are assumed to be 256 color levels to calculate easily on Windows software Visual Basic. These membership functions are shown in Fig. 3 and Fig. 4.

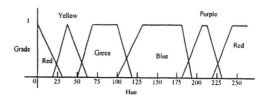

Fig. 3 Membership functions of hue

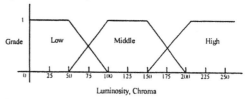

Fig. 4 Membership functions of luminosity and chroma

In Fig. 3 and Fig. 4, the membership functions are shown as continuous functions. However the values of three attributes are actually not continuous in our application soft. Then the membership functions concerning color are considered as dispersive functions. For example, the membership function of "red" (Re) is shown as following:

$$Re(u_i) = \{1/0, 0.97/1, \ldots, 0/100, \ldots, 1/255\},$$

where 0.97/1 means that the value of membership function Re is 0.97 when the value of hue u_i is equal to 1.

5.3 Set-up of consequent fuzzy sets

Using previous fuzzy sets, IF-THEN rules that decide the value of hue are constructed as Table 1. Ye is yellow, Re is red, Bl is blue, Ge is green, and Pu is purple. Here, Ga is gray which the value of chroma is 0 and the value hue is arbitrary (assume 1). Furthermore the IF-THEN rules that decide luminosity and chroma are shown below. Although the existence of the fuzzy IF-THEN rules which give the optimal control to the fuzzy system has been guaranteed (Shidama *et al.*, 1996; Mitsuishi *et al.*, 2000; Mitsuishi *et al.*, 2002), since the concept of optimal control in this color construction system is not clear now, the authors constituted this rules.

Table 1 IF-THEN rules of hue

		Age			
		VY	Y	O	VO
Temp.	W	Ye	Re	Bl	Ge
	C	Bl	Ge	Pu	Ga

Table 2 IF-THEN rules of luminosity

		Age			
		VY	Y	O	VO
Temp.	W	H	H	M	M
	C	M	M	L	L

Table 3 IF-THEN rules of chroma

		Age			
		VY	Y	O	VO
Temp.	W	H	M	M	L
	C	M	M	L	L

For example in Table 1, Ye (1 line, 1 sequence) means following IF-THEN rule:

IF x_1 is W and x_2 is VY THEN u_1 is Ye.

(IF Temperature is Warm and Age is Very Young THEN Hue is Yellow).

5.4 Numerical example

In this part, the numerical example of color construction system described above is shown. Given temperature $x_1 = 18$ and age $x_2 = 32$, we have:

$$HU(u_1)$$
$$= W(18) \cdot VY(32) \cdot Ye(u_1) + W(18) \cdot Y(32) \cdot Re(u_1)$$
$$+ W(18) \cdot O(32) \cdot Bl(u_1) + W(18) \cdot VO(32) \cdot Ye(u_1)$$
$$+ C(18) \cdot VY(32) \cdot Bl(u_1) + C(18) \cdot Y(32) \cdot Ge(u_1)$$
$$+ C(18) \cdot O(32) \cdot Pu(u_1) + C(18) \cdot VO(32) \cdot Ga(u_1)$$

then the value of hue is

$$u_1{}^* = \sum_{u_1=0}^{255} u_1 HU(u_1) \bigg/ \sum_{u_1=0}^{255} HU(u_1) = 230$$

In the process which the center of gravity is calculated before, the sigma is used instead of the integral because the function HU inferred by premise part is not continuous. Therefore it is not needed to use the integral. The values of luminosity and chroma are calculated similarly. We have luminosity $u_2{}^* = 109$ and chroma $u_2{}^* = 109$ as the final outputs of first system. The coloration composed of these three attributes is near purple. The proposed coloration system was constructed using Microsoft Visual Basic 6.0 and built on fuzzy coloration IF-THEN rules. It was confirmed that the colorations decided by this system based on fuzzy inference method are adapted to the opinion of a coloration specialist K. Saigusa.

6. CONCLUSION

The color decision support system analyzed by the dual fuzzy control system is proposed. The 1st system has been already constructed, and we are designing the membership functions and constructing the fuzzy production rules of 2nd system on the field of sensibility engineering and color engineering. The valid oration refracted the opinion of coloration specialist were proposed in the prototype built on IF-THEN rules. It was confirmed that the proposed coloration was availableness for each user. Hence, it is possible to propose coloration pattern considering the influence of a psychological and sensibility using our system proposed in this paper. Our goal is to make more useful application for the field of the architecture.

REFERENCES

Chijiiwa, H. (2001). *An outline of color science*, Chap. 2. University of Tokyo Press, Japan.

Mamdani, E. H. (1974). Application of fuzzy algorithms for control of simple dynamic plant. *Proc. IEEE121*, No.12, pp1585-1588.

Mitsuishi, T., Kawabe, J. and Shidama, Y. (2000). Continuity of fuzzy controller. *Mechanized Mathematics and Its Applications*, Vol. 1, No.1, pp31-38.

Mitsuishi, T., Kayaki, N. and Saigusa, K. (2003) Color construction using dual fuzzy system. *Proc. of IEEE International Symposium on Computational Intelligence for Measurement Systems and Applications*, pp136-139.

Mitsuishi, T. and Shidama, Y. (2002). Minimization of quadratic performance function in T-S fuzzy model. *Proc. of International Conference on Fuzzy Systems (FUZZ-IEEE2002)*, pp75-79.

Mizumoto, M. (1990). Improvement of fuzzy control (IV) Case by product-sum-gravity method. *Proc. 6th Fuzzy System Symposium*, 9-13.

Nakanishi, S., Takagi, T. and Nishiyama, T. (1922). Color planning by fuzzy set theory. *IEEE I.C. on Fuzzy Systems*, pp5-12.

Oda, Y. (1992). Experimental research on the measuring the attribute function - Comparison of a reaction time method and a consultation scale method -. *Proc. of 2nd Non Engineering Fuzzy Workshop*, pp68-72.

Sagawa, K. (1999). Visual comfort to colored images evaluated by saturation distribution. *Color Res. Appl.*, 24(5), pp313-321.

Sagawa, K. (2000). Visual comfort evaluated by number of categorical colors in a colored image. *Color Res. Appl.*, 25(3), pp193-199.

Shidama, Y.,Yang, Y., Eguchi, M. and Yamaura, H. (1996). The compactness of a set of membership functions and its application to fuzzy optimal control. *The Japan Society for Industrial and Applied Mathematics*, No. 1, 1-13.

Terano, T., Masui, S., Sugiura, H. and Yamauchi, K. (1991). Linguistical expression of image by fuzzy logic. *IFES '91*, pp995-1002.

Zadeh, L. A. (1965). Fuzzy Sets. *Information and control*, 8, pp338-353.

CASE STUDY FOR USING ADVANCED CONTROL STRATEGIES FOR LEVEL CONTROL IN A TWO TANK SYSTEM

Zs. Preitl*, R. Bars*, R. Haber**

**Department of Automation and Applied Informatics, Budapest University of Technology and Economics,
MTA-BME Control Research Group
H-1111, Budapest, Goldmann Gy. tér 3., Hungary; fax : +36-1-463-2871
e-mail : preitl@aut.bme.hu, bars@aut.bme.hu, robert.haber@fh-koeln.de
**Department of Plant and Process Engineering, University of Applied Science Cologne, Germany*

Abstract: The paper presents advanced control algorithms exemplified on a case study of a two tank system, suitable for laboratory exercise in teaching Control Engineering. Adaptive PID control is presented based on Modulus Optimum tuning criterion, underlining the problem of bump transfer when switching between manual and automatic modes or between two controllers. Solution to this consists in bumpless transfer, one realization of it being implemented here. Adaptive General Predictive Control is presented and combined with static linearization using the inverse static characteristics of the plant. Finally Internal Model Control (IMC) is used, implemented for the non-linear plant. In the conclusion an evaluation of the algorithms is given. *Copyright © 2004 IFAC*

Keywords: adaptive control, PID control, nonlinear control, Internal Model Control, adaptive predictive control.

1. INTRODUCTION

Advanced control strategies include the most different types of control algorithms. In order to make them more familiar it is necessary to be able to implement, test and somehow compare algorithms, if possible on a real, complex system. The algorithms which are presented incorporate various design techniques: adaptive PI(D) control with bumpless transfer, Internal Model Control (IMC), General Predictive Control (GPC). The plant that was chosen is the classical two tank system.

First the system, its model and its working principles are presented; then the identification of the system parameters is briefly introduced. The system is nonlinear. Its static characteristics is calculated taking into account the physical limitations (defined by the maximal accepted level in the tanks), and the inverse static characteristics is deduced and used in the controller part of the IMC structure. Problems such as bumpless transfer with PI(D) control and application of GPC are emphasized on the two tank system.

2. THE TWO TANK SYSTEM. MODELING, IDENTIFICATION

The simplified scheme of the considered two tank system (2TS) is depicted in Figure 1. The mathematical model is based on the balance of potential energy and motion energy, out of which the velocity of the outlet water is obtained:

$$v_1 = \sqrt{2gh_1} \qquad (1)$$

$$y_1 = \mu a v_1 = \mu a \sqrt{2g}\sqrt{h_1} = k_1\sqrt{h_1} \qquad (2)$$

where μ is the viscosity factor. The changes in the levels can be written as:

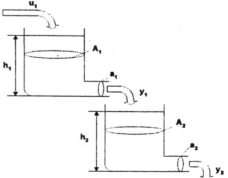

Fig.1 Scheme of a two tank system

$$\frac{dh_1}{dt} = \frac{u_1 - y_1}{A_1} = \frac{1}{A_1}u_1 - \frac{k_1}{A_1}\sqrt{h_1} \qquad (3)$$

$$\frac{dh_2}{dt} = \frac{1}{A_2}(k_1\sqrt{h_1} - k_2\sqrt{h_2}) \qquad (4)$$

Introducing relative units and taking into account that $A_1 = A_2$ and $h_{1max} = h_{2max}$, it results

$$\frac{dh_{1rel}}{dt} = \beta_{1rel}u_{1rel} - \gamma_{1rel}\sqrt{h_{1rel}} \qquad (5)$$

$$\frac{dh_{2rel}}{dt} = \beta_{2rel}\sqrt{h_{1rel}} - \gamma_{2rel}\sqrt{h_{2rel}} \qquad (6)$$

where the plant parameters regarding the above equations result from measurements on the system (filling and emptying the tanks): $u_0 = 0.6$, $\beta_{1rel} = 0.0033$, $\gamma_{1rel} = 0.003$, $\beta_{2rel} = \gamma_{1rel} = 0.003$, $\gamma_{21rel} = 0.00378$, and *rel* refers to relative units..

According to Eq.(5) and (6) the plant is nonlinear. The main task is to control the level of the water in the second tank h_2 through controlled modification of the water input flow $q_1 = u_1$. The behaviour of the system can be modelled in Matlab/Simulink.

Linear control algorithms can be applied if linearization is executed on the system model. There are a lot of possibilities in case of this system: (1) - linearization around a certain working point, (2) - feedback linearization, (3) - static linearization based on the insertion the inverse static characteristics of the plant.

An important step in identification of the 2TS is the determination of its static characteristics. Defining as the input of the system the inflowing water amount u_1, and the output being the level h_2 (both in relative units), and not forgetting the presence of the first tank, where h_1 represents its water level (the first state variable), with its natural limitation to $h_{1max} = 1$, then the range of usable input values is reduced to the interval $u_1 \in [0 \ldots 0.9]$.

Furthermore, if the technique of inserting the inverse static characteristics is used, this "usable range" is even more limited to the interval where the inverse characteristics is defined, see Figure 2. In this case the 2TS model can be identified as a statically linear system. The linear dynamic part of the system can be approximated by a second-order lag element, the time constants of the measurement setup are between 100 and 500 seconds, depending on the working point.

3. CONTROL STRATEGIES APPLIED TO THE CONTROL OF THE TWO TANK SYSTEM

Based on linearization around a working point, and static linearization, three control strategies applied and tested on the 2TS are presented.

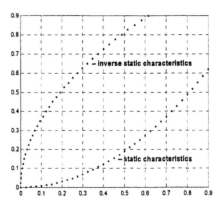

Fig. 2. Static- and inverse static characteristics of the 2TS

3.1. ADAPTIVE PI(D) CONTROL, BUMPLESS TRANSFER

The experiment on the 2TS has the following plan: first a working point is chosen, then identification is performed (using rectangular periodic signal of large period, the identification is open loop). At a certain time there is a switch from pure identification to adaptive PI control, using the RLS continually. This technique is widely used in practical applications, see [Kiong, 1999].

For exemplification let the working point be $u=0.2$, and the according output be $h_2=0.0305$. Identification is performed using rectangular signal (see Figure3). Simulations have been run in Matlab/Simulink. It can be noticed that in the beginning the plant output and the identified model output differ significantly, but after some periods they tend to be the same.

An adaptive PI controller was designed on basis of the plant model obtained through RLS on-line identification. Starting from the theoretical model, the experimental model is considered to be of second order. At each sampling period the continuous transfer function is calculated, and the PI controller is tuned according to the Modulus Optimum method (MO-method) [Åström and Haegglund, 1995], [Föllinger 1978]. Mainly, if the continuous transfer function is:

$$H_P(s) = \frac{A}{(1 + T_1 s)(1 + T_2 s)} \qquad (7)$$

and the PI controller is:

$$H_{C-PI}(s) = k_C \frac{T_I s + 1}{s} \qquad (8)$$

(one pole is cancelled). The closed loop transfer function results in form of:

$$H_r(s) = \frac{a_0}{a_2 s^2 + a_1 s + a_0} \qquad (9)$$

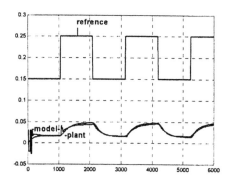

Fig.3. RLS identification of the 2TS

The method is based on the fulfilment of certain conditions in frequency domain imposed to the closed loop system ($H_r(j\omega)$ being the closed loop transfer function):

$$M(\omega) = |H_r(j\omega)| = 1 \qquad (10)$$

for an ω-Domain as big as possible.

This condition can be fulfilled if between the coefficients of the characteristic equation of the closed loop the following condition is insured [Åström and Hägglund, 1995]:

$$2\,a_0 a_2 = a_1^2 \qquad (11)$$

Applying this relation an "optimal" behaviour is obtained and the parameters of the PI controller are:

$$T_I = T_1, \quad k_C = \frac{1}{2AT_2} \qquad (12)$$

The system performances are considered "optimal", representing a damping coefficient of $\zeta \cong 0.707$, over-shoot $\sigma_1 \cong 4{,}3$ %, settling time $t_s \cong 9{,}6 \cdot T_2$ and phase margin $\varphi_m = 60°$. The controller is implemented as a digital one, so proceeding from form (8), its pulse transfer function is calculated. The adaptive tuning of the controller can be realised in Simulink by a separate block that is connected to the identified parameters, see Figure 4.

There is a switch between the reference signal and the control signal, which means that in the beginning there is only open loop RLS identification, followed by closed loop adaptive control, combined with identification.

A problem that must not be neglected is the bumpless transfer between two working modes. Generally, if there is a switch at time k between manual and automatic mode or between two control algorithms, even if suppose $e=0$ at switch time (particularly if $e_\infty=0$, steady state of the system), there will be a significant jump caused by the high value of the integral part, which leads to a very long settling time [Åström and Hägglund, 1995]. In some cases it can even lead to instability.

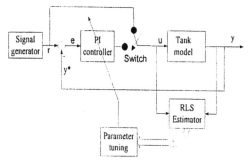

Fig.4. Scheme of adaptive PI control

An example of such bump transfer is presented in Figure 5, where at moment $t=10000$ sec there is a switch between open loop identification regime and closed loop adaptive control. The effects of the bump transfer can be well observed.

In order to reduce this effect, the following simple condition has to be insured:

$$u_{ID} = u_{CONT} = u_k \qquad (13)$$

where u_{ID} is the reference signal (used at identification), u_{CONT} is the control signal and u_K represents the control signal at switch time k. Explicitly, if the controller transfer function is

$$H_C(z^{-1}) = \frac{q_0 + q_1 z^{-1}}{1 - z^{-1}} = \frac{u(z^{-1})}{e(z^{-1})} \qquad (14)$$

it means that the control signal would be:

$$u_k = q_0 e_k + q_1 e_{k-1} + u_{k-1}\quad . \qquad (15)$$

One solution to implement bumpless transfer consists in considering e_{k-1} and u_{k-1} known signal values (open loop functioning), imposing condition (13) to the control signal, calculating a "virtual" value for e_k (eq. (16)) out of which the control signal in the next moment u_{k+1} is calculated (e_k being shifted to e_{k-1}) based on eq. (15) and imposed.

$$e_k = \frac{1}{q_0}u_k - \frac{q_1}{q_0}e_{k-1} + \frac{1}{q_0}u_{k-1} \qquad (16)$$

Further, the adaptive PI controller is used with no modifications. Implementing and simulating the bumpless transfer as defined above, the behaviour of the system is depicted in Figure 6.

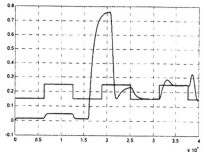

Fig.5. Bump transfer example

157

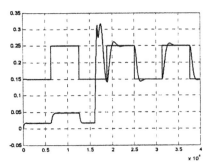

Fig. 6. Adaptive control with bumpless transfer

As expected, the bumpless transfer does not attenuate all oscillations at the switching. It is because, generally, at every point of the identification a new controller is designed, and at early moments of identification, when the model is poor, a good controller is designed for a bad model that can not control the real plant. At the switch time a controller is calculated, after the bumpless switching. In time, the model becomes more and more accurate, and the control performance accordingly, will be improved.

For the non-adaptive case of a PI controller behaviour of the bumpless transfer algorithm is calmer. In this case, at time $t=10000$ there is a switch from open loop to closed loop, calculating a PI controller for that working point with the transfer function:

$$H_{PI}(z) = \frac{0.25z - 0.225}{z - 1} \qquad (17)$$

The behaviour in this case is presented in Figure 7.

3.2 ADAPTIVE GPC CONTROL

In principle, predictive control is based on the idea of calculating the control sequence from the actual time point minimizing the deviation of the reference signal and output signal in the future horizon, based on a given cost function [Camacho and Bordons, 1998]. Let the cost function be:

$$J = \sum_{n_e=n_{u_1}}^{n_{e2}} \gamma_{yn_e} [y_r(k+d+n_e) - \hat{y}(k+d+n_e \mid k)]^2 +$$

$$+ \sum_{j=1}^{n_u} \gamma_{u,j-1} \Delta u^2(k+j-1) \Rightarrow \underset{\Delta u}{MIN}$$

(18)

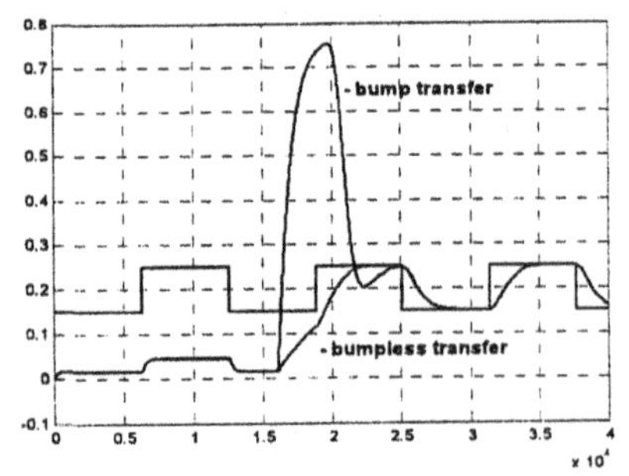

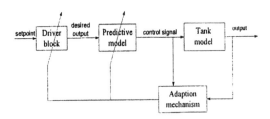

Fig. 7. Bumpless transfer of a non-adaptive PI control structure

where $n_{p2} - n_{p1}$ is the prediction (optimization) horizon, n_u is the control horizon (the number of the supposed consecutive changes in the control signal), $\gamma_{yn_{p1}}, ..., \gamma_{yn_{p2}}$ are weighting factors of the control error, and $\gamma_{u_0}, ..., \gamma_{n_u-1}$ are weighting factors of the control increments. The role of these factors is that on one hand the weighting of the control increments decreases the control effort on cost of a slower control with overshoots, on the other hand weighting the control errors accelerates the control and decreases the overshoots. All these are the tuning parameters of the controller, the quality of the control is depending on their choice.

By minimising the cost function a series of control increments are calculated, and according to different strategies, a certain number of them is applied as control input. One version of GPC is Adaptive GPC, where an on-line identification is performed and the model of the plant is changed accordingly. The tuning parameters (n_{p1}, n_{p2}, n_u, weighting factors) remain the same in this case. The scheme of an Adaptive GPC is depicted in Figure 8.

The driver block generates the desired output trajectory that will guide the process output, the predictive model calculates the control signal which insures that the plant output is close to the desired trajectory. The adaptation algorithm adjusts the predictive model parameters and also informs the driver block about the difference of the output to the reference trajectory, so that the reference trajectory may be redefined if needed [Sanchez et.al., 1996].

As application let the tuning parameters of the controller be: $n_{p1}=1$, $n_{p2}=10$, $n_u=2$, $\lambda=0.01$, $d=0$ (dead time). In the beginning there is a "learning period", where the system has PRMS (Pseudo Random Multilevel Signal) excitation with three levels. The PRMS is an alternative to other test signals that can be very useful in case of non-linear systems. First RLS identification has been executed with PRMS signal in open loop, then adaptive predictive control was used in closed loop tracking a rectangular reference signal. comparison is made based on the identification technique: identification of the non-linear plant without any linearization vs. static linearization (inserting the inverse static characteristics), the results are presented in Figure 9

Fig.8. Adaptive GPC control scheme A

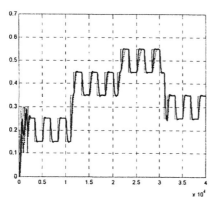

Fig.9. Adaptive GPC simulation results

(line: with inverse static characteristics, dot: without inverse static characteristics). It can be noticed that in both cases after the learning period the control is good. With the inverse static characteristics the performance is better.

A second comparison can be performed to analyze the effect of controller tuning parameters. The control scheme containing the inverse static characteristics is used, the prediction horizon is reduced:

(1) $n_{e1}=1$, $n_{e2}=10$, $n_u=2$;
(2) $n_{e1}=1$, $n_{e2}=5$, $n_u=2$.

The results are depicted in Figure 10.

It is seen that with shorter prediction horizon the transient behaviour is worse. The influence of other parameters can be analyzed in the same manner.

3.3 INTERNAL MODEL CONTROL WITH INVERSE STATIC CHARACTERISTICS

Internal Model Control (IMC) [Morari and Zafiriou, 1989], [Lunze, 1997] is a widely used control strategy. The main idea is that the controller itself already incorporates the model of the plant, as seen in Figure 11.

As presented in the literature (see for example, [Lunze, 1997], [Datta 1998] et. al.), if the plant

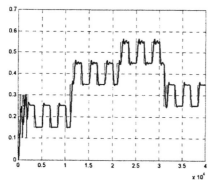

Fig.10. Adaptive GPC using different prediction horizons

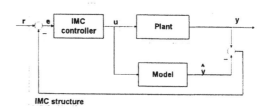

IMC structure

Fig.11. Structure of the IMC control system

is open loop stable then these control structures can be very efficient for some applications. If there is no mismatch between the plant and the model (see [Farkas et.al. 2002]), then the control is actually open loop, the best controller is the realizable (approximate) inverse of the plant. If there is mismatch, in the closed loop a feedback signal appears which works against this effect. Stability of the closed loop has to be ensured. A filter inserted in the feedback may reduce further the effect of mismatch.

In the case study of the 2TS the static non-linearity must be compensated. For this reason the IMC controller includes the inverse static characteristics. It has to be underlined again, that this method can be applied only where the inverse static characteristics is defined, in other regions different control algorithms have to be considered.

The limitation that characterizes the maximum flow that can be introduced into the plant (in order to avoid the command signal to step out of the physically possible range $0<u<0.9$ – at $u=0.9$ the water from the first tank starts to flow out) in this case will be placed inside the IMC structure. This way the control scheme will be as presented in Figure 12.

To see how the structure works, first let the simulation run for perfect match. The chosen working points have to be in the range where the inverse static characteristics is defined. The result is presented in Figure 13.

In this case the control system works actually in open loop. On the other hand, when there appears plant-model mismatch, let the identified model be for the working point 0.2. The continuous linearized model is a second order lag with time constants $T_1=450$ and $T_2=400$ approximately. If the model mismatch is drastic ($T_1=400$ and $T_2=100$) (this situation could

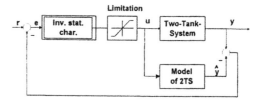

Fig.12. IMC control of the 2TS

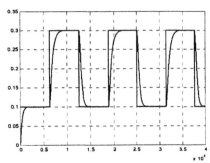

Fig.13. IMC control in case of perfect match

appear e.g. if online identification is performed and the model is changed accordingly), the modified pulse transfer function of the plant will be:

$$H_M(z) = \frac{0.0070z + 0.0064}{z^2 - 1.7182z + 0.7316} \quad (19)$$

In this case, if the obtained linear model is inserted into the IMC structure, preceded by the static characteristics of the 2TS, the following behaviour is obtained (Figure 14). It can be seen that even though there is significant mismatch between the plant and the model, the control still gives good performance.

4. CONCLUSIONS

In the paper different advanced control strategies have been applied to a model of a laboratory setup of a Two-Tank-System. The system is non-linear.

The presented study shows the effect of three control strategies. All steps of designing automatic control are incorporated: mathematical modelling, linearization, identification, classical conventional control with controller tuned according to "traditional" method (Modulus Optimum), adaptation of parameters in the working point, and also more advanced control techniques, as GPC and IMC.

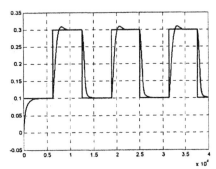

Fig.14. IMC structure for plant-model mismatch

The analysis of control strategies on the same process model is favourable since it allows the comparison of efficiency of different control structures. The presence of non-linearity brings the

didactical model closer to a real one. The control system is analysed both in simulation and on the real laboratory equipment.

The whole activity was accompanied by the use of Matlab/Simulink environment. The study of different control structures integrated into one case study can be welcome with much interest from students.

ACKNOWLEDGMENT

The authors' work from BUTE was supported by the fund of the Hungarian Academy of Sciences for control research and partly by the OTKA fund T042741. The research was also supported by the bilateral EU-Socrates-Erasmus cooperation between Budapest University of Technology and Economics and the University of Applied Science Cologne.
The author's work from the University of Applied Science Cologne has been supported earlier by the Ministry of Science and Research of NRW (FRG) in the program „Support of the European Contacts of the Universities / Förderung der Europafähigkeit der Hochschulen" and now is supported by the University of Applied Science Cologne in the program "Advanced Process Identification for Predictive Control" and by the program of EU-Socrates.
All supports are kindly acknowledged.

REFERENCES

Åström, K.J., Hägglund, T., (2000), *Benchmark systems for PID control*, PID'00 IFAC Workshop on Digital Control, Preprints, Terrassa (Spain), pp.181-182.

Åström, K.J., Hägglund, T., (1995), PID *Controller Theory, Design ant Tuning*, Instrument Society of America, Research Triangle Park.

Camacho, E.F., Bordons, C., (1998), *Model Predictive Control*, Springer Verlag, London

Datta, A., (1998), *Adaptive Internal Model Control*, Springer Verlag.

Farkas, I., Vajk, I., (2002), *Internal Model-Based Controller for a Solar Plant*, IFAC 15th Triennial World Congress, Barcelona, Spain

Föllinger, O., (1978), *Regelungstechnik*, Elitera Verlag, Berlin.

Kiong Tan KoK, et. al., (1999), *Advances in PID Control,(Advances in Industrial Control Series)* Springer Verlag, London, Berlin, Heidelberg

Lunze, J., (1997), *Regelungstechnik - 2* Springer Verlag.

Morari, M., Zafiriou, E., (1989), *Robust Process Control*, Prentice-Hall, Englewood Cliffs.

Sanchez, J.M.M., Rodellar, J., (1996), *Adaptive Predictive Control*, Prentice Hall, London-New York.

ELSEVIER

IFAC
PUBLICATIONS
www.elsevier.com/locate/ifac

ADAPTIVE MECHANISM FOR CONTROL OF SOCIAL STABILITY IN THE PROCESS OF TECHNOLOGY DEVELOPMENT

Tsyganov V.V.

*The Institute of Control Sciences, Russian Academy of Sciences,
Profsoyuznaya st., 65, Moscow, 117997, Russia
e-mail: shoubine@ipu.rssi.ru*

Bagamaev R.A.

*Commercial bank "VITAS',
Eniseyskaja, 22, Moscow, 108432, Russia,
e-mail: R.Bagamaev@vitasbank.ru*

Scherbyna N.N.

*Council of Minister of the Ukraine,
Institutskaja, 4, Kiev, 52020, Ukraine,
e-mail: scherbyna@kmu.gov.ua*

Abstract: The liberalization and centralization cycles of the societies, states and the world community are influenced by the technological changes rises gap between rich and poor countries. Big part of the world has no possibility for adaptation to these changes and need the support. Mechanism of such support should be adaptive, corresponding to the level of this gap in a way used in adaptive control. But "active" bureaucracy may predict the results of the adaptive procedures and to reach own aims. In many cases support given to poor countries had no effect because of the failure of the mechanism used (corruption, capital and brain drain etc.). Methodology of creating progressive adaptive and intelligent mechanisms for the stability of the societies, states and the world community is described. *Copyright © 2004 IFAC*

Keywords: stability, active element, adaptive, intelligent, control.

1. INTRODUCTION

Social stability of the society and the state is the result of changing the liberalization and centralization mechanisms and periods. Technological changes are applied to the liberalization and centralization cycles and provide the progress. But these changes produce also gap between rich and poor countries. Many poor countries of the third world have no possibility for learning and adaptation to these changes and need the support. Developed countries provide some efforts to diminish this gap and to improve international stability. Chestnut and Kopacek (1989) indicate that to reduce the risks and likelihood of the international instability it is possible to apply adaptive control principles for international conflict resolution. Adequate adaptive procedures provide identification of controlled object structure and parameters of environment and, finally, generating the controlling actions making use of current information, obtained from the elements, in order to achieve the optimal state of the system as a whole. The third world support should correspond to the level of this gap between rich and poor countries in a way used in adaptive control systems. From the other side, systemic investigation and the design of the adaptive systems to control organizations taking into account human factor effect. It should be understood as an activity manifestation of the people or collectives (elements of control systems) is caused by the availability of their own aims, not necessary coinciding with the goal of the system in entirely (Burkov and Tsyganov, 1986). Such elements may utilize available information channels connected with the system control center in order to improve the current or future state. From the other side, "active" people deals with the process of this support may predict the results of the adaptive control procedures and to use that knowledge to reach their own aims. In fact, in many cases support given from developed to poor countries had no effect because of the failure of the mechanism used (corruption, capital and brain drain, distortion of information etc.). For example, one of the important feature of such failure is the activity of the bureaucracy used their possibilities to manipulate supporting resources. Progressive adaptive and intelligent mechanisms for the international regimes functioning are intended for eliminate this activity. Methodology of creating such mechanisms is described below.

2. THE SOCIAL CYCLES AND THE TECHNOLOGICAL CHANGES

Changing the liberalization and centralization mechanisms of the society and state functioning are influenced by technological changes created by high technologies. Advanced technology produces both positive and negative effects of the societies, states and the world community stability. The technically advanced systems that are in existence, as well as those being built, can be used effectively to help people to realize the benefits that are possible with the present-day high technology. From the other side, potential of high-tech now is accumulated in a few major leading countries. Multinational corporations from these countries supply high-tech all over the world. To provide new high-tech it is necessary to realize R&D, know-how etc. That needs appropriate intellectual and financial resources. The prices of high-tech goods and services are mainly not competitive prices because of the monopoly on the results of R&D and know-how. From the other side, price of traditional goods, provided by developing countries, are under strong competitions. For this reason investigations and investments in high-tech are in most of cases preferable. Financial and intellectual resources drain from developing to leading countries. The result of the capital and brain drain is the lack of financial and intellectual potentials in developing countries. It becomes more and more difficult to provide R&D, know-how and new high-tech. Therefore technological development produces economical and social rupture between leading and developing countries. That provides likelihood of social crisis and instability, and as a consequence tensions between developed countries and the third world. Corresponding to the well-known business magazine "Euromoney", edited in UK, in 1999 almost one half of 180 countries in the world (85 countries in March and 87 in September) had credit rating, equal to zero in the same periods. More the one half of all countries (92 from 180) had no access to international bank finance. For example, up to year 2015 total capital flow from Russia is estimated as $600 billions. But Russian budget at the same period is estimated only as $800 billions. This provides obstacles to the realization of a more peaceful and stable set of domestic and international relations.

3. LIBERALISM AND THE CAPITAL DRAIN

Global economic liberty produces capital and brain drain from the third world. One of the important ways to diminish a gap between rich and poor countries is to minimize capital and brain drain. Let us consider is the problem of capital drain. The concept of adaptive control mechanism of international cooperation derived by Tsyganov (1990a) is applied to manage capital flows between leading and developing countries. The mathematical approach used the modified model of the developing active element described by Burkov and Tsyganov (1986). Under consideration is the owner provides two businesses take place in two countries. Potential of each business rises in accordance with the equation:

$$q_{it+1} = C_i q_{it} + B_i u_{it}, \qquad q_{i0} = q^*_i$$

where q_{it} – potential of i-th business, i=1,2, u_{it} –investments in i-th business in period t, C_i and B_i - positive coefficients, t - number of period, t = 0,1,....
The profit of each business is:

$$z_{it} = A_i q_{it}, \qquad A_i > 0$$

The total profit of the owner is:

$$z_t = z_{1t} + z_{2t} = A_1 q_{1t} + A_2 q_{2t}$$

Total investments of the owner u_t are equal to the total profit:

$$u_t = u_{1t} + u_{2t} = z_t$$

Tsyganov, Shubin, and Kulba (2003) considered the situation when the purpose of allocation of these investments is to maximize present value. Let us consider the case when the purpose of allocation of the investments is to maximize total profit in the period t:

$$V_T = \sum_{i=1}^{2} u_{iT} \quad \underset{u_t}{\rightarrow} \quad max$$

where ρ is a discount rate, $\rho < 1$, T - horizont for business planning (farseeing) of the owner.

Statement. The country 1 is called as the capital attractor if all the investments are allocated in the business takes place in this country:

$$u_{2t} = 0, \qquad t = 1,...,T-1$$

Theorem. The country 1 is the capital attractor if and only if

$$I_{1t} > I_{2t}, \quad I_{it} = A_i B_i [1 - (\rho C_i)^t]/(1 - \rho C_i),$$
$$i = \overline{1,2}, \quad t = \overline{1,T}$$

This theorem gives necessary and sufficient conditions about of the definite direction of the capital flow between two countries. With the aid of this theorem measurement means and criteria for monitoring can be found. In this approach $I_i = (I_{i1},...,I_{iT})$ is the vectors indicator of the investment climate. If $I_1 \gtreqless I_2$ then it should be income of the investments in country 1. Let us consider owner operating in different branches of economy. Then in branches in country 1, were investment climate is better, income of investments takes place, etc. Another important indicator (Q_i) is the total quantity of capital drain from the country should be calculate as a difference between outcome and income of the investments in different branches of economy. These two indicators are the functions of the parameters both the domestic and the international liberal mechanisms such as prices, efficiency of investment, taxes, trade and customs rules etc. Information about I_i and Q_i may provide indication of leading events for various nations that may describe conditions of normal, alert, and emergency operation for various potential trouble spots in the world. Possible courses should be developed for action to resolve situations corresponding to alert and emergency operation. Various possible alternative actions can be explored. Through the use of models, simulations, discussions with knowledgeable experts from both the countries involved, as well as with third-party experts, it should be possible to get various impressions of what may be the possible outcome of alternative actions. In the adaptive mechanism there are special procedures used for adjusting of parameters both international and domestic mechanisms to control capital flow. Simulation of progressive adaptive mechanisms of multistage negotiation based on the new information technologies considered by Tsyganov (1990b). Incentives and motivations for cooperation may be developed. In each of the countries involved the people responsible for the decision-making that causes a capital flow can be provided with various incentives for keeping the capital from the flight. In real situations capital and intellectual flows are managed by the special international procedures and mechanisms. They are called by Young (1982) and Krasner (1983) as international regimes. Coates and Seamen (1989) indicate that international regimes are closely linked with domestic political and economical procedures and mechanisms. Total mechanism to control capital and intellectual flows includes both international and domestic regimes. For example, detailed description both these regimes, dealing with capital flows through joint ventures, had been given by Tsyganov (1991b).

4. CENTRALISM AND THE ADAPTIVE MECHANISM OF SOCIAL STABILITY

Adjustment of the states and the international mechanisms to prevent capital flight discussed in the previous item may be considered as an indirect way of support to avoid instability. The other way is a straight support of poor country with the aid of international organization such as International Monetary Fund etc. Concept for designing of the adaptive control mechanisms for international cooperation (including rates, incentives, taxes, norms, etc.) had been considered by Tsyganov (1990a). This approach is based on the analysis of the problem dealing with designing the adaptive mechanism of functioning (AMF) of the two-level organization included the Center on the upper level and the Agent as the farseeing active element on lower level. The role of Center is played by international organization. The role of Agent is played by the government of the supported country. The AMF includes both adaptive procedure **A** for parameter estimation and procedures of decision making: planning **P**, control **C** and stimulation **S** (see fig. 1).

A method to avoid distortion of information consists of designing the so-called progressive AMF wherein the present value of the Agent long-term goal function corresponding to the solution of the game with the Center increase with the growth of the efficiency of the Agent functioning (Tsyganov, 1986).

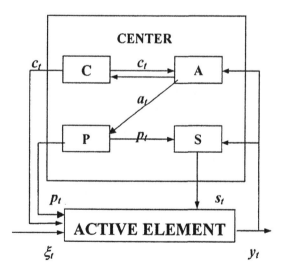

Fig. 1. AMF structure. Here a_t – adaptive parameter, p_t – plan, c_t – control, s_t – stimuli, ξ_t – noise, y_t – output, t – time period.

The designing of AMF caused the development of the approach based on the obtained problem solutions of progressive AMF synthesis. Burkov and Tsyganov (1987) made detail consideration of two AMF main types. The first one is intended for maintaining the processes of the state forecasting, planning and control of the Agent. Under consideration are the adaptive procedures of time series forecasting and designing of regressive model. The second type i.e. rank AMF, is designed to provide information for learning of decision making (classification and pattern recognition). They are used mainly for adaptive estimation of the parameters of the decision making procedure, control and stimulation of the Agent. In some cases both expert adaptive mechanisms derived by Tsyganov (1991b) and intelligent functioning mechanisms considered by Tsyganov, Borodin and Shishkin (2004) can be used.

5. INTELLIGENT LIBERAL AND CENRILIZED MECHANISMS OF SOCIAL STABILITY

Developed liberal and centralized mechanisms may be used for diminishing rupture between and poor countries. Real state and international regimes are much more comprehensive and need intelligent control systems. Drawing from experience gained in implementing intelligent control to a varied range of large scale systems, Tsyganov, Borodin and Shishkin (2004) highlights the need for a multilevel self-learning and self-organized systems. Particular attention is directed toward adaptations of the widely used self-learning algorithms in an attempt to increase the effective applicability, range of self-organizing control with the aid of artificial intelligence methodology. On the other hand, as a rule the possibility to control of a complex organization in dynamics with no complete information, is based on intelligent information systems (IIS) derived by Burkov and Tsyganov (1987). They realizes model, identification of control objects structure and outside parameters, predictions, and forms the base of control actions on the basis of current information, received from the elements, to attain the systems aim on the whole. To avoid information distortion, passed by the elements to the center of the system, it is necessary to consider the problem of information system designing in a total problem of procedures synthesis such as planning, regulation and stimulation accepted in this control mechanism. In recent years the newly appeared direction in the theory rapidly develops as well as the practice control for hierarchic organizations with a stochastic structure. It has been connected with the designing of the intelligent functioning mechanisms (IFM). The IFM includes IIS, and procedures of planning, regulating and stimulating. In the IFM information received in the process of element functioning is used by the center for decision-making and achievement of the systems aim. These IFM ensure possibility to identify the internal structure of elements and their parameters as well as the utilization of internal elements resources in accordance with the center goal. Tsyganov and Borodin and Shishkin (2004) indicate that the main types of IFM are:

- Learning functioning mechanisms,
- Self-organizing mechanisms,
- Expert intelligent mechanisms.

Learning functioning mechanisms (LFM) provide the possibility of estimating the parameters of the organizational potential in its dynamics, supplying more information to plan the organization output indices at the account of learning processes. Self-organizing mechanisms should combine learning and planning for output organization indices (the way it is done in LFM) with the control of organization inputs, i.e. direct influence on the potential of an organization. In expert intelligent mechanisms (EIM) the knowledge base is the part of IIS. EIM combines learning with indistinct and qualitative commands from the center and control on these commands basis. To design such mechanisms it is necessary to create hierarchic computerized systems with such intelligent possibility as multi-level learning. The knowledge base consists of:

- main knowledge base includes well-known dependences, any accurate data and the results of individual and collective expertise,
- system of knowledge acquisition functioning in an interactive mode with the decision makers who are responsible for a problem solution and answering the question: "what may actually happened, if…?".

With respect to EIM synthesis the idea of theoretical results consists in the fact that with the sufficiently flexible information usage to solve problems of planning, control and incentives the optimum will be reached. As a result – there appears the possibility of potential identifying on extra basis of received information and gradual slow output to the required level of development. The developed approach is directed to creation of EIM, including procedures of analysis and forecasting of economic elements potential with a high degree of approximation and procedure of decision-making. The approach suggested to the solution of the problem of adaptive control for the international regimes implemented by Grishutkin and Tsyganov (2001) to global management of the intangible technologies.

6. CONCLUSION

Both liberal and centralized intellectual mechanisms may be used to diminish rupture between rich and poor countries. Many poor countries of the third world have no possibility for adaptation and need the support. Mechanism of such support should be adaptive. Usually provision of ability to control complex international organization in its dynamics with incomplete information is, as a rule, based on application of adaptive control. From the other hand, control of hierarchic international systems implies the consideration of a special type of human factor - elements activity connected with the availability of the elements own goals. The center obtains information from the active elements in the course of their functioning and uses it for estimating their states to reach the aim of the control. But the farseeing active element may predict the center controlling action and chooses its states in such a way that its effects on the results of state estimation and adaptive control to maximize its own goal function. For this reason the problem of the designing of progressive adaptive functioning mechanism (including estimation, planning, control and stimulating procedures) for the hierarchical international regimes should be taken into account.

REFERENCES

Burkov, V.N. and V.V. Tsyganov (1986). Stochastic Mechanisms of the Active System Functioning. In: *Preprints of the 2ⁿᵈ IFAC Symposium on Stochastic Control*. vol.1, pp.259-263. Nauka, Moscow.

Burkov, V.N. and V.V. Tsyganov (1987). Adaptive information system to control organization activity. In: *Preprints of IFIP TC 8 Conference on Governmental and Municipal Information Systems*. pp.85-95. Budapest.

Chestnut, H. and P. Kopacek (1989). Supplemental Ways for Improving International Stability. In: *Report on the IFAC/EPCOM Working Group (WG 7.2) "Control Engineering and International Conflict Resolution"*. Vienna.

Coates, J.F. and S. Seamen (1989). The Relation of Management to Control Technology – Futher Applied Studies in Creative Management of Potential Conflict at International Levels. In: *Preprints of the IFAC SWIIS Workshop "International Conflict Resolution Using Systems Engineering"*. pp.47-54. Budapest.

Grishutkin, A.N. and V.V. Tsyganov (2001). Progressive Adaptive Mechanisms of Globalization. In: *Proceedings of the International Conference on Systems Cognition*. Vol. 2, pp.101-107. IPU RAN, Moscow.

Krasner S. (1983). *International Regimes*. Cornell University Press, N.-Y.

Tsyganov V.V. (1986). Adaptive control of Hierarchical Socio-Economic Systems. In: *Preprints of the 4ᵗʰ IFAC/IFORS Symposium @Large Scale Systems. Theory and Applications*. Vol. 2, *pp. 694-698*. Pergamon Press, Zurich.

Tsyganov V.V. (1990a). Modelling of Adaptive Control Mechanism for International Cooperation. In: *Report on the Research Workshop "Models and concepts of interdependence between nations". "Soviet-American Dialogue in the Social Sciences"*, p. 74. National Academy Press, Washington, D.C.

Tsyganov, V.V. (1990b). Simulation of Progressive Adaptive Mechanisms of Multistage Negotiations and the New Information Technologies. In: *Preprints of the 11ᵗʰ IFAC World Congress*. Vol.1, pp. 202-206. Tallinn.

Tsyganov V.V. (1991a). *Joint venturing in the USSR*. Radianska osvita, Kiev.

Tsyganov, V.V. (1991b) Expert adaptive mechanisms. In: *Proceedings of the 8ᵗʰ Conference on Systems Engineering*. Vol.1, pp. 92-95. Coventry.

Tsyganov, V.V., Borodin, V.A., and Shishkin, G.B. (2004). Intelligent Enterprize: Mastering Capital and Power. Logos, Moscow.

Tsyganov V.V., Shubin A.N., and Kulba V.V. (2003). Adaptive Mechanisms for Improving International Stability at the Technological Changes. In: *Preprints of the 10ᵗʰ IFAC Conference on Technology and International Stability*. pp. 75-79. Waterford, Ireland.

Young O.R (1982). Regime Dynamics: The Rise and Fall of International Regimes. *International Organizations*, **36**, pp.43-52.

EFFICIENT NON-LINEAR AND NONPARAMETRIC PREDICTION

Zaka Ratsimalahelo

University of Franche-Comté
U.F.R. Science Economique, CRESE
45D, Av., de l'Observatoire
25 030 Besançon-France
Phone:+33(0)381 666 759
Fax: +33(0)381 666 737
E-Mail: Zaka.Ratsimalahelo@univ-fcomte.fr

Abstract: In this paper, we propose a class of measures of predictability that are based on the nonlinear measures of multivariate dependence for random vectors. We consider a class of measures of dependence based on the Fisher Information matrix which is related to the predictability. These measures capture both linear and nonlinear dependence. Moreover, these measures of predictability are conceptually very general as they can be used for a set of variables which can be a mixture of continuous, ordinal-categorical and nominal-categorical variables. The approach is purely non-parametric in the sense that it is based only on data and makes no assumption on the nature of predictive models.
Its properties, including its relationship to the notion of linear predictability and to the notion of the mutual information are presented. An application to financial time series is also given. *Copyright © 2004 IFAC*

Keywords: Forecasting; Fisher's information matrix; Non-linear dependence; Predictability measures; Nonparametric prediction; Mutual information; Dirichlet distribution.

1. INTRODUCTION

One of the most important challenges in business, finance and economic activity is the prediction of future events. Forecasting has been dominated by linear statistical methods for several decades. Although linear models possess many advantages in implementation and interpretation, they have serious limitations in that they cannot capture nonlinear relationships in the data which are common in many complex real world problems (see Granger and Terasvirta (1993)). The approximation of linear models to complicated nonlinear forecasting is often not satisfactory.

During the past two decades, a number of nonlinear time-series models have been developed such as:

the threshold autoregressive (TAR) model first proposed by Tong (1978), (see also Tong (1995)),

the self-exciting threshold autoregressive (SETAR) model of Kräger and Kugler (1993),

the smoothing transition autoregressive (STAR) model of Chan and Tong (1986),

the autoregressive conditional heteroscedastic (ARCH) model of Engel (1982) and its extensions,

the generalized autoregressive conditional heteroscedastic (GARCH) model of Bollerselv (1986),

the Exponential GARCH (EGARCH) model of Nelson (1991).

While these models can be useful for a particular problem and data, they do not have a general appeal for other applications. The a priori specification of the model form restricts the usefulness of these parametric nonlinear models since there are too many possible patterns. In fact, the formulation of an appropriate nonlinear model to a particular data set is a very difficult task compared to linear model building because there are more possibilities, many more parameters and thus more mistakes can be made, Granger (1993).

Furthermore, one particular nonlinear specification may not be general enough to capture all non-linearities in the data. As opposed to the model-based nonlinear methods, we investigate in this paper a nonparametric approach which can capture nonlinear data structures without prior assumption about the underlying relationship in a particular problem. This approach is a more general and flexible modelling and analysis tool for forecasting applications in that it not only can find nonlinear structures, but also model linear processes.

As the prediction is based on the assumption of the dependence between the exogenous (input) and the endogenous (output) variables. Darbellay (1998) used the measures of dependence as measures of predictability. He developed the mutual information of two random vectors as measure of predictability. The advantage of his approach is that the mutual information takes into account the full dependence structure of both linear and non-linear dependences. Unfortunately, his measure of predictability is not always larger than a measure of linear predictability. In this paper, we propose a more general approach where the class of measures of practicability is larger than a measure of linear predictability. We consider a class of measures of dependence based on the Fisher Information matrix which is related to the predictability. These measures of dependence capture both the linear and the non-linear dependence. Moreover the Fisher information has another advantage because it is related to Shannon's form of entropy and to Kullback-Leibler entropy see (Frieden, (1998)).

The paper is organized as follows. In Section 2 the concept of Fisher information matrix and the Cramer-Rao bound are developed. We develop a class of measures of dependence based on the Fisher Information matrix which is related to the predictability. In Section 3 it is shown that if the exogenous and the endogenous variables are normal distribution, then particular functional forms of the class of measures of predictability are nothing else than the measure of linear predictability. Moreover, its relationship to the notion of mutual information is proven. In Section 4, using bivariate inverted Dirichlet distribution, we show that the measure of dependence exists even if the correlation coefficient does not exist. In Section 5 an application of finance time series and simulation is given.

2. PREDICTABILITY AND FISHER INFORMATION

2.1 Fisher information matrix

In this section, we first define the Fisher information matrix. Let $X = (X_1,...,X_p)$ and $Y = (Y_1,...,Y_q)$ be two vectors of continuous variables taking real values with a joint probability density function $f(x,y)$. The vector X is the exogenous vector and Y is the endogenous vector. Each X_i can be understood as a predictor variable and each Y_j is to be predicted. The above notation is more compact, indeed, in multivariate time series

analysis, we consider a process $\{y_t\}$ where y_t is a m-dimensional vector constituted of endogenous variables and a process $\{x_t\}$ where x_t is a n-dimensional vector of exogenous variables. Thus, we define the vector Y as the future endogenous vector by $Y = \{y_t', y_{t+1}',...,y_{t+f}'\}'$ where Y is a $q = m(f+1)$-dimensional vector and where f can be chosen freely according to some defined forecasting horizon. Defined further, vector X contains a vector of past endogenous variables and past and present exogenous variables; that is $X = \{y_{t-1}',...,y_{t-l}', x_t', x_{t-1}',...,x_{t-h}'\}$ where X is a $p = (l+h+1)$-dimensional vector. Note that we may consider particular cases (i) $X = \{y_{t-1}',...,y_{t-l}'\}$ which contain only the past of endogenous vectors and (ii) $X = \{x_t', x_{t-1}',...,x_{t-h}'\}'$ without the past of endogenous vectors The procedure developed below can be used for both one and multiperiod-step prediction.

The derivative of the log likelihood function with respect to x and y are defined by

$$S_x = \frac{\partial log(x,y)}{\partial x} \qquad (1)$$

and $$S_y = \frac{\partial \log f(x,y)}{\partial y} \qquad (2)$$

respectively, that is the score function vectors of order p and q respectively.

We have the following property that the expectation of the score function vectors is zero:
$$E(S_x) = 0 = E(S_y).$$

Let us denote $S' = (S_x', S_y')$, the Fisher Information matrix $J(X,Y)$ of order $(p+q) \times (p+q)$ be the matrix variance-covariance of the score functions.

$$E(SS') = J(X,Y) = \begin{bmatrix} J_{11} & J_{12} \\ J_{21} & J_{22} \end{bmatrix} \qquad (3)$$

where the matrix J_{11} of order $(p \times p)$ is defined by

$$J_{11} = \iint (\frac{\partial \log f(xy)}{\partial x_i} \frac{\partial \log f(x,y)}{\partial x_j}) f(x,y) dx dy$$
$$\text{for } i,j = 1,...p,$$

the matrix J_{22} of order $(q \times q)$

$$J_{22} = \iint (\frac{\partial \log f(x,y)}{\partial y_i} \frac{\partial \log f(x,y)}{\partial y_j}) f(x,y) dx dy$$
$$\text{for } i,j = 1,...,q,$$

and the matrix J_{12} of order $(p \times q)$

$$J_{12} = \iint (\frac{\partial \log f(x,y)}{\partial x_i} \frac{\partial \log f(x,y)}{\partial y_j}) f(x,y) dx dy$$

for $i = 1, ..., p$ and $j = 1, ..., q$.

$J_{21} = J_{12}'$ (Prime denotes the transpose of matrix and vector).

2.2 The global Cramer-Rao lower bound.

In this section, we will show some relationships between the Fisher information matrix and the global Cramer-Rao bound (C.R.B.).
Mayer-Wolf (1990) and Kagan and Landsman (1999) established the relation between the covariance of a random vector and the Fisher information matrix.
The variance-covariance matrix of Y and X is defined as

$$Cov(X, Y) = \Sigma(X, Y).$$

So $\Sigma(X, Y) \geq J(X, Y)^{-1}$ where $J(X, Y)$ is the Fisher information matrix defined in (3).

It is interesting to forecast the endogenous variable Y based on an information set X (exogenous variable). The best prediction is the conditional expectation defined by $\hat{Y} = E(y_{t+f} / X) = \varphi(X)$, where $\varphi(X)$ is a linear or non-linear function, the MSE is bounded by the following relation:

$$E\left[(Y - \hat{Y})(Y - \hat{Y})'\right] \geq J(X, Y)^{-1} \qquad (4)$$

This relation is called the global Cramer-Rao bound (see Bobrovsky, B. Z., Mayer-Wolf E., and Zakai, M. (1987)).

2.3 A class of predictability measures.

We consider a class of predictability measures in terms of the Fisher information matrix $J(X, Y)$.

Definition. Let $h(w)$ be a non-negative increasing or decreasing one to one function of $\omega(\in \Re)$ and possesses continuous derivatives at least up to the first order such that $h(0) = 0$ and $h'(0) = 1$.

Particular examples of functions $h(w)$ satisfying the conditions of definition are $h(\omega) = \left[\exp(\mu\omega) - 1\right] / \mu$, $\mu \geq 0$, and the Box-Cox transformation $h(w) = \left[(1 + \omega)^{\mu} - 1\right] / \mu$, $\mu \geq 0$. Other familiar examples are the logarithmic function $h(\omega) = -\log(1 - \omega)$, the identity function $h(\omega) = \omega$ and the function $h(\omega) = \omega / 1 - \omega$. These functional forms are considered further later in this paper.

We construct a class of measures of predictability as a measure of dependence $\delta(X, Y)$ between X and Y defined by:

$$\delta(X, Y) = \sum_{i=1}^{n} h(\lambda_i^2) \qquad (5)$$

where $\lambda_1^2, ..., \lambda_n^2$ are the non-zero eigenvalues of $J_{11}^{-1} J_{12} J_{22}^{-1} J_{21}$.

The number $\delta(X, Y)$ captures the full dependence, both linear and non-linear between X and Y and can be interpreted as the predictability of Y by X. This measure of predictability does not need any particular model to predict Y from X.

Note that the above definition characterizes, a class C of measures of predictability between X and Y. We now give the properties of a class C of measures of predictability. One can verify that $\delta(X, Y)$ have the following properties:

(1) $\delta(X, Y) = 0$ if X and Y are independent or equivalently X contains no information on Y, which means that X cannot be predicted by means of Y. Hence, we have $\delta(X, Y) = 0$ if and only if $J_{12} = E(S_x S_y') = 0$. This result is a new condition for the independence of the random vectors X and Y, which is equivalent to

$$\frac{\partial^2 \log f(x, y)}{\partial x \partial y} = 0$$

(2) $\delta(X, Y) = 1$ if there exists a one to one relationship between X and Y.
(3) It is invariant under non-singular linear transformation for the variables X and Y.
(4) It is unchanged by the interchange of X and Y, that is $\delta(X, Y) = \delta(Y, X)$.

A natural postulate for an appropriate measure of dependence states that the value of a measure of dependence, when it is applied in a bivariate normal model, must be equal to the absolute value of the correlation coefficient of the model. This postulate is investigated in the following example for the measure $\delta(X, Y)$.

3. NORMALITY

3.1. Multivariate normal distribution

Let (X, Y) be a vector of $(p + q)$ normal random variable with mean μ and covariance Σ.

$$(X, Y) \sim N(\mu, \Sigma),$$

where $\mu = \begin{pmatrix} \mu_x \\ \mu_y \end{pmatrix}$ is the $(p+q)$ vector of expectation of random vectors (X,Y) and $\Sigma = \begin{bmatrix} \Sigma_{11} & \Sigma_{12} \\ \Sigma_{21} & \Sigma_{22} \end{bmatrix}$ is the $(p+q) \times (p+q)$ variance-covariance matrix of (X,Y).

According to Mayer-Wolf (1990) and Kagan and Landsman (1999), if (X,Y) is a multivariate normal, then the variance-covariance of (X,Y) is equal to the inverse of the Fisher Information matrix

$$Cov(X,Y) = J(X,Y)^{-1}$$

This is equivalent to

$$J(X,Y) = [Cov(X,Y)]^{-1}$$

Thus we have

$$J(X,Y) = \begin{bmatrix} J_{11} & J_{12} \\ J_{21} & J_{22} \end{bmatrix} = \begin{bmatrix} \Sigma_{11} & \Sigma_{12} \\ \Sigma_{21} & \Sigma_{22} \end{bmatrix}^{-1}$$

Proposition.

If (X,Y) is a multivariate normal, then we have the following equalities of matrices $\Sigma_{12}\Sigma_{22}^{-1}\Sigma_{21}\Sigma_{11}^{-1} = J_{11}^{-1}J_{12}J_{22}^{-1}J_{21}$ or equivalently $\Sigma_{21}\Sigma_{11}^{-1}\Sigma_{12}\Sigma_{22}^{-1} = J_{22}^{-1}J_{21}J_{11}^{-1}J_{12}$ and let $\lambda_i^2 \ \ i = 1,...,k$ be the non zero eigenvalues of $J_{11}^{-1}J_{12}J_{22}^{-1}J_{21}$ or $J_{22}^{-1}J_{21}J_{11}^{-1}J_{12}$, where $k = rank(J_{12})$. Hence, the λ_i^2 are the non zero canonical correlations between X and Y ■

Let $r_i^2 \ \ i = 1,...,k$ be the non zero eigenvalues of $\Sigma_{12}\Sigma_{22}^{-1}\Sigma_{21}\Sigma_{11}^{-1}$ or $(\Sigma_{21}\Sigma_{11}^{-1}\Sigma_{12}\Sigma_{22}^{-1})$ where $k = rank(\Sigma_{12})$. They are also the canonical correlations as defined in the multivariate statistical analysis, (Anderson (1984)). According to the equalities of matrices, we have also the equalities of the non zero eigenvalues $r_i^2 = \lambda_i^2$ for $i = 1,...,k$. Hence the λ_i^2 for $i = 1,...,k$ which are the non zero eigenvalues of $J_{11}^{-1}J_{12}J_{22}^{-1}J_{21}$ or $J_{22}^{-1}J_{21}J_{11}^{-1}J_{12}$ can be interpreted as the canonical correlations between X and Y.

Then for the logarithmic function $h(\omega) = -\log(1-\omega)$, we have

$$\delta(X,Y) = 1 - \prod_{i=1}^{k}(1-\lambda_i^2) \qquad (6)$$

Measure $\delta(X,Y)$ given by equation (6) is a function of the canonical correlations.

Using a different approach, the same result was obtained by Darbellay (1998). He used the mutual information of (X,Y) denoted by $I(X,Y)$. Consequently, when the random vector (X,Y) is normally distributed, the measure of predictability $\delta(X,Y)$ is equal to the mutual information $I(X,Y)$. The canonical correlations are frequently used to assess the degrees of linear dependence between X and Y.

3.2. Particular cases

(i) For $q = 1$, the measure of predictability is equal to the square of the multiple correlation coefficient $\delta(X,Y) = R^2(X,Y)$.

(ii) For $p = 1$ and $q = 1$, the measure of predictability is equal to the square of the correlation coefficient. $\delta(X,Y) = \rho^2(X,Y)$.

The covariance matrix is often related to the accuracy of linear prediction while the inverse of the Fisher information is related to the accuracy of the best (usually) highly non-linear prediction. We have shown that if the vector (X,Y) is multivariate normal then $\delta(X,Y)$ is equivalent to linear predictability.

Darbellay (1998) also obtained the same result. He used a measure of dependence based on mutual information. Moreover, Darbellay's measure of dependence is not always greater than linear predictability.

In the general case, the relation between the predicted Y and the predictor vectors X is not always linear in which case the correlation coefficient linear does not exist. We now show that the measure of dependence exists even if the correlation coefficient does not exist.

4. BIVARIATE INVERTED DIRICHLET DISTRIBUTION

In this section we will show that the measure of dependence defined above can capture the relation between the exogenous and endogenous variables even if the correlation coefficient does not exist. We suppose that the random vector (X,Y) follows the bivariate inverted Dirichlet distribution with joint density,

$$f(x,y) = \frac{\Gamma(2m+\alpha)x^{m-1}y^{m-1}}{\Gamma^2(m)\Gamma(\alpha)(1+x+y)} \quad x \geq 0, y \geq 0 \text{ and}$$

$m \succ 0, \alpha \succ 0$, where m and α are the parameters of joint density and $\Gamma(.)$ is the gamma function.

Thus the Fisher type information matrix is

$$J(X,Y) = \frac{\alpha(\alpha+1)}{2m+a+1} \begin{bmatrix} \dfrac{m+\alpha+3}{m-2} & -1 \\ -1 & \dfrac{m+\alpha+3}{m-2} \end{bmatrix}$$

$m \succ 2$ and $\alpha \succ 0$.

Hence for $h(\omega) = \omega$, the measure of dependence is equal to

$$\delta(X,Y) = \left(\frac{m-2}{m+\alpha+3}\right)^2. \qquad (7)$$

Moreover the correlation coefficient of X and Y is given by

$$\rho(X,Y) = \frac{m}{m+\alpha-1} \text{ for } \alpha \succ 2. \qquad (8)$$

We can see that both $\delta(X,Y)$ and $\rho(X,Y)$ decrease with α, and increase with m. According to equation (8), the correlation coefficient of X and Y does not exist for $\alpha \prec 2$.

5. APPLICATION TO FINANCIAL TIME SERIES

5.1 The data.

We use a $20-$year sample period from 1/1/1980 to 30/6/2000, or 5,348 daily observations, to compute log return time series on the following financial assets: (*i*) daily stock returns on the Dow Jones Industrial Average (DJIA), the NASDAQ, FTSE-All Share, and Nikkei indices, (*ii*) daily returns on the JPY/USD, GBP/USD and SFR/USD exchange rates, and (*iii*) daily returns on the 3-month Eurodeposit rates for the US dollar, pound sterling and deutschemark[1]. Using the ADF tests all the series in log returns are stationary, so we can look at correlations.

5.2 Estimating δ for financial returns: whole sample.

In general, we consider $X = (X_1,...,X_p)$ exogenous (explanatory) variables and $Y = (Y_1,...,Y_p)$ endogenous (dependent) variables. Recall from Section 3.2 that if $p = q = 1$ then $\delta(X,Y) = \rho^2(X,Y)$, that is the square of the correlation coefficient equivalent to the R^2 of an OLS regression of Y on X. However, if $p \succ 1$

___

[1] All data is from Datastream

then $\delta(X,Y) \succ \rho^2(X,Y)$, captures the underlying non-linear dependence.

Estimating $\delta(X,Y)$ under the assumption of multivariate normality is done in three stages. First, the Fisher information matrix $J(X,Y)$ is computed as the inverse of the variance--covariance matrix. At this stage we also want the standard errors of Σ to translate into standard errors of J. Then, the canonical correlations are computed as the eigenvalues of $J_{11}^{-1}J_{12}J_{22}^{-1}J_{21}$.

Over the whole sample ($N = 5,348$), the correlations between pairs of assets returns are shown in Table 1:

Table 1: R^2 and δ : whole sample

GBP/USD → FTSE	DJIA → NASDAQ
$R^2 = 0.002$	$R^2 = 0.44$
$\delta(r_{US}) = 0.0087$	$\delta(r_{US}) = 0.4402$
$\delta(r_{US}, DJIA) = 0.1130$	$\delta(r_{US}, r_{GER}) = 0.4403$
$\delta(r_{US}, DJIA, NASDAQ) = 0.1549$	$\delta(r_{US}, r_{GER}, JPY/USD) = 0.440$

On the one hand, Table 1 makes it clear that $\delta(X,Y)$ can increase dramatically over R^2 if the latter is very small, that is if the two return series are linearly unrelated over the sample. This is the case for the GBP/USD and FTSE returns. On the other hand, if linear dependence is already very significant as is the case for DJIA and NASDAQ returns over the $20-$year period then the non-linear measure will contribute very little.

5.3 Estimating δ for financial returns: rolling windows.

Although the data over the whole sample is asymptotically normal, in practical applications, the presence of many structural breaks implies that practitioners often have to rely on correlations based on the most recent observations, in particular those over the last month. We therefore also compute $22-$day (1 working month) non-overlapping correlations for pairs of assets and examine the evolution of linear (ρ) and non-linear $\delta(X,Y)$ dependence over the whole sample period.

Over our $20-$year sample period, this yields 242 non-overlapping periods of returns for the various assets. We consider the case of the correlation between GBP/USD returns defined w log to be the exogenous, or explanatory variable and FTSE returns, defined to be the endogenous, or dependent variable. For δ, we additionally consider Eurodollar interest rate returns and JPY/USD returns as exogenous variables helping to explain FTSE returns. Using our earlier terminology, we therefore have $p = 3$ and $q = 1$.

171

Next, we consider the same set of variables, except that we now use DJIA returns rather than JPY/USD returns as an exogenous variable. The means and standard deviations of linear dependence (R^2) over the rolling windows then are the same, but those for δ are different. They are reported in the following table 2 .

Table 2: R^2 and δ : rolling window statistics

$$\mu_{R^2}(GBP/USD, FTSE) = 0.10$$
$$\sigma_{R^2}(GBP/USD, FTSE) = 0.11$$

δ :JPY/USD	δ :JPY/USD, DJIA
$\mu_\delta = 0.21$	$\mu_\delta = 0.25$
$\sigma_\delta = 0.14$	$\sigma_\delta = 0.15$
$\mu_{\delta-R^2} = 0.11$	$\mu_{\delta-R^2} = 0.16$
$\sigma_{\delta-R^2} = 0.10$	$\sigma_{\delta-R^2} = 0.12$

Clearly, the average δ of a 1−month rolling window of data increases as we substitute a return series (GBP/USD) which is less relevant to the one that we are trying to forecast (FTSE) with a more relevant one (DJIA). The average standard error of the δ over the 1−month rolling window also goes up, but less than above.

6. CONCLUSION

In this paper, we have proposed a class of predictability measures based on the Fisher Information matrix . It presents the advantage of capturing both linear and nonlinear dependences. Moreover, the nonlinear predictability measure $\delta(X,Y)$ exists and can be used in cases where the correlation coefficient does not exist.

At present, we are applying theses measures to real-world data, including financial time series. Our objective is to understand the type of problems for which it works well. It appears that the measures are particularly well suited for relative large data sets with nonlinear structure.

Finally, the proposed method is useful not only for the analysis of the general predictability but also for that of causality. We can apply the proposed method to the detection of causality between two time series. We can study linear and non-linear causality separately.

REFERENCES

Anderson, T. W. (1984). *An introduction to multivariate statistical analysis*, 2nd ed. Wiley, New-York.

Bobrovsky, B. Z., Mayer-Wolf E., and Zakai, M. (1987). Some classes of global Cramer-Rao Bounds, *Annals of Statistics*, **15, 4**, 1421-1438.

Bollerslev, T. (1986). Generalized Autoregressive Conditional Heteroskadasticity, *Journal of Econometrics*, **31**, 307-327.

Chan W. S. and Tong H. (1986). On tests for non-linearity in time series analysis. *Journal of Forecasting*, **5**, 217-228.

Darbellay, G. A. (1998). Predictability: an information-theoretic perspective, In: *Signal Analysis and Prediction* Prochazka, A. et al. eds. Birkhauser, Boston.

Diebold F. X. and Kilian L. (2000). Measuring Predictability: Theory and Macroeconomic Applications, Working Paper

Engel R. F. (1982). Autoregressive conditional heteroscedasticity with estimates of the variance of U.K inflation, *Econometrica*, **50**, 987-1008.

Granger CWJ (1993). Strategies for modelling nonlinear time-series relationships, *The Economic Record*, **69**, 233-238.

Granger CWJ, and Terasvirta T. (1993). *Modelling nonlinear economic relationships*, Oxford University Press, Oxford.

Frieden R. B. (1998). *Physics from Fisher Information*, Cambridge University Press.

Kagan A. and Landsman Z. (1999). Relation between the covariance and Fisher information matrices. *Statistics and Probability Letters*, **42**, 7-13.

Kräger H. and Kugler P. (1993). Non-linearities in foreign exchange markets: a different perspective. *Journal of International Money and Finance*, **12**, 195-208.

Mayer-Wolf, E. (1990). The Cramer Rao functional and limiting laws. *Annals of Probability*, **18**, 840-850.

Nelson, D. B. (1991). Conditional Heteroskedasticity in Asset Returns: A New Approach, *Econometrica*, **59**, 347-370.

Ratsimalahelo, Z. (2002) Efficient Non-linear and Nonparametric Prediction, Working Paper, A second version (2003).

Tong, H. (1978). On a threshold model. In: *Pattern Recognition and Signal Processing*, C. H. Chen, ed. Sijhoff and Noordoff, Amsterdam.

Tong H. (1995). *Non-linear Time series. A Dynamical System Approach*. Clarendon Press, Oxford.

Zografos K. (1998). On measure of dependence based on Fisher's information matrix, *Communication in Statistics Theory and Methods*, **27, 7**, 1715-1728.

ELSEVIER

IFAC

PUBLICATIONS
www.elsevier.com/locate/ifac

DECISION MAKING PROCESS IN VIRTUAL MANUFACTURING REGARDING HUMAN RESOURCES

Mihael Debevec, Tomaz Perme and Dragica Noe

Laboratory for Handling, Assembly and Pneumatics,
Faculty of Mechanical Engineering, University of Ljubljana
Aškerceva 6, 1000 Ljubljana, Slovenia
e-mail: miha.debevec/tomaz.perme/dragica.noe@fs.uni-lj.si

Abstract: One of the core parts of virtual manufacturing system (VMS) is a Model of Human Resources (MHR) where also human's decision-making processes has to be modelled. A part of MHR is a Model of Decision Making Processes (MDMP) where every separate decision-point in manufacturing has to be modelled and all feasible decisions at that particular decision-point have to be considered. The tasks which have to be executed by human workers and operators in manufacturing are mostly defined with a technological plan and time schedule, but that is not enough for modelling of the decision-making processes. There are tasks and decisions that are executed and made by human operators in everyday manufacturing processes and that are not exactly defined or described in any manufacturing documentation. The execution of these tasks and decisions depends mostly on knowledge, skills, and even intuition of every particular operator and worker. The problem is however to collect and model all these tasks and decisions. In the paper, the problems of virtual manufacturing regarding a decision-making process are pointed out and a possible solution for effective modelling of worker's tasks and decisions that appear in small and medium size tool-making companies is presented. *Copyright ©2004 IFAC*

Keywords: decision making process, human resources, modelling, virtual manufacturing, tool-making company.

1. INTRODUCTION

Virtual Manufacturing Systems (VMS) are currently more and more indispensable for obtaining the information about future properties and the state of manufacturing in any production company. Available computer programmes for Virtual Manufacturing (VM) are still islands that solve some specific problems and cover some activities in different phases of the product and production system life cycle. Besides that, they are mostly very expensive and so appropriate almost only for large companies, particularly in the automotive industry and in the mass production industry.

Tool-making companies are a typical representative of small and medium- sized companies (SME) with typical individual methods of production. For these companies the virtual manufacturing could be an important tool that could help them to improve or at least to preserve their position on the market.

The goal of the presented work is to establish the concept of virtual manufacturing for SME, especially adapted for the tool-making companies. One of their basic characteristics, which have to be incorporated in the concept of virtual manufacturing, is the high dependency between the manufacturing processes, the work of operators and the decisions that are made by operators.

The tasks which have to be executed by human workers and operators in manufacturing are mostly defined with a technological plan and time schedule, but that is not enough for the modelling of the decision-making processes. There are tasks and decisions that are executed and made by human operators in everyday manufacturing processes and that are not exactly defined or described in any manufacturing documentation. There are also situations where the worker has the ability to choose between several different possibilities. The execution of these tasks and decisions depends mostly on knowledge, skills and even intuition of every particular operator and worker. The problem is, however, to collect and model all these tasks and decisions.

The present state in companies shows that manufacturing systems are in most cases developed and built without considering the special influences on manufacturing like interactions between workers and their working environment are (Hadfield, *et al.*, 2000). Those factors are not taking into consideration in the classical simulation execution of planned manufacturing system.

In a modern approach, human operators have to be considered not just in a manner of capacity, but also with all their properties that have influence on the execution of tasks and making of decisions. It is obvious that decisions depend on the instantaneous combination of other workers influences, instantaneous working tasks, and the state of the manufacturing system.

The aim of this paper is to introduce the concept of virtual manufacturing for small and medium-sized companies, and discuss the human resources aspect. An appropriate solution for the modelling of human resources is introduced and some essential features that have to be considered in human resource model are presented.

2. SOME BACKGROUND ABOUT VIRTUAL MANUFACTURING

Virtual manufacturing (VM), introduced by Onosato and Iwata (Onosato and Iwata, 1993), is a concept that can be an imperative for obtaining a new tool for automation and control engineering. Several authors in the different branches of knowledge discussed the theme Virtual Manufacturing (Banerjee, 2001; Dai, 1998; Lederer, 1995; Virtual MUW, 1994).

The term "virtual" means that the activities within the information and manufacturing process are accomplished by and within the computer, and the resources and material are the objects described and represented by the information (Perme 1995). According to the VM concept, the manufacturing system can be either real or virtual and can be presented in the form of a real physical subsystem (RPS), a real informational subsystem (RIS), a virtual

physical subsystem (VPS) and a virtual information subsystem (VIS)(Fig. 1).

From these four basics interpretations of manufacturing systems, combinations can be set up that have significance in a real environment. The combination of a VPS and a RIS is sensible for establishing a virtual manufacturing environment (VME). In effect, VPS-RIS means a virtual manufacturing system (physically) where a virtual product is manufactured by a computer on the basis of real information.

Important elements of VM include the simulation and presentations of the results. The "reality" of the VM system is closely connected with the accuracy of the simulation results and the assumption of reality in the presentation. The results of the simulation must be accurate and generated in real time; the designer must find a balance between these conditions within the given computational capacity.

From the various possible combinations, a VME can be set up containing three basic models of real manufacturing systems (RMS): the product, the process, and the factory model. These models can be analysed by simulation and the results can be presented in a form approaching reality.

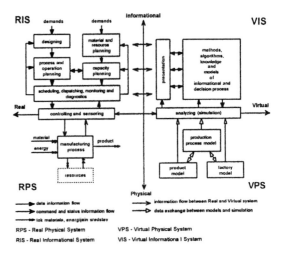

Fig. 1. Physical and virtual parts of the information and manufacturing processes.

The product model contains the representation of all instances of products. In the process model, all feasible processes that can be performed in the shop-floor are listed in detail. The factory model is actually the layout of shop-floor with all the resources, and contains several sub models, which are indispensable for performing a virtual manufacturing. The sub models of factory model represent the following resources:

– machining centres,
– measuring machine(s),
– cutting-tools,

- special equipment,
- clamping devices,
- transportation devices and
- human resources.

The most complex and demanding of all aforementioned sub models is the Human Resource Model (the operator or worker model) (HRM). In the HRM, all possible conditions and activities regarded as the operator's work have to be included, as well as all properties and features that characterise the particular human operator. The most important features, which are directly connected with operator's work, are:

- some basic properties,
- operator priorities,
- operator decisions if more than one feasibility of prosecution is possible,
- operator's decision making if prosecution activity is not defined but required,
- modelling the worker's tasks that are not precisely defined, but have to be considered as possible and inevitable.

2.1. Features of the human operator

Basic operator properties are basic descriptions of the operator states. Operator can be in following states:

- INACTIVE (available for a job),
- ACTIVE I (performing an ordinary job),
- ACTIVE II(performing other jobs until no ordinary job is scheduled),
- IN SHORT BREAK (not available for a job for some period),

The operator's state is considered INACTIVE when the operator is available for a next job. This type of state is generally named *waiting for a job*. The operator's state is considered ACTIVE when the operator is performing a job. Operator's job is divided into two different types of the operator's job: in the first type the operator is performing an ordinary job, and in the second type no ordinary job is scheduled – the operator is performing less momentous tasks like his primary tasks are. The operator's state is considered as ON A SHORT BREAK when the operator is momentarily not available to perform a job.

Operator priorities in the HRM depend on the operator's knowledge, and for him/her the assigned list of authorized tasks. Each operator has an exactly defined set of tasks that can or is qualified to perform.

In the cases, when the operator could choose between many possible tasks at the moment, the HRM has to incorporate the mechanism that prevents the virtual system deadlock because of the "WHAT TO DO NOW" occurrence. This mechanism has to "respond" in the same way, as the operator would make a decision.

The operator's decision-making process depends on the instantaneous combination of individual and environmental factors on the operator (Table 1).

In decision-making process operator's performance factors are also important and bring out the full description of every particular operator. Performance of the operator can be characterized with a combination of the following factors: dependability, error rate, absenteeism, staff turnover, and accident rate (Hadfield, *et al.*, 2000).

Table 1: Influental factors in operator's decision making process (Hadfield, *et al.*, 2000)

Individual factors:	
– general intelligence factor ('g')	– skills
– IQ	– work-related attitudes, beliefs, values
– conscientiousness	– work ethic
– extroversion	– goals
– neuroticism	– locus of control
– agreeableness	– organizational commitment
– openness	
– age	– job satisfaction
– gender	

Environmental factors	
physical:	organizational:
– noise level	– shift patterns
– air temperature	– work teams
– light level	– maintenance
– humidity	– training
– ventilation	– hierarchical structure
	– diversity
	– job rotation
	– communication
	– organizational climate

It is required that every detail of the real manufacturing system must be considered in the model. This is the basic rule that assures successful performing of virtual manufacturing.

3.   THE BASICS OF HUMAN RESOURCE MODELLING

Generally, the modelling of human resources can be divided in three directions as follows:

- human resource modelling (physical description) (Baudet, *et al.*, 1996; Hadfield, *et al.*, 2000; Letizia, 2000; Plekhanova, 2000),

- human behaviour modelling (Biddle, *et al.*, 2003; Kenyon, 2002; Luscombe, *et al.*, 2002; Quispel, *et al.*, 2001; Tolk, 2002; Wise, *et al.*, 2001; Wray and Laird, 2003), and
- human decision modelling.

Human resource modelling mostly considers common and physical characteristics of workers. This means a description of those human characteristics that describe all rough attributes of humans, like the determination of the number of workers, shift patterns, and routines. There is also included all information from manufacturing documentation that are directly connected to humans.

The field of human behaviour modelling has nowadays the most research interest in the human resource modelling area, and becomes one of the most important and difficult emerging disciplines in virtual worlds. A worker behaviour is influenced by various aspects of the surrounding environment: physical factors (noise, temperature, illumination, etc.), organisational factors (work patterns, incentives and supervision, etc) and by individual socio-cultural factors (demography, attitudes, values, and beliefs). To varying degrees these people-oriented issues have an important effect on the performance of a manufacturing system and, clearly, if not considered and resolved at the strategic design stage they may be difficult and expensive to correct once a factory is operational. Also human-machine and human-human interactions are complex, non-linear, adaptive, strongly coupled, and evolving. Complex patterns emerge from these interactions. Effective characterization of human dynamic behaviour has been lacking. That this characterization is essential for engineering design, in its broadest sense, is witnessed by the failure of countless engineered systems, with consequences from commercial failure to loss of life. The major number of research in the field of human behaviour modelling has been done for military applications.

Human decision modelling is connected very closely to human behaviour modelling. The decision-making of a single person is namely dependent on her/his behaviour. Authors present several possibilities for human reactions modelling, which are carried out like a consequence of giving momentary conditions. The major number of research in the field of human decision modelling has been done for the military and passenger evacuation applications.

Human resource modelling is mainly discussed on the following areas:

- military and aviation (Luscombe, *et al.*, 2002; Tolk, 2002; Wray and Laird, 2003),
- manufacturing (mostly human resource management and operational research) (Baudet, *et al.*, 1996; Hadfield, *et al.*, 2000; Plekhanova, 2000),

- ship evacuation (Evi, 2004; Letizia, 2000),
- computer games (Biddle, *et al.*, 2003),
- space research (Wise, *et al.*, 2001),
- driving vehicles (Quispel, *et al.*, 2001),
- fire protection,
- medicine (nursing).

From the available literature it is evident that the major number of research has been done in the field of military and aviation. Authors mostly discuss human behaviour modelling on the basis of game theory.

Several software packages were developed for the simulation of human behaviour, mostly for military (for a simulation of soldier's behaviour on battlefields and pilot's behaviour in action), and passenger evacuation in the marine environment.

An example of human resource modelling, especially appropriate for human behaviour modelling, presented by Hadfield (Hadfield, *et al.*, 2000), shows the basic theory in this field. It is established as a relationship model that says that the behaviour of a person is a function of their personal characteristics and the environment within which they exist (Fig. 2).

This model will be used as the basis for human resource behaviour modelling and human decision-making modelling in our research area.

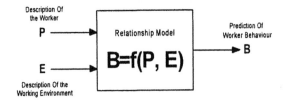

Fig. 2. Lewin's Relationship Model for predicting worker's behaviour (Hadfield, *et al.*, 2000).

4. HUMAN RESOURCE MODELLING IN VIRTUAL TOOL-MAKING COMPANY

The human resource model for a virtual tool-making company will be presented on the basis of theoretical assumptions available in the literature. It will incorporate all the characteristics of a real human that are necessary for a successful performing of manufacturing processes. In the model all human decisions have to be described for instances of given situations. Models will include specific characteristics significant for Slovenian tool-making companies.

In comparison to serial production where the production plan is worked out in detail and jobs are assigned to the machines in advance in a small batch or individual production, like the production in tool-making companies the scheduling is normally carried out on the shop floor level simultaneously. This allows greater flexibility and quicker responsiveness but demands from operator more self-dependency and also more responsibility.

The production in tool-making companies is planned roughly. The main document in a workshop is a technological plan that flows through the shop floor together with a work piece. A technological plan determines all required manufacturing processes on machines and other treatments for every work piece, but doesn't define exact procedures like clamping, cutting tool setting, etc. So operators in machining centres have to decide what, when, and how to prepare the work piece, clamping devices, and even cutting tools. Although the machining parameters are set in the NC programme, it is usual that the operator changes them and so optimises the cutting processes.

There are different situations that require from the operator to make a decision, and can be described as follows:

– More than one near equal solution is possible, but just one could be selected for prosecution (choosing between identical cutting tools or functional equal clamping devices). The solution is trivial, but has to be made because of the mutual exclusion problem.
– The Operator has to choose between several but different solutions. An example for that is a selection of an appropriate cutting tool in the case when a defined one is worn-out or broken, and there is no identical one. This also causes the definition of new cutting parameters.
– The Operator has to decide how to schedule an activity. An example for that is the clamping of a work piece in a machining centre. The technological plan set just the machining operations, but the complete clamping procedure is left to the operator. The Operator had to select proper clamping devices, choose a proper clamping method, and proper clamping forces.
– Operator has to also define the technology for manufacturing of a work piece. These are cases when the jobs are not scheduled or even planned in advance. An example for this is an intervention, i.e. urgent repair of a wear-out or broken tool.

A decision-making process depends on subsequent planned tasks, possible solutions, and the actual state of the manufacturing system. Operator's decisions are under the influence of individual operator's characteristics and work team characteristics. Regarding the goal, knowledge, and experiences, an operator can produce a sequence of activities that have to be fulfilled (Fig. 3).

All stated operator's characteristics have to be incorporated in virtual manufacturing system because human factor has a great influence on production

results. The human resource model has to describe the physical characteristics of workers (i.e. availability) as well as the ability to make a decision in all possible situations.

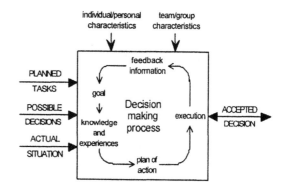

Fig. 3. Impact factors on the decision-making process.

4.1. Example of VMS

A first prototype of VMS for tool-making companies has been developed on the basis of a commercially available, integral informational system (IIS) for planning and control of production in such companies.

In addition to the programmes that enable a virtual environment, i.e. models and simulation engine, integrated in IIS, a special program for human decision execution has been developed and integrated into IIS. This programme consists of four modules:

– EQ_DECS that models the decision-making process when more than one near equal solution is possible. On the basis of demands from production documentation this module copies out from database (IIS) all corresponding solutions and from the list select first one.
– SEV_DECS models the decision-making process when several but different solutions are possible. It acts as previous module, but it needs additional data from IIS. These additional data are attributes of every single resource.
– ACT_DECS models the decision-making process when a decision has to be made how to do a scheduled activity. This method copies out from the database all suitable solutions that are selected on the basis of work piece dimensions and the demanded clamping method. All clamping devices data contain attributes that define all demanded clamping activities regarding a clamping method. Attributes also include clamping forces, which depend on work piece size ranks.
– A TECH_DECS program performs the decision-making process when a decision has to be made how to do the technology for the manufacturing of a work piece. On the basis of dimensions of a

work piece, the program copies out from the database all suitable technologies that are selected on the basis of work piece dimensions and demanded technology. The database includes a list of typical and simple technologies (surface milling, milling of rounds). Every predefined technology contains the sequence of manufacturing activities, data of demanded resources, and cutting parameters.

The decisions "how to do the technology for manufacturing of a work piece" are very complicated to model because these decisions depend on operator experiences. Modelling operator decisions on the basis of experiences is important research area for future work.

5. CONCLUSION

The goal of presented research work is to propose the concept of Virtual Manufacturing System for tool-making companies. The important part of that concept is dedicated to human resources. The presented concept gives a framework for modelling of human operators and integration of these models into the VMS and offers the possibility to configure rather than model the particular tool-making company.

At the moment the simple models of human resources have been developed that make possible an implementation of virtual environment for small for small and medium-sized tool-making companies. First tests show the suitability of the concept and developed tools, and direction of further work.

REFERENCES

Banerjee, P., (2001). *Virtual Manufacturing*, John Wiley & Sons, Inc.

Baudet, P., Azzaro, C., Pibouleau, L., Domenech, S., (1996). *A Genetic Algorithm for Batch Chemical Plant Scheduling*, International Congress of Chemical and Process Engineering CHISA'96, Prague.

Biddle, E. S., Stretton, M. L., Burns J., (2003). *PC Games: A Testbed for Human Behavior Representation Research and Evaluation*, Proceedings of the Twelfth Conference on Behavior Representation in Modeling & Simulation Conference (BRIMS 2003), Scottsdale, Arizona.

Dai, F., (1998). *Virtual reality for Industrial Applications*, Springer - Verlag.

Evi, A Designer and Operator Tool for Effective Passenger Evacuation in the Marine Environment. (2003). Safety At Sea Ltd, Glasgow, UK.

Hadfield, L., Mason, P., Fletcher, S., Mason, S., Siebers, P.-O., (2000). *Human Performance Modelling as an Aid in the Process of Manufacturing System Design*, Research Project, Manufacturing Systems Department, School of Industrial & Manufacturing Sciences, Cranfield University, Cranfield.

Kenyon H. S., (2002). *Controlling Cybernetic Crowds*, SIGNAL Magazine, December 2002, AFCEA's Journal for Communications, Electronics, Intelligence, and Information Systems Professionals.

Lederer, G., (1995). *Making Virtual Manufacturing a Reality*, Industrial Robot, Vol 22/4/1995, MCB University Press.

Letizia, L., (2000). *Developments in Evacuation Systems and Techniques*, SAFER EURORO Workshop, Madrid.

Luscombe, R., Mitchard, H., Gill, A., (2002). *Using Agent Based Distillations to Model Human Intangibles for Dismounted Infantry Combat*, Land Warfare Conference 2002, Brisbane.

Onosato, M., Iwata, K., (1993). *Development of a Virtual Manufacturing System by Integrating Product Models and Factory Models*, Annals of the CIRP, Vol. 42/1/1993, pp. 475-478.

Perme, T., (1995), *A Virtual Manufacturing System in Education*, Proceeding of Tempus Workshop on Automation and Control Engineering in Higher Education, July 5-7 1995, Vienna, Austria, pp. 7-14.

Plekhanova, V., (2000). *Applications of the Profile Theory to Software Engineering and Knowledge Engineering*, Research paper, Conference on Software Engineering and Knowledge Engineering (SEKE'2000), Chicago, Illinois, pp. 133-141.

Quispel, L., Warris, S., Heemskerk, M., Mulder, L. J. M., Brookhuis, K. A., Van Wolffelaar, P. C., (2001). *Automan, a psychologically based model of a human driver*, Clinical Asessment, Computerized Methods and Instrumentation, Lisse, The Netherlands: Swets & Zeitlinger.

Tolk, A., (2002). *Human Behaviour Represenations – Recent Developments*, Virginia Modeling Analysis & Simulation Center (VMASC), Suffolk, Virginia.

Virtual Manufacturing User Workshop, (1994). *Technical Report*, Lawrence Associates Inc., Dayton, Ohio, 12-13 July.

Wise, B. P., McDonald, M., Reuss L. M., Aronson, J., (2001). *ATM Human Behavior Modeling Approach Study, Task Order (TO) 69*, Technical Research in Advanced Air Transportation Concepts & Technologies (AATT), Science Applications International Corporation (SAIC), Arlington, Virginia.

Wray, R. E., Laird, J. E., (2003). *Variability in Human Behavior Modeling for Military Simulations*, Proceedings of the Twelfth Conference on Behavior Representation in Modeling & Simulation Conference (BRIMS 2003), Scottsdale, Arizona.

ELSEVIER
IFAC
PUBLICATIONS
www.elsevier.com/locate/ifac

REHABILITATION AND CARE - A PROMISING APPLICATION AREA FOR ROBOTICS?

Gernot Kronreif[1], Martin Fürst[1], Andreas Hochgatterer[2]

1) *ARC Seibersdorf research GmbH,*
Mechatronic Automation Systems – Robotics Lab,
A-2444 Seibersdorf, AUSTRIA
e-mail: gernot.kronreif@arcs.ac.at

2) *ARC Seibersdorf research GmbH,*
Rehabilitation and Integration,
A-2700 Wiener Neustadt, AUSTRIA,
e-mail: andreas.hochgatterer@arcsmed.at

Abstract: Due to the continuously increasing life expectancy of people in most countries, the percentage of people with motor impairments or other disabilities is constantly increasing. One main goal to handle this situation is to keep the older people independent as long as possible. The use of new technologies as "living aid" could be a suitable approach to offer support for "activities of daily living" (ADLs) and/or rehabilitation measurements in a cost-efficient manner. Especially robotics applications could play an important role in enhancing the quality of live for disabled or older members of the society. This paper gives a short overview on the "state-of-the-art" of these systems, asking also about key factors for a more significant market penetration of rehabilitation robot systems.

Keywords: Rehabilitation Robot, Assistive Systems, Service Robotics, Augmented Manipulators

1. INTRODUCTION

Encouraged by the unquestionable opportunities for this very special topic, rehabilitation was one of the very first non-industrial application areas for robotics. Although first approaches for rehabilitation robots are going back to the 1960s, focused research has been started about twenty years ago. First prototypes resulting from these efforts mostly can be seen as tailored systems for one particular user – fitting to her/his requirements and thus mostly lacking of generality. Encouraged by these very "personalized" developments, an avalanche of similar R&D projects was triggered, most of which were financed by national as well as international funding organisations. Especially the European

Union's TIDE (Technological initiative for the socioeconomic Integration of the Disabled and Elderly) funding initiative should be mentioned at this place (CEC, 1999; CEC, 1998; CEC, 1994). TIDE was set up as a pre-competitive technology research and development initiative specifically aimed at stimulating the creation of a single market in rehabilitation technology in Europe. The TIDE pilot phase started in 1991 and was succeeded by a bridge phase in 1993 – having robotics research and development represented in each phase. The main phase finally started in 1995 and was contained within the 4th EU research framework under the general heading of Telematics Applications.

For the purposes of TIDE, rehabilitation technologies was defined as those provided directly to elderly and/or disabled people, to enable them to live more independent lives and become integrated in the social and economic activity of their communities, preferably outside institutional care – this definition also perfectly characterizes the scope of rehabilitation robotics and is therefore used in this paper as well.

Beside the above mentioned initiatives, development of rehabilitation robotics also has been largely considered in other more recent R&D activities of CEC joint research, e.g. at the IST initiative. Also the huge research efforts undertaken in the USA (with lots of projects funded by the Veterans Administration) and in many Asian countries (e.g. HWRS-ERC (Human-friendly Welfare Robot System) at KAIST/Korea) with very spectacular results have to be noted at this place.

However, now - after more than two decades of research – some questions must be asked: What is the market penetration of rehabilitation robots? Are those systems finding their way to the end-users? And what is the current state of rehabilitation robots? It is the aim of this paper to give a short overview on major results in this field and also to critically discuss their market situation.

2. WHY TO USE ROBOTS FOR REHABILITATION?

In essence the main goal of each rehabilitation robot is to provide handicapped people with a similar degree of autonomy and mobility as other "non-handicapped" people have. There are two aspects of paramount importance for the usability of rehabilitation robots: the interaction between user and machine (human-machine-interface, HMI) as well as the interaction between robot and each object to be manipulated. Especially for the design of an adequate HMI the developers have to bear in mind, that the user neither is getting overstrained by operation of the system nor being unchallenged. A well-designed interface has to account the specific needs of the operator and thus – for the ideal case – should be tailored to the abilities and the working-style of the particular user.

What are the main reasons to use robotic systems in rehabilitation and care? Generally speaking, robotic systems could be a key component for solving problems with individual ADLs (activities of daily living) for handicapped persons – a user group which also includes older people in need of care as well as people with chronic diseases. With the aid of a robotic assistant, lost functionalities could be (partly) replaced and/or reduced ones could be amplified. Despite of the obvious very different robot tasks, there are some considerable analogies between this area and classical (industrial) automation. In general,

both application fields are benefiting from the key features of robotic systems: Reliability, accuracy and cost-benefit ratio.

Further promotive factors for the use of robotic systems in rehabilitation and care are:

- Qualified nursing staff is only available to some extend. A reasonable discharging from routine jobs by means of a robot system helps to move real nursing activities to the fore.
- Further development of more general service robots and of their subcomponents could support an extended use of new ICT components (also) in the areas prosthetics and (tele)support systems.
- Approval of rehabilitation aid by health insurance funds and similar organisations may lead to a better market penetration and thus to a significant reduction of system costs.

What is a „robot"?

Before describing selected application examples, definition of the term „robot" in the context of rehabilitation and care should be discussed in short. According to established (there is no real standard yet) definitions of the term robot, where such a device is an "automatically controlled, reprogrammable multipurpose manipulator programmable in three or more axes" (ISO 8373) most of the known assistive systems would not fall into the category of a "robot system". More helpful might be the definition given by the International Federation on Robotics (IFR), where a service robot is a system "which operates semi or fully autonomously to perform services useful to the well being of humans and equipment, excluding manufacturing operations". Another less restrictive definition for rehabilitation robots could be "a programmable device for augmentation of manipulation and mobility" where programmable stands for "adequate input of user commands causing robot activity" (Kronreif, 2001).

3. ROBOTS FOR REHABILITATION – SELECTED APPLICATION EXAMPLES

If one tries to categorize the activities in rehabilitation robotics, three major application fields can be identified. First of all, robotic systems may be used for providing mobility to disabled users. This group of course includes all the different approaches for wheelchair systems as well as any kind of mobile assistance platforms. The second group of research includes manipulators for assistance functions. Finally, there is a growing field of applications for therapy purposes.

3.1. Assistive Systems with Mobile Platforms

Johann Borenstein's "Nursing Robot" (Borenstein and Koren, 1985) - completed in 1986 as a DSc.

thesis - was one of the first fully functioning mobile robots equipped with a manipulator arm. The Nursing Robot system comprises three major components: a self-propelled vehicle, a robotic arm mounted on it, and a communications workstation next to the disabled person's bed. Onboard the mobile robot low-cost microcomputers are interconnected as a hierarchical network, in order to control a variety of activities: Sensor data processing, motion control, path-planning, communication, and others. With this equipment the vehicle was enabled to move autonomously in a room with unexpected obstacles. Main idea for the robot was to perform simple services, such as fetching a glass of water, operating electrical appliances or replacing a cassette in a video-recorder.

Another more recent system – but for the same application scenario – is Care-O-bot ® from Fraunhofer IPA in Stuttgart. Generally speaking, the robot is aimed to facilitate daily activities, especially for people in need of care. For example, the robot is designed to fetch and carry objects, to support the user for stand up and walking or to serve as a mobile communication platform (Hans, et al., 2002).

Similar functionalities are implemented in ARC Seibersdorf's mobile communication platform prototype ARChy. Here the main focus is directed to the development of a dedicated HMI which can be easily adapted to the special requirements of the particular user. Another aspect of research includes the replacement of typical sensor equipment known for navigation of mobile platforms by "low-cost" sensor systems, which also are suitable for practical use. Main feature of ARChy is the connection to ARC's tele-support system MoniCare (Fugger, et al., 2003) which opens access to any installed home automation system and which also connects the mobile platform to any kind of communication systems, like WWW or telephone.

3.2. Wheelchair Systems

Grounded on the intensive research and commercial application of "Autonomous Guided Vehicles" (AGVs) during the last years, a number of projects dealing with the development of "smart" wheelchairs - incorporating features such as obstacle avoidance, tracking along a wall, track or manoeuvring through a door - entered the scene.

The OMNI project, for example, provided an important step forward in natural wheelchair control by persons with a severe physical or multiple disabilities by introducing omni-directional mobility of the wheelchair. This allows the chair to move in any direction and the linear motion to be combined with a rotation around any given point, including the centre of the wheelchair. In addition to the omni-directionality, which facilitates intuitive user control

and increases mobility, ultra-sonic and infrared sensors for environment analysis with an obstacle-avoidance system are provided. This enables high-level control of the wheelchair navigation, reducing the complexity of the control task for the user while guaranteeing safety (Hoyer, et al., 1997).

Another successful implementation of technology from mobile robotics is demonstrated by the Bremen Autonomous Wheelchair "Rolland". Here, a commercial available powered wheelchair system (Meyra Genius 1.522) is equipped with different sensor systems and a PC (figure 1). In addition to a theoretical examination of the basics of navigation, robust techniques are developed which enable the wheelchair to learn even complex behaviours by combining several elementary behaviours such as wall-following or doorway-passage. As first application, a driving assistant was implemented that adapts the speed ordered by the user via joystick to the current obstacle situation. In addition, the driving assistant supports the user by passing door-frames and by turning round (Lankenau and Röfer, 2000).

3.3. Robotic Systems for Manipulation Assistance

Looking back in history, one of the first systems for manipulation assistance was developed at Stanford and Palo Alto VA Hospital. Several generations of the DeVAR (Desktop Vocational Assistive Robot) workstation were aimed at a vocational environment (Van der Loos, et al., 1989). Main part of the system was a PUMA 260 industrial robot mounted upside-down on an overhead track in order to increase its working envelope. For manipulation the robot was equipped with a modified prosthesis from Otto-Bock as gripper. DeVAR was used in a real working environment on an 8 hours a day basis at the Pacific Gas and Electric Company - the high cost of about USD 100.000,-- were justified by saving on the cost of a human assistant for this particular installation.

Fig. 1. Bremen Autonomous Wheelchair "Rolland"

A similar application was developed around the RT robot series (produced by OxIM, Oxford, UK) at Boeing in Seattle for one of their own disabled programmers. More recently, as part of the European TIDE funded RAID (Robot for Assisting the Integration of the Disabled) project (Jones, 1999), an RT200 robot - the arm is of SCARA geometry and has dimensions approximating those of an adult - was built into an extended workstation for office-based tasks (figure 2). At the end of the RAID project the workstation was commercialised by the manufacturers of the RT robots.

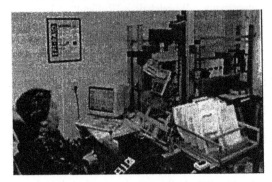

Fig. 2. System RAID for vocational assistance

Another important application for robotic augmented manipulation is for feeding assistance. This field of application includes one of the probably most commercially successful rehabilitation robot systems known until now – HANDY 1 (Topping, 2001). The development started in 1987 as a Masters research project, where M. Topping set himself the task of helping his neighbour - a 12 year old boy with spastic paraplegia - to feed himself. HANDY 1 consists of a 5DOF (degree of freedom) robotic arm and a food tray mounted on a wheeled base unit. The arm is controlled by a single switch input device in scanning mode. Up to now, over 250 units have been sold world-wide with very positive feedback of its effectiveness (Hillman, 2003). Originally designed for eating assistance only, the system in the meantime has been extended for other activities including drinking, make-up, shaving and teeth cleaning (figure 3).

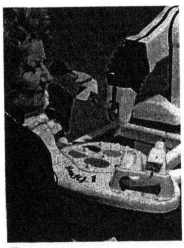

Fig. 3. System HANDY 1, equipped with make-up tray

Another very successful rehabilitation robot is the MANUS arm which is being now sold commercially by Exact Dynamics. MANUS was a collaborative Dutch project started in 1984 and centred at the Institute for Rehabilitation Research IRV. The robot arm was built on the conclusions of the Spartacus project by developing a wheelchair-mounted manipulator (Kwee, et al., 1989). The original MANUS manipulator consists of a 5 degree of freedom (DOF) arm mounted on a rotating and telescoping base unit which could be attached to a variety of electric wheelchairs. The use of slip couplings and a firmware watchdog increase the safety of this manipulator. Despite of the relatively high costs of the robot arm (about USD 35.000,--) more than 150 units were already sold world-wide (Hillman, 2003).

3.4. Robot Augmented Therapy

During the last 5 years the application of robotic systems for therapeutic issues – especially for post-stroke rehabilitation – became one of the major themes in rehabilitation robotics. Up to now, the main focus clearly is on therapy of the upper extremities.

In a co-operation between Stanford University and VA Palo Alto a robot-assisted device was developed which is capable of moving an upper limb in simple predetermined trajectories by directly controlling the position and orientation of the forearm (Shor, et al., 2001). In the current setup of MIME (Mirror Image Movement Enabler), a PUMA-560 robot is mounted beside the patient. The robot is attached to a wrist-forearm orthosis (splint) via a 6-axis force transducer, a pneumatic breakaway overload sensor set to 20 Nm torque, and a quick-release coupling mechanism. The subject's arm is strapped into the splint with the wrist in neutral position. Robot/forearm interaction force and torque measurements from the transducer are recorded and archived by a personal computer. A 6-axis position digitizer is mounted on the other side of the table. This device is attached to a splint on the other forearm. When the position digitizer is attached to the paretic limb, it can be used to quantify voluntary movement kinematics. In a series of clinical trials, the therapeutic efficacy of MIME is being evaluated in chronic stroke subjects by comparing functional and motor-control improvements in subjects receiving robot-aided exercise or conventional treatment.

Another robot-aided therapy system for stroke rehabilitation is being developed by the team of H.I.Krebs at MIT. For MIT-Manus (Krebs, et al., 2003) the subject sitting at a table puts the lower arm and wrist into a brace attached to the arm of the robot. A video screen prompts the person to perform an arm exercise such as connecting the dots. If movement does not occur, MIT-Manus moves the person's arm. If the person starts to move on his own, the robot provides adjustable levels of guidance and assistance to facilitate the person's arm movement. The robot system could be successfully evaluated in a series of clinical trials and is now available as a commercial product.

Other similar – and partly commercial - projects for robot augmented therapy are ARM, GENTLE-S (both of which are dedicated to rehabilitation of upper extremities) as well as LOKOMAT, TEM and others for the lower extremities.

3.5. Robot Assisted Playing

The use of robots as learning aid for disabled children has been discussed for a long time. Robots indeed seem to be very attractive for children and a major part of research in this context is focussed on autistic children. The team of Dautenhahn at the University of Hertfordshire, for example, started their AuRoRA project (Autonomous Robotic platform as a Remedial tool for children with Autism) in 1998 which involves the use of a mobile robotic platform as a teaching aid for autistic children (Werry, et al., 2001). The AuRoRA project differed from more 'traditional' robotic applications in the use of interaction and expression. AuRoRA is having no specific task to perform and the emphasis has been shifted to the expression of the robot. As a primary concern the children should enjoy interacting with the robot.

Based on the results from the above mentioned project a group at the Universite de Sherbrooke has designed a set of different robotic toys (Michaud and Theberge-Turmel, 2002) that can take into account the interests, strengths and weaknesses of each child, generate various levels of predictability, and create a more tailored approach for personalized treatment. Target group again is the one of autistic children. With these robotic systems the possible factors that might influence the child's interests in interacting with the toy - like shape, colours, sounds, music, voice, movements, dancing, trajectory, special devices, and others – can be observed.

In a recent project of ARC Seibersdorf Research a prototype of a system for "robot assisted playing" is being developed in order to give assistance to severe physically handicapped children. The main goal is to feature autonomous playing activities for this user group. Thus, the chosen approach puts the robot back to a supporting role; not the robot is the toy – but the robot helps to use the toy. Using the functionality of the robot system, the user is now in the position to manipulate real objects (toys) in the real world, despite of her/his handicap. In a first feasibility study, a dedicated custom-made robot system has been designed and realized for manipulation of small LEGO™ bricks (Kronreif, et al., 2003). The system consists of a storage device for different types of

bricks, a 4DOF robot system in Cartesian configuration, a special gripper device and a playground (figure 4). Using a dedicated 5-key input device (or a single-switch input device in scanning mode) the user can chose a particular brick type and define the inserting position. The prototype is available since summer 2003. A series of user tests evaluated the concept and did show a very high acceptance by the users. Further tests evaluating learning effects are currently in progress.

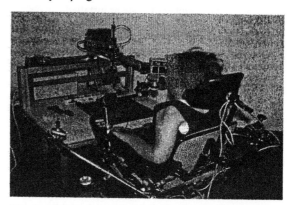

Fig. 4. ARC "robot assisted playing" during user trial

4. MARKET IMPACT

Aside from technical fascination all the work done in the field of rehabilitation robotics will only benefit people with disabilities if it is made available to them. Thus, "success" has to be measured in the commercial availability of devices.

Most of the past research projects have to be judged as technology-driven. Thus, the resulting prototypes show technical feasibility but are far away from practical use because the "reality" of the market had been simply ignored. Existing prototype wheelchair systems, for example, are sometimes degenerating the wheelchair to a fully autonomous mobile platform equipped with sensors and controllers but with almost no free space to transport a person in an aesthetically acceptable way. Prototypes available now are showing lots of fancy features which sometimes are not addressing real and essential user needs. Sensors used in the most cases are not practicable for daily use – either because of their measurement principle or because of very high costs. In more detail, the following barriers to a successful market penetration have to be dealt with:

- Most existing solutions show limited versatility and are lacking of modularity to cover different user requirements.
- No sufficient concern for aesthetics.
- Human-machine interface is not suitable for practical use.
- The narrow-purpose design and the high specialisation of state of the art devices are not economic at all.

- The danger exists of getting stuck to the image of the 'ever to remain' future technology.
- There is little co-operation among manufacturers and no trade association exists; there is no forum where manufacturers, researchers, rehabilitation centres and user representatives meet.
- The field has a relatively low profile with respect to support from EU and national funds.
- The current state of the market is very much in an initial phase:
  - small number of systems installed
  - systems are still rather expensive
  - little data on user experience and cost-effectiveness, slow diffusion of experience
  - considerable difficulty to have solutions accepted by health care providers and insurance companies for reimbursement
  - lack of collective insight into user experiences, and into options and issues involved in distribution, support and service

5. CONCLUSION

During the last 20 years lot of research was accomplished in the field of rehabilitation robotics. Much has been achieved to date and current developments indicate promising possibilities for the future. Dozens of prototypes can be found for several application fields – like mobility and/or manipulation assistance, robot-augmented therapy and assisted playing; some selected examples are given in this paper.

As far as commercial exploitation is concerned, many of the systems are delivered as prototypes only and do not make it to the market. Some reasons for this under-exploitation have been indicated above. They have to do with the structure of the (European) rehabilitation robotics market, in terms of national fragmentation, lack of distribution and service channels, financing, information flow, and co-operation and competition amongst actors. An extended range of solutions would lower the barriers to product acceptance, and the increased exposure of solutions to the public and purchasers would stimulate diffusion.

REFERENCES

Hillman, M. (2003). Rehabilitation Robotics from Past to Present – a Historical Perspective. In: *Proc. ICORR 2003*, pp. 1-4. Taejon, Korea.

Van der Loos M., et al.(1989). Design and Evaluation of a Vocational Desktop Robot. In: *Proc. RESNA 12th Annual Conf.* New Orleans.

Jones T. (1999). RAID – Towards greater independence in the office and home environment. In: *Proc ICORR 1999*, pp201-206. Stanford, USA.

Fugger E., et al. (2003). Integrated Alert & Communication System for independent Living of older Adults. In: *Proc. AAATE 2003*, pp. 803-807. Dublin, Ireland. IOS Press.

Topping, M. (2001). Handy 1, A robotic aid to independence for severely disabled people. In: *The Information Age* (Ed M. Mokhtari), pp. 142-147. IOS, Netherlands.

Kwee, H.H., et al. (1989). The MANUS Wheelchair-Borne Manipulator: System Review and First Results. In: *Proc. IARP Workshop on Domestic and Medical & Healthcare Robotics*, Newcastle.

Hoyer, H., at al. (1997). An Omnidirectional wheel-chair with enhanced comfort features. In: *Proc ICORR 1997*, pp. 31-34. Bath, UK.

Krebs, H.I., et al. (2003). Robotic applications in neuromotor rehabilitation. *Robotica 21*, **Vol 1**, pp 3-12.

Borenstein, J., Y. Koren (1985). A Mobile Platform for Nursing Robots. *IEEE Transactions on Industrial Electronics*, pp. 158-165.

Shor, P.C., et al. (2001). The effect of Robotic-Aided Therapy on Upper Extremity Joint Passive Range of Motion and Pain. *Integration of Assistive Technology in the Information Age* (Ed M. Mokhtari), IOS, pp 79-83. Netherlands.

Kronreif, G., et al. (2003). Robot Toys for Severe Physically Handicapped Children. In: *Proc ICORR 2003*, pp. 282-286. Taejon, Korea.

CEC (1999). Telematics for the integration of the disabled and elderly. In: *Synopses of Projects January 1999*, European Commission.

CEC (1998). High TIDE – A review of the pilot phase of the TIDE projects from 1991-1994, European Commission.

CEC (1994). TIDE – Bridge phase synopses, European Commission.

Werry, I., K. Dautenhahn, W. Harwin (2001). Investigating a Robot as a Therapy Partner for Children with Autism. In: *Proc AAATE 2001*, pp. 379-383, IOS Press. Ljubljana, Slovenia.

Michaud, F., C. Theberge-Turmel (2002). Mobile Robotic Toys and Autism. In: *Socially Intelligent Agents - Creating Relationships with Computers and Robots*, Kluwer Academic Publishers, Boston.

Kronreif, G. (2001). Roboter in Rehabilitation und Pflege (in German). In: *Proc Independent Living: Medizinisch-technische Hilfen für ein selbständiges Leben im Wohnbereich für alte, kranke und behinderte Menschen*, OCG Schriftenreihe, Wien.

Hans, M., B. Graf, R.D. Schraft (2002). Robotics Home Assistant Care-O-bot: Past -- Present -- Future. In: *Proc IEEE ROMAN 2002*, pp. 380-385. Berlin, Germany.

Lankenau, A., T. Röfer (2000). The Role of Shared Control in Service Robots - The Bremen Autonomous Wheelchair as an Example. In: *Service Robotics - Applications and Safety Issues in an Emerging Market*, Workshop Notes ECAI 2000, pp.27-31.

E-TRAINING USING REMOTE ASSEMBLY SYSTEMS

Tomaž Perme and Dragica Noe

*Laboratory for Handling, Assembly, and Pneumatics,
Faculty of Mechanical Engineering, University of Ljubljana
Aškerèeva 6, 1000 Ljubljana, Slovenia
e-mail: tomaz.perme/dragica.noe@fs.uni-lj.si*

Abstract: The growing complexity of modern manufacturing and assembly systems demands engineers that can plane, build, and operate these systems, and have a more comprehensive and deeper knowledge about the operation and control of components and complete systems. During the education and training processes, students and engineers have to get this knowledge, as well as some practical experience. The concept of Remote Assembly System (RAS) can help lecturers to give students more on-spot examples, and students as well as engineers to get more practical experience with programming and control of real assembly systems. In this paper, the RAS concept, including examples and practical experience are presented. The example of RAS consists of assembly systems and a personal computer with a remote control system that enables programming and control of assembly systems over the internet. Students have used these systems in a practical course and the results are very promising – they spent less time in the laboratory but they got more practical experience by working at home. In spite of that there are still some problems from a technological point of view that are presented and discussed. *Copyright © 2004 IFAC.*

Keywords: assembly, remote control, e-training, practical course, internet, remote laboratory, remote panel, LabVIEW.

1. INTRODUCTION

A modern engineer needs a continuous education process in order to work efficiently. The responsibility of universities is to support this with efficient undergraduate and postgraduate study, including workshops and seminars for engineers. An advanced educational concept and highly evolved support tools are needed which can enhance and strengthen learning in a variety of disciplines. Especially for delivering knowledge of manufacturing and assembling technologies and automation, an interdisciplinary and flexible curriculum is required, where obtaining practical experience is very important.

The particular subjects have to include recent developments and consider new concepts, as well as different approaches and trends in various fields. The concept of Virtual Assembly (VA), which is based on an idea of Virtual Manufacturing (VM), introduced by Onosato and Iwata (1993), was an imperative for obtaining a new tool for automation and control engineering. The purpose of VA is to integrate existing manufacturing models with analysing techniques and representation forms in a coherent system in such a way that new planning and control methods and techniques can be tested and verified without disturbing or having a real assembly system. The concept of VA can also be efficiently used in education (Perme 1995) and for training purposes.

The main advantage of VA is that the user doesn't need a real assembly system, and that he can use the VA anywhere any time. But no matter how detailed the models are, how precise the simulation is, and how realistic the presentation form is, the VA can just supplement real assembly systems, because in education and also in training process it is still necessary to get a tangible feeling of reality.

The Remote Assembly System (RAS) is a tool based on the concept of Remote Laboratories (RL) that gives the operator the possibility of programming and controlling real assembly systems over the internet, and get feed-back information about the behaviour of the real system. The RAS can also be the basis for a VA system, and in some cases for education and training both systems that can be used in combination.

Many examples of remote laboratories can be already found on the Internet or in different articles and papers. The examples are mostly from the field of control and electrical engineering, and just few of them also include on-line feed-back information. These examples are experiments rather than systems that can be also found in industry. It is also hard to find any information about the practical applicability of remote laboratories and response from the users, i.e. students and engineers that have used them for practical courses and training.

The RAS is proposed in addition to the VAS as an efficient concept for lectures and practical courses on handling and assembly. In the paper, the concept, an example of RAS as well as practical experience will be presented. The example consists of four assembly stations, an automatic transfer system, a programmable logical controller (PLC), and a personal computer with a remote control system which enables programming and controlling the RAS over the Internet. Students have already used the assembly system from the example in a practical course and the results are very promising – they spent less time in the laboratory but they got more practical experience by working at home.

2. REMOTE LABORATORY

The original idea of the remote assembly system (RAS) comes from the idea of remote laboratories, i.e. laboratories that the user can reach via internet. There are several levels of remote laboratories, but all of them want to bring an experiment from laboratories over the internet to the user (Fig. 1.).

The level of the remote laboratory depends on the ability of the user to prepare and control the experiment over internet. The usual course of the work in a laboratory is as follows:
– set-up an experiment,
– execution of experiments,
– analyse the results of the experiment.

An experiment is usually set-up via the Internet in cases when the set-up comprises some selections of predefined algorithms or methods, or setting some parameters. The setting up of an experiment

that requires some (re-)programming is more exceptional, but can also be done over the Internet. In both cases the physical part of the experiment is set-up in advance. Although the physical reconfiguring of the experiments at a distance could have practical meaning, it will not be considered as very necessary for the education and training.

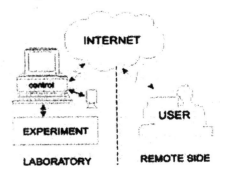

Fig. 1. Concept of remote laboratory

Execution of physical experiments always run in a real laboratory. Computer simulation and experimentation with virtual systems (in a virtual environment) are not denoted as physical systems as they usually just consist of computers, software, and human machine interfaces (HMI). Physical experiments are conducted on real systems or on physical models which are installed in laboratories, but they can be controlled remotely, so there is no need for an operator to be in the laboratory. In that way an experiment could be executed anytime and from anywhere.

Remote control of an experiment can be done in the following ways:
– the experiment can be initialised and started from a distance and then executed automatically,
– the execution needs some intervention from the operator, and
– the interaction between control of the experiment and operator is needed.

In the first case no on-line feed-back information is needed during the execution of the experiment, so we cannot really talk about the control of the experiment. In the case where the operator's intervention is needed, the operator needs some feed-back information during the execution of the experiment. In the last case the operator interacts with the control of the experiment on-line and real-time communication is required. The most communication intensive is the visual feed-back from the camera (this is normally the case of tele-operation).

For educational and training purposes just initialisation and control to start and stop the experiment are usually sufficient. In that case the control of an experiment is located in the laboratory. The examples that exist on the Internet mainly use this kind of experiment control. The intervention and also tele-operation could be interesting for education, but they are normally too cost intensive for remote laboratories. The exception to this is visual feed-back of the experiment execution (visual monitoring), that is appropriate for education. It is not necessary to be in real-time so also some "low-cost" solutions like web cameras can be efficiently implemented.

The results of an experiment can be in different forms from tables, charts, diagrams or plots. The data can be collected during the experiment, stored locally, and at the end of the experiment transferred to a remote site where they can be analysed. One possible way of showing the execution of the experiment is also in the form of an off-line movie. In that way and with on-line visual feed-back the user can see (and/or hear) the experiment execution. This is very useful for education and training because the user gets a more realistic and tangible representation of the experiment.

3. REMOTE ASSEMBLY SYSTEM

The Internet-based and remote systems also have the advantage in cases where the cost intensive software and hardware are needed. An assembly system is an example of such case, so the purpose was to establish remote assembly system (RAS).

The RAS is based on a concept of remote laboratories. The users have access to the remote laboratory i.e. the experiments, instruments, and equipment from outside over the Internet. The important characteristic of the RAS is that the user can control an experiment with a standard Web browser.

3.1. Actual assembly system

An actual assembly system in a real laboratory consists of four assembly stations, an automatic transfer system, a programmable logical controller (PLC), and a personal computer (Fig. 2).

Assembly stations are equipped with two automatic screwing devices and a manipulator for part handling. All except a vibration feeder and turning table are pneumatic devices. The controller needs to control 11 actuators (18 outputs and 12 inputs), so there are two PLC needed which are equipped with an operator panel, interconnected via CANbus and connected through RS 485 to the personal computer. The software on the personal computer allows users to programme both PLCs and monitor the programme execution. The operator panels are designed for the everyday interaction of operators with the assembly system, when this is running in real production.

Fig. 2. Actual assembly system

3.2. Demands on remote assembly system

There are two options for remoteness of the actual assembly system. First, consider just the remote operator panel. In that case the user can just start and stop the programme already loaded into the PLC. The user can also change some operational parameters, for example waiting times and counter values.

Besides the remote operator panel, the second option also includes remote programming and monitoring, so the user can also write a programme, load it on the PLC, start it, and monitor the execution of the programme on the actual assembly system.

The visual feedback is for both options important feedback information about the operation of the actual assembly system. The operator panel and also monitoring are just the tools that help operator and programmer with their work. They still need to see the operations of assembly devices, and the same is with the remote user.

From these considerations some demands on a remote assembly system were pointed out. The RAS has to comprehend:
– visual feedback,
– a remote operator panel, and
– a remote programming and monitoring device.

3.2. Developed remote assembly system

The core equipment for the development of remote assembly system consists of a low-cost Web-camera and graphical programming environment LabVIEW with a remote panel. The picture from the camera installed in the laboratory is used as visual feedback information, and to access the remote assembly system over the internet a remote panel from LabVIEW is implemented. The concept of RAS using a LabVIEW remote panel and Web-camera is presented on Fig.3.

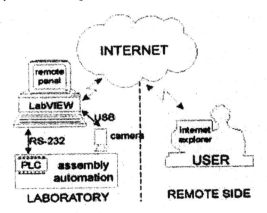

Fig. 3. Concept of RAS with a LabVIEW remote panel

Using the remote panel feature in LabVIEW, someone can remotely control the LabVIEW application over the Web with no additional programming. The developed application can be embedded in the front panel in a standard Web browser. Then, designated users can control the application remotely from a standard Web browser without installing LabVIEW.

Part of the application developed in LabVIEW can also be a Web-camera display. So the visual feedback information can be part of the application that is remotely controlled using a standard Web browser. In that way the user can control the application and get visual feedback information.

On the basis of the concept of RAS with a Web-camera and a LabVIEW programming environment two applications have been developed:
 - an application with a Web camera and a remote operator panel, and
 - an application with a Web camera and remote programming of PLC.

3.2.1. RAS with remote operator panel

An application has been developed that displays the operator panel and image from Web camera on the screen of the personal computer (Fig. 4). The Web camera is connected to the personal computer via USB interface. The application grabs the digital picture frequently from the Web camera and displays it on the screen. In that way the actual assembly system can be seen on the screen.

The actual operator panel is connected to the PLC via serial interface. With a digital input/output card installed in personal computer, the application is also connected to that communication and so the operator panel is emulated in the application. In this way the user can work with the operator panel on the screen as he works normally with the actual one.

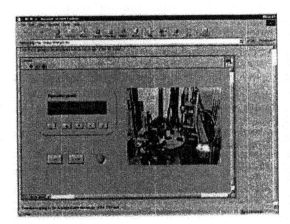

Fig. 4. Snapshot of Web browser with remote operator panel and visual feedback

When this application is published on the Web (LabVIEW remote panel feature) and the PLC is properly programmed, the actual assembly system can be controlled and observed with Web browser on remote computer.

3.2.2. RAS with remote programming of PLC

The PLC programming languages (there are five types) for regulatory, discrete, and sequential control applications are standardised (IEC 1131). This means that the syntax and semantics of these languages is defined, but implementation is left to the particular developer and producer of PLCs, and so each producer of PLCs has its own programming system. Practically this means that a particular PLC programme written in ladder diagram format on a programming system from producer A cannot be used on the programming system of producer B.

The existing programming system for an actual PLC cannot be integrated into the LabVIEW application and so published on the Web. There are also other internet applications that allow remote control of computers, but they are not appropriate for the remote laboratory.

These two problems lead us to the development of our own application for programming of PLCs. To solve the first problem an industrial digital I/O card integrated into a personal computer was used instead of PLC. Using the LabVIEW for development of the application solved the second problem. So an application has been developed that allows the user to write a programme and execute it. When this application is published on the Web, the remote user can programme and control the actual assembly system over the Internet (Fig. 5).

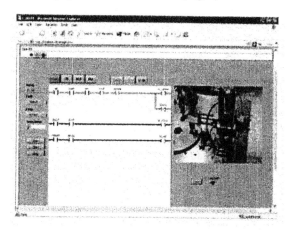

Fig. 5. Snapshot of Web browser with remote programming and visual feedback

The number of inputs and outputs of the implemented I/O card is limited, so at the time the remote assembly system with remote programming consists of only one manipulator.

In addition to the RAS, the application has been developed, that allows programming and simulation of one manipulator (Fig. 6) on a local computer. In this way the user can make a programme and test it with the simulation.

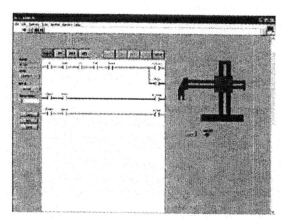

Fig. 6. Snapshot of programming and simulation of PLC

4. USING RAS FOR EDUCATION

The main purpose of education in manufacturing systems is to deliver knowledge about topics such as methods, techniques, processes, and resources needed for designing, planning, scheduling, monitoring, and controlling the manufacturing of the product comprehended in different subjects.

Assembly and handling are very important topics in the education of manufacturing engineer. The subject of *handling and assembly* includes a lecture and a practical course. The lecture gives students:
- an overview of the organisation of assembly systems focusing on assembly and logistics,
- a fundamental knowledge of the resources needed for transport, handling, and assembly, and
- a fundamental knowledge of planning and control techniques of transport and handling devices and assembly.

The practical course is divided into a laboratory course and a project. In the laboratory course the students get practical experience with handling and transport devices, robotized assembly cells, and shop-floor production systems. The projects are focused on solving problems of organisation, or automation of assembly and handling and have to be carried out by the students. With the project work, the students obtain some additional and advanced knowledge of particular topics.

Some problems that have impact on the quality of the practical course on handling and assembly can be summarised in the following points:
- planning, configuration and especially programming of the assembly system cannot be done by every student,
- effects of different organisational concepts, scheduling and control techniques in shop-floor production systems cannot be exposed without disturbing a real shop-floor production system, and
- in most cases the results of the projects cannot be tested or implemented in practice, so their practical value cannot be verified.

3.1. Implementation of RAS in courses on handling and assembly

The RAS can be implemented in lectures as well as in practical courses. In lectures the RAS with a remote operation panel can help the lecturer to show the structure and operation of a real assembly system. With the changing of operation parameters, different working modes can be demonstrated on the spot.

For the practical courses, the RAS with remote programming can be very useful. In the first place it can help the lecturer not just to explain the principles of writing a PLC programme, but also to show in each case the operation of a particular command and/or programme on the real assembly system.

After the students get some basic knowledge about programming of PLCs, they can use the programme for programming and simulation. With this programme they learn how to make an error-free and functional PLC programme. With the obtained knowledge, the students can use the RAS with remote programming to make a programme and to verify it on the real assembly system from a distance.

5. PRACTICAL EXPERIENCE WITH RAS

The experiences with first prototype of remote assembly system were obtained during the last semester on a practical course of assembly. Part of that course is also the programming of assembly systems, i.e. programming of PLCs. Previously, the programming methods and programming language were explained and students have to write a programme at home, and then test it in a laboratory on the actual assembly system. The problem was, that the students could write with the existing programming tools an error-free PLC programme, but in all cases the programme didn't work at all, or at least properly.

In the last semester the developed application for RAS with remote programming was used to explain to students the programming principles. Some students used the developed prototype for programming and simulation, as well as the developed prototype of RAS with remote programming.

The results of the first tests proved the concept of RAS. From one side it was simpler to explain and from another easier to understand the programming principles and implemented programming language. The questions were clarified immediately with practical examples.

There was also a very positive response from the students that used RAS for practical work. They managed to write error-free and functional programmes. The simulation was more useful for learning the programming language than the RAS with remote programming, because there was no need for Internet connection.

There are also some shortcomings of the developed applications. The students that used RAS over the Internet had some difficulties with responsiveness of the remote application. The obvious problem was also the visual feedback, which requires more intensive data transfer. But these problems were not so significant, when the RAS was used from the computers connected to campus network.

6. CONCLUSION

The use of the remote assembly systems concept can result in an increased interest in particular lectures and better practical courses. The students can learn more and easier, because of the possibility of permanent testing and verification of results of their practical work.

The first prototype of a remote assembly system with remote programming was implemented on simple assembly device (manipulator) so the implementation of simulation instead of RAS was more convenient for learning PLC programming. When the actual assembly system is more complex, like a robotized assembly cell is, than also a more complex simulation is needed, which cannot entirely substitute practical work with a real system.

The RAS can be useful also for education at a distance. In e-education, the students can get some "practical" experience with actual assembly systems using RAS over Internet. In the same time their practical work can also be documented. For example if a particular student makes a PLC programme and tests it, he can save the programme and he can record the execution of that programme. In that way he has a piece of evidence that he accomplished his practical work.

The RAS can be used also for e-training. If there is an actual assembly system in testing phase, the future users can use RAS to make training at a distance.

The presented work shows some prospectives of remote assembly systems. The experience with first prototype will help to develop a RAS for learning of programming and handling of robotized assembly cell for education and also for e-training.

REFERENCES

Gustavsson I. (2002). Remote Laboratory Experiments in Electrical Engineering Education. *Proceedings of the Fourth IEEE International Caracas Conference on Devices, Circuits and Systems*, 17-19 April 2002, p I025-1-5.

Miele D.A., Potsaid B. and Wen J.T. An Internet-based Remote Laboratory for Control Education. *Proceedings of the 2001 American Control Conference*, Jun 25-27 2001, Arlington, VA p 1151-1152.

Perme, T., (1995), *A Virtual Manufacturing System in Education, Proceeding of Tempus Workshop on* Automation and Control Engineering in Higher Education, July 5-7 1995, Vienna, Austria, pp. 7-14.

Onosato, M. and K. Iwata (1993). Development of a Virtual Manufacturing System by Integrating Product Models and Factory Models. *Annals of the CIRP*, Vol. **42**/1/1993, pp. 475-478.

Schmid C. A Remote Laboratory Using Virtual Reality on the Web. *Simulation*, v 73, n 1, July 1999, p 13-21.

ELSEVIER
IFAC
PUBLICATIONS
www.elsevier.com/locate/ifac

PETRI NET WEB SIMULATOR FOR INTERNET-BASED TRAINING IN DISCRETE EVENT SYSTEMS

C. Mahulea, C. Lefter, M.H. Matcovschi, O. Pastravanu

Department of Automatic Control and Industrial Informatics
Technical University "Gh. Asachi" of Iasi
Blvd. Mangeron 53A, Iasi, 700050, ROMANIA
Phone/Fax: +40-232-230751, E-mail: opastrav@delta.ac.tuiasi.ro

Abstract: The paper presents the construction and exploitation of a Petri Net Web Simulator (abbreviated as PNWS), which provides Internet-based training services for the simulation and analysis of event-driven dynamical systems. The PNWS was designed and implemented as a client-server application, relying on web technologies such as http, soap, xml, jaxm. The main objectives envisaged by the PNWS are: the simulation of Petri nets within a familiar framework (using any Java-supported Internet browser), the independence of the operation system, the integration with any http server, the convenient development of further facilities. Moreover, the PNWS was meant to ensure full compatibility with the MATLAB software, as representing a standard for the scientific computation in many fields of engineering. *Copyright © 2004 IFAC*

Keywords: control education, Internet-based training, distance learning, discrete-event dynamic systems, Petri nets, client - server applications, web technologies, MATLAB.

1. INTRODUCTION

Nowadays, the Petri net models are intensively exploited in the analysis and design of various types of discrete-event dynamic systems. Therefore the training in this field requires modern instruments, offering a large access to all those interested in acquiring adequate knowledge. Within such a context, we have developed a Petri Net Web Simulator (abbreviated as PNWS) that is able to answer the basic education needs in handling Petri net models.

The PNWS has been constructed so as to ensure full compatibility with the MATLAB software, as representing a standard for the scientific computation in many fields of engineering. The PNWS has been designed and implemented as a client - server application, relying on web technologies such as http, soap, xml, jaxm, and envisaging the following objectives: (i) the simulation of Petri nets within a

familiar framework (using any Java-supported Internet browser), (ii) the independence of the operation system, (iii) the integration with any http server, (iv) the convenient development of further facilities. The PNWS has been tested on a large number of examples that cover the usual area of problems encountered in control engineering education. In these tests, we have also involved a number of students who prepared laboratory experiments, homeworks and projects under PNWS.

Our paper is structured in six sections, as follows: Section 2 gives a brief description of the PNWS facilities; Section 3 discusses the organization of the client - server application; Section 4 focuses on the implementation and programming techniques; Section 5 illustrates the PNWS usage by several case studies; Section 6 formulates some concluding remarks on the general benefits resulting from the PNWS exploitation.

2. BRIEF DESCRIPTION OF THE PNWS FACILITIES

The PNWS allows the simulation of four classes of Petri net models, namely: untimed, transition-timed, place-timed and stochastic Petri nets. The last class includes stochastic, generalized stochastic and stochastically rewarded Petri nets. The firing rule ensuring the simulation progress acts in accordance with the specificity of each class. The compatibility with MATLAB allows creating the following supplementary facilities: (i) the usage of the symbol **Inf** for operating with infinite-capacity Petri nets, (ii) the usage of the distribution functions available in the Statistics Toolbox for assigning priorities to the concurrent transitions.

The graphical interface of the PNWS was built in Java, by using the libraries **awt** and **swing** so as to ensure the user's full control of all the options. In its design the authors derived a full benefit from their previous experience aquired during the implementation of the Petri Net Toolbox for MATLAB (Mahulea *et al.*, 2001), (Matcovschi *et al.*, 2003), (Pastravanu *et al.*, 2004). Figure 1 presents the main window that appears whenever PNWS is started by an http browser. This figure permits a quick visualization of the following six elements whose role is detailed below: (1) Menu Bar, (2) Toolbar, (3) Drawing and simulation panel, (4) Sim/Draw button, (5) Drawing area, (6) Status Bar.

The **Menu bar** (1) is placed horizontally, in the upper side of the window and displays a set of seven menus by means of which the user can select all the facilities available in PNWS. These menus are briefly described in the following: The **File** menu (including the sub-menus **New, Open, Save** and **Exit**) provides functions for handling files. The **Options** menu allows the user to personalize the visualization and simulation conditions, by choosing the graphical theme of the entire application ("Look and feel"

principle), the initial color(s) of the Petri net nodes, the animation speed when the simulation is run in the slow mode. The **Draw** menu offers tools for graphical editing (places, transitions, arcs, markings and labels) in the drawing area. The **Petri Net Type** menu permits assigning the net type to a new model, or changing the net type of an already existing model. The **Simulation** menu gives the user the possibility to control the simulation progress, to record the simulation results and to specify the stop condition when the simulation is run in the fast mode. The **Analysis** menu provides tools for studying dynamical properties of the net. The **Help** menu offers on-line information on the exploitation of PNWS.

The **Toolbar** (2) is equipped with a number of buttons mapping the most frequently used facilities available in the menus displayed by the **Menu Bar**. The **Drawing and Simulation Panel** (3) maps either the drawing or the simulation commands in accordance with the user's option selected via the **Sim/Draw Button** (4). Within the **Draw** mode, the **Edit** button enables the access to each graphical element of the net displayed in the **Drawing Area** (5), by means of specific dialog boxes. As an illustration, figure 2 shows the dialog boxes associated with the places, transitions and arcs of a P-timed Petri net with infinite token capacity. Within the **Simulation** mode, the **Drawing Area** (5) accommodates the performance of the simulation, for which the user can choose the running type: step-by step (accompanied by animation), slow (accompanied by animation) and fast (without animation). The **Status Bar** (6) displays information typical to the current mode of the PNWS exploitation, selected by the **Sim/Draw Button** (4). The information corresponding to the **Simulation** mode comprises the simulation time elapsed and the number of events occurred since the beginning of the simulation.

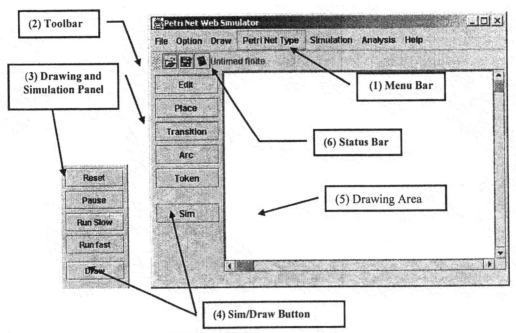

Fig. 1. The main window of the PNWS opened by an http browser.

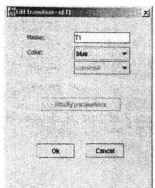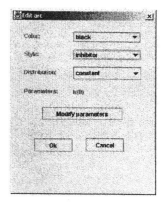

Fig. 2. The dialog boxes used for editing the characteristics of the places, transitions and arcs in a P-timed Petri net with infinite token capacity.

3. ORGANIZATION OF THE CLIENT - SERVER APPLICATION

The general organization of the client-server application can be synthetically presented by means of the diagram in figure 3 that points out the interaction between the main software modules.

3.1. The Client Application

The key functionality of PNWS can be visualized only from the client side, by the help of a Java frame. The choice of Java (Sun Microsystems, Inc., 2002b) as the programming language for developing this application resulted in two remarkable advantages: (i) the independence of the platform, meaning that the application runs identically under different operating systems; (ii) the usage of the same code by a Java applet, a web page, or a standalone application.

The application can be integrated with an html page and can be started by the Internet browser as a regular applet, which, thus, imposes its security requirements for running. In the definition of the applet, the identifier of a Petri net model can be introduced as a parameter, and, consequently, the PNWS is initialized with the net corresponding to that identifier. Since the implicit configuration

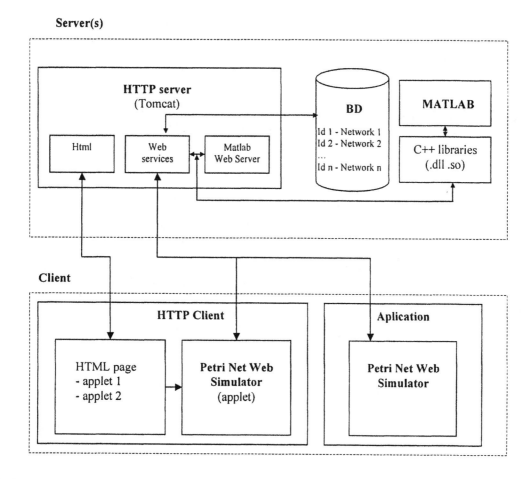

Fig. 3. Schematic presentation of the interactions between the modules of the client - server application

prevents the Java code from accessing the local file system, the **Load/Save** option of the **File** menu is activated only when such operations become possible. It is worth mentioning that javapolicy (Sun Microsystems, Inc., 2002a) can be used to personalize the users' rights for file handling.

Immediately after loading an html page, only a button is visible, which allows displaying a Petri net in a dedicated window, in accordance with a given template. PNWS communicates with the http server via the soap protocol (World Wide Web Consortium, 2001a) and requires the Petri net corresponding to the given identifier. Figure 4(a) is a hardcopy of the screen reproducing that part of the html document displayed before starting the application. Afterwards, when the application is started, the user is permitted to visualize / modify the net, or to run the simulation. Figure 4(b) is a hardcopy of the PNWS window that becomes available on the screen after the application has been started. When editing, additional communication might be initiated with the server for the validation of certain input data. When running a step-by step simulation, for speed reasons and for minimizing the network traffic, the request to the server should refer to a predefined number of steps.

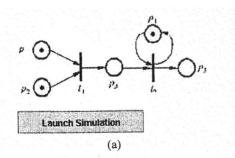

(a)

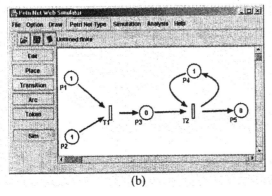

(b)

Fig. 4. Hardcopies of the screen (a) before starting the application; (b) after starting the application.

3.2. The Server Application

The server application is organized modularly. It contains three independent components (module) that can operate on the same computer, or in different locations (case when the http protocol is used for communication). The only advantage resulting from the usage of a single machine for hosting the three components is the increased speed in responding to the PNWS requests.

The *http Server* represents the first component of the server application and it can be any http service provider ensuring compatibility with the specifications of the Servlet 2.3 technology. It stores the application code in a compressed form that is sent to the client when the client loads an html page containing one or several applets of the application. Once an adequate html page is accessed, the **Launch Simulation Button** (see figure 4(a)) appears on the screen only after all application files are loaded in the cache.

The location of the application code on the hhtp Server means that the user needs no specific configuration before launching the PNWS and, moreover, he/she has always access to the latest version of the application. When the http Server receives, from a client, a request for the web services typical to PNWS, specific procedures are started to communicate either with the database of the application or with the MATLAB via C++ libraries (The MathWorks Inc., 2001a) or the MATLAB Web Server (The MathWorks Inc., 2001b). Eventually, the results are returned to the corresponding client. The modular organization of the application allows data to be processed on any machine that is able to communicate with the http Server. The locations of the PNWS database, C++ libraries and MATLAB Web Server are given only in the configuration files of the http Server (i.e. ensuring total transparency from the client's point of view).

MySQL represents the second component of the server application and it administrates the following types of stored information about the Petri net models: (i) predefined models, (ii) intermediate steps requested by the simulation of the predefined models so as to reduce the number of MATLAB calls; (iii) a local cache of the simulated models (other than the predefined ones); (iv) models saved by the users.

The *MATLAB connecting module* represents the third component of the server application and its role consists in controlling the MATLAB calls. This module has been implemented by two different strategies: (i) the integration of the MATLAB Web Server with the http Server by means of cgi scripts; (ii) the usage of the C++ libraries via api interfaces.

In strategy (i), the MATLAB is started by the MATLAB Web Server and keeps running in the background no matter if there exist requests from clients. This strategy is preferred whenever the MATLAB and the http Server run on different machines. The server includes the called MATLAB functions and their input parameters into a post-type request, and the results are available from the response page of the MATLAB Web Server. In strategy (ii), the C++ libraries operate as a gateway between Java and MATLAB, which, in the background, opens a MATLAB session for the called functions, and their results are returned to the Java programs. This strategy is preferred whenever the MATLAB and the http Server run on the same machine.

4. IMPLEMENTATION AND PROGRAMMING TECHNIQUES

In the implementation of PNWS, the Java language was the main instrument used for developing the graphical interface, the web services and the procedures running on the server. The object-oriented programming ensures the flexibility and modularity of the code, as well as various possibilities for its further extension. The MATLAB language was used for the implementation of the procedures requested by simulation and analysis. The algorithms have been designed to permit different implementations for the transition firing rules, in accordance with the specificity of the Petri net: finite or infinite capacity, untimed, *T-timed* or *P-timed*. The progress of simulation is driven by an asynchronous clock corresponding to the occurrence of events (e.g. (Cassandras, 1993)). In the untimed case, the sequencing of the events is reduced to simply ordering their occurrence, without any temporal significance, unlike the timed case when simulation requires a continuous correlation with physical time (e.g. (Cassandras, 1993; David and Alla, 1992)).

Requests and replies within the framework of the application are conveyed on the world wide web. For storing the information about the Petri nets, xml files (World Wide Web Consortium, 2001b), are used. This technique allows the exploitation of the soap protocol for the client-server communication (operating as an exchange of structured information within a decentralized and distributed environment).

5. CASE STUDIES

We illustrate the exploitation of PNWS for two examples regarded as very relevant, because they refer to physical systems frequently encountered in the structure of digital equipment.

Example 1. Consider a sending - receiving system, adapted from (Murata, 1989) and (Desrocheres and Al-Jaar, 1993), whose Petri net model is introduced in the **Drawing Area** of PNWS as shown by the screen capture given in figure 5.

The simulation of the untimed Petri net provides qualitative information, proving the cyclic operation induced by the send - acknowledge mechanism. If a T-timed Petri net model is used, the simulation also supplies quantitative information that permits the analysis of the dynamics in terms of the efficient exploitation of the resources. Assume that to the six transitions of the Petri net model we assign exponential distributions with the following parameters: 3 ms for t1, 4 ms for t2, 2 ms for t3, 3 ms for t4, 5 ms for t5, 1 ms for t6. The steady-state functioning is characterized by the performance indices: a period of 15.6 ms, a utilization degree of over 40% for the communication channel; a utilization degree of 30% for the equipment preparing data to be sent; a utilization degree of 70% for the equipment processing the received data.

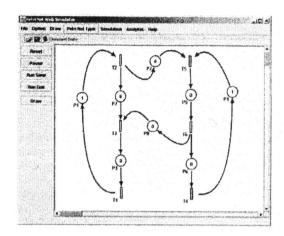

Fig. 5. Screen capture of the PNWS Drawing Area containing the Petri net model discussed by Example 1

Example 2. Consider a biprocessor system with two parallely shared resources, adapted from (Dijkstra, 1968) and (Zhou and DiCesare, 1993), whose Petri net model is introduced in the **Drawing Area** of PNWS as shown by the screen capture given in figure 6.

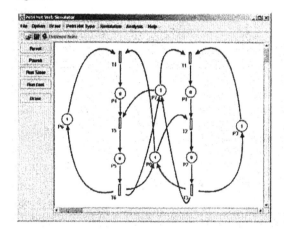

Fig. 6. Screen capture of the PNWS Drawing Area containing the Petri net model discussed by Example 2.

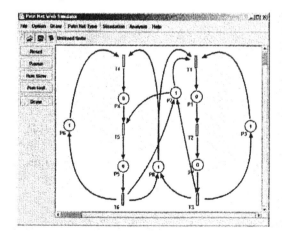

Fig. 7. Screen capture of the PNWS Drawing Area containing the Petri net model with deadlock prevention discussed by Example 2.

The simulation of the untimed Petri net model draws attention on the deadlock occurrence, resulting from the circular blocking of the shared resources. The parallel mutual exclusion shown by the screen capture given in figure 7 is able to guarantee the deadlock-free operation of the system, fact that can be simply demonstrated by the simulation of the new model revealing the cyclic functioning. If a T-timed Petri net model is used, the simulation also supplies quantitative information that permits the analysis of the dynamics in terms of the efficient exploitation of the resources. Assume that to the transitions t6 and t3 of the Petri net model we assign exponential distributions with parameter 20 ms and 15 ms, respectively, such that our future simulation uses a stochastically T-timed Petri net model (with the value 0 assigned to the other four transitions). The simulation results show that utilization degrees of the two processors are 57% for the processor modeled by p6, and 60% for the processor modeled by p3.

6. CONCLUSIONS

The paper has presented the design, implementation and testing of the client – server application PNWS, elaborated for Internet-based training in discrete-event systems. The PNWS offers a very efficient instrument for computer-aided education and deserves a special attention from the institutions interested in the organization of distance learning programs for Control Enginnering. The whole architecture of PNWS was meant to ensure a large flexibility and to permit further development relying on the same implementation techniques. Besides its numerous facilities for studying the dyanamics of the systems modeled by Petri nets, the PNWS can serve as an example for the construction of modern academic software, focused on the extensive exploitation of the Internet resources.

REFERENCES

Cassandras, C.G. (1993). *Discrete Event Systems: Modeling and Performance Analysis*, Irwin, Boston.

David, R. and H. Alla (1992). *Du Grafcet aux réseaux de Petri* (2e édition), Hermes, Paris.

Desrocheres, A.A. and R.Y. Al-Jaar (1993). *Modeling and control of automated manufacturing systems*. IEEE Computer Society Press, Rensselaer, Troy, New-York.

Dijkstra, E. W. (1968). Co-operating sequential processes. In *Programming Languages*, F. Genyus (ed.), New-York: Academic, pp. 43-112.

Mahulea, C., L. Bârsan, and O. Păstrăvanu, (2001). MATLAB tools for Petri-net-based approaches to flexible manufacturing sytems. In: *Prep. of the 9th IFAC Symposium on Large Scale Systems LSS 2001* (Filip, F.G., Dumitrache, I. and Iliescu, S.S., Eds.), pp.184-189, Bucharest.

Matcovschi, M., C. Mahulea, and O. Păstrăvanu (2003). Petri Net Toolbox for MATLAB, *Proc. of the 11th IEEE Mediterranean Conference on Control and Automation MED'03*, Rhodes, Greece, 18-20 June, 2003, Abstracts pp. 71, CD-ROM (6pag).

Murata, T. (1989). Petri nets: properties, analysis and application. In: *Proc. of the IEEE*, **77**, pp.541-580.

Pastravanu, O., M. Matcovschi and C. Mahulea (2004). Petri Net Toolbox – teaching discrete event systems under Matlab. In (Voicu, M., Ed.), *Advances in Automatic Control*, pp. 247-256, 2004, Kluwer Academic Publishers, Boston, ISBN: 1-4020-7607-X.

Sun Microsystems, Inc. (2002a). *Java 2 Platform Security* (http://java.sun.com/j2se/1.4/docs/guide/security/index.html)

Sun Microsystems, Inc. (2002b). *Java Technology and Web Services* (http://java.sun.com/webservices/docs.html)

The MathWorks Inc. (2001a). *MATLAB C/C++ Math Library*. Natick, Massachusets.

The MathWorks Inc. (2001b). *MATLAB Web Server*. Natick, Massachusets.

World Wide Web Consortium (2001a). *SOAP Version 1.2 Part 1: Messaging Framework* (http://www.w3.org/TR/2001/WD-soap12-part1-20011217/)

World Wide Web Consortium (2001b). *XML Fragment Interchange*. (http://www.w3.org/TR/xml-fragment)

Zhou, M.C. and F. DiCesare (1993). *Petri Net Synthesis for Discrete Event Control of Manufacturing Systems*, Kluwer, Boston.

EDUCATION FOR NEW AUTOMATION TECHNOLOGIES

P. Kopacek

Institute of Handling Devices and Robotics,
Vienna University of Technology,
Favoritenstr. 9-11, A-1040 Vienna
Tel.: +43 – 1 – 58801 – 318 01, FAX: +43 – 1 – 58801 – 318 99,
Email: kopacek@.ihrt.tuwien.ac.at

Abstract: "New Automation Technologies" are introduced more and more in industrial automation. Therefore "advanced" education concepts are necessary for this fast changing topics and subjects. In this contribution some ideas based on the experience of the author from education and postgraduate education mostly in the field of automation and management are shortly described.

Keywords: Automation, education, postgraduate education, virtual university.

1. INTRODUCTION

Currently we are in the "Third industrial revolution – the information age". Information is available for most of the companies and private individuals via modern communication facilities e.g. Internet .... Today's markets require flexible enterprises because the customer demands are more and increasing and are changing in shorter times. Headlines are cooperation, globalization, automation and information. Many companies were faced to reengineer their organizations, making them more flexible and adaptable to changes in the market.

Modern enterprises have to be
- Forced to have a strategic ability to create and select opportunities
- Able to combine global strategies with local implementation
- able to work processes leveraged across business
- able to rapidly deploy functional skills doing the work which best exploits the business opportunities.

Such enterprises which adapted fastest will be able to capture most of the market share. As a result they will become industry leaders.

Another most important fact is the early knowledge of "New Technologies" especially in automation and the introduction in the company. These must be accompanied by a longtime investment and business plan.

2. "NEW – AUTOMATION - TECHNOLOGIES"

Automation technology mainly deals with all theoretical and practical methods which enable a working process to progress partially or completely automatically.

Automation devices must function to guarantee certain conditions or production parameters; to act according to programs or allow approximations of optimum performance – i.e. for energy, material and information flow.

Moreover the operating and monitoring personal is to be activated assisted and relieved. In the classical way automation consists of the sub areas of
- Measurement
- Control
- Computing

In "modern" automation some new technologies from our point of view are necessary and in the following listed. We were confronted with these in education as well as scientific and industrial research.
- Manufacturing
  Water jet cutting, Hydro forming, "symmetrical" welding by robots, soldering by laser, cutting of plastics, disassembly operations.......
- Mechatronics
  Smart sensors, Embedded systems, Micro-, Nano-, Femtotechnology, Integrated systems
- Information Technology
  Hardware: Miniaturization, PLCs ....

Software: Speech programming, high speed image processing ....

- Communication Technology
  WLAN, mobile communication, Bluetooth technology ....
- Automation
  Process automation: AI; fuzzy, neuro, neurofuzzy, intelligent, digital control, genetic algorithms   ....
  Manufacturing automation: Robotics – mobile, intelligent and cooperative robots, CIM, ims, agile manufacturing systems ....

We have recognized – all these technologies require because of their complexity an interdisciplinary approach – a team of especially educated people.

3. EDUCATION FOR NEW AUTOMATION TECHNOLOGIES

Until now, in "High Tech" Engineering Education we have the following classical headlines:

- Mathematics and science are primary
- technical problems have technical solutions
- engineering is not only computing
- Knowledge can be divided in separate blocks. Curriculum design is a matter of choosing the right blocks and arranging them in a right sequence.

Until now the Austrian educational system in technical fields was a little bit complicated (Fig. 1) but it will harmonized with the rest of Europe according to the Bologna process.

Disadvantages of the current education system:

- "Classical" technical education institutions (HTL, TU, ...) have usually very "stiff" curricula's and it's very difficult to adapt these continuously.
- The lecturers are usually not in favor to include the latest developments in their courses.
- Laboratory equipment for "High tech" is usually very expensive – donations from companies – manpower for operation and maintenance

Advantages:

- Free of charge or reasonable cheap

In the nearest future the education in this field have to become more and more interdisciplinary. The topics and there contents have to be "exchangeable" very fast to teach always the latest "state of the art". Some times it could be difficult to estimate which new automation technologies will "survive" and which will be important for the industry in the future. A continuing education of the lecturers (teach the teachers) is therefore absolutely necessary.

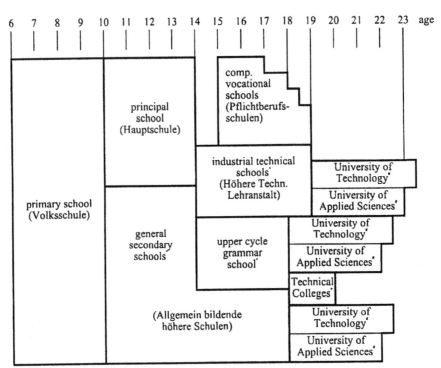

. maturaty examination
. technical maturaty examination
. diploma examination

Fig. 1 Educational possibilities in technical fields in Austria, revised version (Kopacek et al 1988)

198

4. POSTGRADUATE EDUCATION

There are a lot of postgraduate education institutions of different quality in Europe and worldwide. They are offering for example more than 1200 so called MBAs studies in mid Europe.

Private Universities were also installed. In Austria we have since 1999 a law for private Universities. Most of them offer courses and studies in non technical fields because technical education requires expensive laboratories, workshops, ...

Advantages of these institutions are:
- Curricula's could be very flexible
- Newest developments in special courses and seminars – length depends from the subject (2 – 40 hours).
- Possibility to hire lecturers familiar with the latest developments from industry as well as from research. The problem is the didactic quality of such lecturers but its usually not so important in technical subjects.

Disadvantages:
- Top education cannot be cheap

5. AN EXAMPLE: THE POSTGRADUATE TRAINING AND EDUCATION PROGRAM IN AUTOMATION

Rapid growth in automation has lead to an increasing demand for engineers and technicians with well based knowledge in this field. Not only large but also a variety of small and medium enterprises need specialists for automation, with a command of the latest technical developments and the ability to apply them in practice.

A survey undertaken on behalf of the Federal Ministry for Science and higher Education showed that automation plays an important role in 49% of all small and medium-sized enterprises. 62% of these enterprises viewed effective staff training as advantageous. Continuing education of specialists for assembly, maintenance, planning and start up of automation facilities should therefore be offered systematically by the available public and private educational institutions. The existing educational possibilities in Austria are, however, as yet incomplete. For this reason, systematic recording of requirements the coordination of existing programs was advisable, thus achieving an integration of existing and newly initiated and promoted education offers. (Kopacek ,1988)

5.1. The Basic Idea – Modularity

The whole program is based on a step-by-step educational concept, allowing entry and completion at various levels. The starting point depends on the existing knowledge of the participant. According to the present vocational education system, it is possible to distinguish the following educational levels:

- E1...completed apprenticeship, compulsory and medium level vocational school
- E2...industrial technical school
- E3...university, technical university

Each of these levels can be reached through the respective course of education and represents a starting point for continuing education. Credits for knowledge acquired by other means (practical work, private studies) can be also counted.

Continuing education falls into 4 general categories (C1 – C4) which leads to 4 final education levels (F I – F IV). Fig. 2 shows the basic structure of these educational possibilities: each of the aspired levels may be reached in varying ways from each of the three starting points.

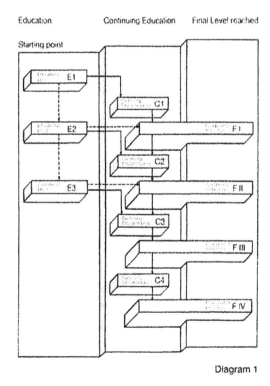

Diagram 1

Fig. 2 The program structure

Final Level F I imparts knowledge and skills required for the assembly and maintenance of automation installations, whereas final level F II provides knowledge for the planning and start-up of these facilities. Final level F III is designed for the basic development of complex automation setups. Those

199

who pass final level F IV will have attained extensive theoretical and practical knowledge, enabling them to make competent decisions both in the technical and in the managerial field of automation.

Minimum Hours per Special Subject Areas

Subject Area	I	II	III	IV	Final Sum
Basic Knowledge and Skills					
1. Fundamentals of process control	15	15	.	.	30
2. Fundamentals of chemical process engineering	15	15	.	.	30
3. Fundamentals of electrical engineering and electronics	30	20	.	.	50
4. Fundamentals of mechanical engineering and precision tool engineering	25	15	.	.	40
Process control					
5. Control engineering I	30	32	40	.	102
6. Control engineering II	30	20	10	.	60
7. Process measurement techniques	30	16	20	.	66
8. Control devices	40	15	.	.	55
9. Control systems, instrumentation and installation	20	10	.	.	30
10. Planning of process control system	.	20	28	20	68
11. Process technology	6	10	30	160	206
Computing					
12. Computing	.	15	30	10	55
13. Microcomputer	.	20	25	20	65
Theory					
14. Control mathematics and system theory	.	15	20	10	45
15. Modeling	.	.	20	10	30
16. Simulation	.	.	15	.	15
General Issues					
17. Economic aspects	4	6	6	10	26
18. Social and political issues	5	6	6	10	27
Final Sum	250	250	250	250	1000

Fig. 3 Credits for the subjects

5.2. The Curriculum

Knowledge available in automation technology has been divided into 18 special subjects areas (Fig. 3). These areas fall into 5 subject categories:

1. Basic Knowledge and Skills
2. Process Control
3. Computing
4. Theory
5. General Issues

Subject category "Basic Knowledge and Skills" consists of
1. Fundamentals of process control
2. Fundamentals of chemical and process engineering
3. Fundamentals of electrical engineering and electronics
4. 4: Fundamentals of mechanical engineering and precision tool engineering

This basic knowledge, acquired from various fields and required for the entire further program, is imparted in the final levels F I and F II. The core of the program is the category "Process Control", consisting of
5. Control engineering I
6. Control engineering II
7. Control equipments and devices
8. Instrumentation and installation

9. Design of control equipment and devices
10. Process technology (applications)

These subjects provide specialized knowledge of automation technology with a strong link to practical application, particularly emphasized in field No. 11, "process technology". A study of at least two out of six branches offered (energy production and distribution, process engineering and related fields, production technology, building control as well as environment and transport systems) is required in the course of the program.

Within the subject category "Computing"
11. Computing and data processing
12. Micro-computers

is designed to further knowledge with respect to future developments in automation technology.

Subject category "Theory" consisting of
13. Mathematics of control and systems theory
14. Modelling
15. Simulation

is of special importance for final levels F II and F IV, as it enables the participant to assess the usefulness of "modern" automation facilities.

In that automation engineers must be increasingly concerned with specific economic, social and also political issues, those were included in the

"General Issues" subject category
16. Economic aspects
17. Social and political issues.

5.3. The structure of the program

The program is conceived as a modular system, modules being for example
- Single lectures
- Courses
- Seminars
- University courses

These events are being offered for a greater number of the 18 special subject areas listed in Fig.4. Each of the four final levels may be reached by "stacking" the above mentioned modules, provided that:

a. The modules selected allow the participant to attain the minimum required number of hours in each field (Fig. 3).
b. The sum of modules completed, given a minimum of 250 hours, will lead to the next level of instruction.
c. The previous level has been completed as a prerequisite for entrance to the next.

Anyone interested in this program has already accumulated more or less "modules" according to his formal education, which may provide a basis for further instruction.

This program is running in Austria since 15 years very successful.

6. VIRTUAL UNIVERSITIES

Information Technology offers new possibilities. By Internet we can reach any person at any time. Therefore group learning, sharing of experiences and access to updated study material changes the knowledge acquisition process from teaching to active learning and from classroom learning to learning on demand.

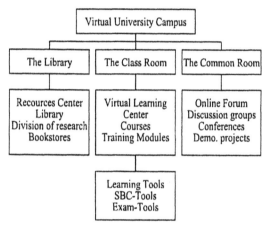

Fig. 4 Basic structure of the Virtual University (Knoth 2000)

The priority customers are:
- companies who have a need to involve employees in collaborate learning processes
- students from business and engineering schools
- other business people, academics and government officials

The basic structure of our "Virtual University" (Fig. 4) consists of
- The classroom
- The common room
- The library

The *classroom* is the formal learning space, the teaching and training environment that facilitates collaborative learning, where courses will be stored.

The *common room*, the informal learning space, allows those interested to join expert driven discussions, explore the content of the formal learning programs and learn about demonstration projects.

The on-line *library* consists of a resource center and an on-line bookstore. It supports the classroom teaching and allows success to materials and data from centers of excellence.

7. SUMMARY

Education in New Automation Technologies requires some changes in the "classical" education mentality. Key points are:
- Ensure a reasonable good basic education (mathematics, physics ...)
- Try to include as early as possible technical subjects in curricula's
- Try to make the curricula's more flexible
- "High tech" requires knowledge in English language.
- Modern technical education has to be interdisciplinary
- "High Tech" is not only computing.
- Follow the concept of the "Virtual University"
- Install and support postgraduate education institutions and technical oriented private Universities

New automation methods and technologies require new educational concepts. Therefore in this contribution two examples realized in Austria are presented and shortly discussed.

8. LITERATURE

Buciarelli, L.L. (1996): Educating the Learning Practitioneer. Proceedings of the *SEFI Annual Conference on "Educating the Engineer for Lifelong Learning"*, Vienna, 1996, p. 13–22.

Knoth,R.; M.Hoffmann; B.Kopacek and P.Kopacek (2000): Virtual University for Sustainability – a new world wide education concept. Proceedings *"Electronics Goes Green 2000+"*, Berlin, Sept. 2000, **Vol.I**, p.845-848.

Kopacek,P.; S.Höllinger and M.Horvat (1988): Training and Continuing Education Program in Automation. Proceedings of the *IFAC Symposium "Trends in Control and Measurement Education"*, Swansea, UK, July 1988, p.163-168.

Michel,Ch. (1996): Continous Learning in an industrial Environment. Proceedings of the SEFI Annual Conference on *"Educating the Engineer for Lifelong Learning"*, Vienna, 1996, p.23 – 32.

Whitwell, J.A. (1996): Education &Training for engineers in the 21st Century. Proceedings of the SEFI Annual Conference on *"Educating the Engineer for Lifelong Learning"*, Vienna, 1996, p.427 – 432.

AUTHOR INDEX